Introduction to Political Analysis in R

Introduction to Political Analysis in R is a comprehensive guide for students and researchers eager to delve into the intersection of data science, statistics, and political science. Aimed at equipping readers with the essential quantitative skills to analyze political data, the book bridges practical coding techniques in R with foundational statistical concepts, emphasizing real-world applications in politics.

The text adopts a progressive structure, beginning with the basics of R and data manipulation before advancing to more complex topics such as data visualization, spatial analysis, text analysis, and modeling. Through accessible language and engaging examples—ranging from U.S. election forecasting to global development trends—it demystifies complex analytical methods. Each chapter integrates coding exercises and real-world datasets to reinforce learning, fostering independent data analysis skills.

Designed for undergraduate political science majors, this book is also a valuable resource for anyone seeking to understand data-driven political analysis, whether for academic research, professional development, or personal curiosity.

Key features include:

- Integrates data science and statistics with a political science focus, offering hands-on coding practice using the R programming language.
- Provides real-world datasets and step-by-step exercises, enabling students to directly apply concepts to political phenomena such as gerrymandering.
- Features a progressive chapter structure, progressing from foundational data handling to advanced methods like text analysis, spatial mapping, and linear modeling.
- Emphasizes accessible coding for beginners, fostering self-sufficiency in data analysis without requiring prior statistical expertise.
- Bridges theory and application with examples that engage students' interest in politics while developing transferable analytical skills.

H. Whitt Kilburn is Associate Professor of Political Science, Grand Valley State University, Allendale, Michigan.

Introduction to Political Analysis in R

H. Whitt Kilburn

CRC Press
Taylor & Francis Group
Boca Raton London New York

CRC Press is an imprint of the
Taylor & Francis Group, an **informa** business
A CHAPMAN & HALL BOOK

Cover: Cheng J, Karambelkar B, Xie Y (2023). *leaflet: Create Interactive Web Maps with the Java Script 'Leaflet' Library.* R package version 2.1.2, https://CRAN.R-project.org/package=leaflet.

First edition published 2026
by CRC Press
2385 Executive Center Drive, Suite 320, Boca Raton, FL 33431, U.S.A.

and by CRC Press
4 Park Square, Milton Park, Abingdon, Oxon, OX14 4RN

CRC Press is an imprint of Taylor & Francis Group, LLC

ISBN: 978-1-032-55451-8 (hbk)
ISBN: 978-1-032-55658-1 (pbk)
ISBN: 978-1-003-43155-8 (ebk)

DOI: 10.1201/9781003431558

Typeset in Latin Modern font
by KnowledgeWorks Global Ltd.

To my beautiful family for their love and support: Megan, Clara, Bea and, of course, Ashla and Odin

Contents

List of Figures

List of Tables

Preface

Organization of the book: a note for instructors

The curriculum in this text represents what I teach political science students about practical skills and foundational concepts in data analysis. It is intended for undergraduate majors in political science and related fields, with widely varying prior preparation from an introductory applied statistics course. Such prior knowledge is not a prerequisite for most of the ideas and **R** tools presented here.

To encourage the students' own self-sufficiency in data analysis, the arrangement of material in the text represents an approach that combines subjects in applied data science and statistics. It is as much about learning practical tools in organizing and cleaning data as it is about constructing tests and models. The text is meant to be accessible and even fun, to develop students' inherent interest in politics and working with data.

The progression of ideas throughout the chapters in the text is unconventional in organization compared to a traditional research methods course that includes a statistics component. Yet it is one I have found in my own teaching to be an organization that engages student interests in different ways, emphasizing varied subject matter and aspects of coding, and hopefully illustrating for students the wide applicability of coding in **R**. Within my own one-semester course students approach each chapter on a roughly weekly basis. With 12 chapters in total, there is 'wiggle room' within a traditional academic semester. Each chapter includes references to resources for the incredible breadth online of applied statistics and data science in **R**, as well as cumulative exercises. Overall, the text provides a broad, applied introduction to essential quantitative data skills in **R** and applicability to the study of politics.

Supporting material and further elaborations of the text are available at https://github.com/whittkilburn/inpolr. All datasets listed in the appendix are hosted at https://faculty.gvsu.edu/kilburnw/inpolr.html

Acknowledgments

The idea for this project originated several years ago at Grand Valley State University (GVSU) when I was appointed a Faculty Fellow of the Pew Center for Teaching and Learning. Supporting the university's digital studies program, I developed workshops and student resources in data analysis, oriented toward students in the humanities and social sciences. I thank former Provost Maria Cimitile and Christine Rener, Pew Center director and Vice Provost for Instructional Development and Innovation, for their support. At the center, the commitment to excellence in teaching demonstrated by my faculty and staff colleagues helped to sustain my interest in bringing to life a curriculum in **R** for my own students in Political Science.

Within my department, I thank chairs Mark Richards and Darren Walhof for their support. Helpful conversations about this project with faculty colleagues John Constantelos, Erika King, and Michelle Miller-Adams influenced its direction, as did Statistics faculty Bradford Dykes, Gerald Shoultz, and John Stephenson. Needless to say, I alone bear responsibility for any errors, oversights, and omissions in this text.

Yet to the extent that the text has any merit, it is partly due to the supportive environment within Political Science and more generally GVSU for taking risks in writing and research. I thank my Political Science colleagues for their camaraderie and encouragement. On that note, I thank my former Dean of the College of Liberal Arts and Sciences Fred Antczak, Interim Dean Don Anderson, and Provost Jennifer Drake, whose vocal support for the idea of faculty being free to engage different parts of our professional selves in our work helped to inspire this project along to completion.

If it were not for the patience and thoughtful observations from students in PLS 300 Political Analysis, this text would never have begun. I thank my students for their reactions to, and suggestions for, early notes and drafts over the years. Students have pointed out incoherent sentences, found coding errors, and offered conceptual suggestions for improvement. In particular I thank Louis Cousino, Trystyn Duke, Isabella Jabbour, Kyle Macfarlane, Sarah Pullins, Hannah Schmidt, and Gavin Shingles. I thank the Office of Undergraduate Research and Scholarship for a grant to fund over various summer breaks research assistance from Kyle Huisman and Alex Schraeger.

Most recently, I am indebted to the assistance of Ellie Klocek and Sophie Gemmen who read various chapter drafts and offered thoughtful suggestions. At CRC Press, I thank Senior Editor David Grubbs, Curtis Hill, Kari Budyk, and Samar Haddad. I thank my family for their patient support and understanding while I completed the manuscript.

Finally, I thank the **R** community and its founders, Robert Gentleman and Ross Ihaka, without whom this book could not be written. I am also grateful to the **R** Core Team, who continue to maintain and advance this extraordinary open-source project. This book depends upon the work of **R** package developers who freely share their expertise. I am especially indebted to Yihui Xie, whose **bookdown** (Xie, 2023a) and **knitr** (Xie, 2023b) packages made it possible to draft the book in an integrated, reproducible format, and to Hadley Wickham and the tidyverse (Wickham, 2023b) team for creating **ggplot2**, **dplyr**, and related packages that appear throughout. Additional packages appear in individual chapters to support specific data analyses. To these developers and the broader **R** community, whose collective work empowers teaching, learning, and research around the world, I am sincerely grateful.

H. Whitt Kilburn

Grand Rapids, January 2025

About the Author

H. Whitt Kilburn is Associate Professor of Political Science at Grand Valley State University, Allendale, Michigan, where he teaches courses on American politics, public opinion, and applied statistics. He holds a Ph.D. in Political Science from the University of North Carolina at Chapel Hill.

1

Introduction

Chapter 1 introduces the importance of data analysis in political science and situates it within the broader context of political science research.

Learning objectives and chapter resources

By the end of Chapter 1, you should be able to (1) explain why data analysis skills are important in the study of politics; (2) explain the practice of data analysis within the broader research endeavor of political science; and (3) explain the purpose of two different examples of coding for data analysis.

Introduction

In 1959 two political scientists, Ithiel De Sola Pool and William McPhee along with a psychologist, Robert Abelson , launched a research project to study the public opinion of American voters, particularly to understand the factors holding together northern African Americans in the Democratic party coalition (De Sola Pool and Abelson, 1961). Of course they needed data. They turned to mass opinion polls, at the time on the cusp of their golden age of widespread use in national elections. The authors pooled together survey data collected by pollsters George Gallup and Elmo Roper, from 1952 to 1958, then merged these responses with separate records of corresponding local election results. Imagined as a spreadsheet, it would consist of hundreds of columns and nearly a hundred thousand rows. From the survey responses, the authors identified 480 distinct demographic "voter types", such as "'Eastern, metropolitan, lower-income, white, Catholic, female Demo-crats[sic]'" (De Sola Pool and Abelson, 1961, 168). The assembly of the data is all the more impressive given the limitations of the era's technology – such data analysis at the time consisted of physical paper 'punch cards' to be sorted slowly by a massive mainframe computer. As explained by historian Jill Lepore, this effort was intended to represent, in the phrase of de Sola Pool, "a kind of Manhattan Project" for public opinion and electoral politics (Lepore, 2021, p. 104).

While not quite the same intellectual achievement nor significance for global history as the Manhattan Project, the event would nonetheless reflect two fundamental changes in data and politics still with us today. One is the beginning of modern campaigns and elections: data intensive with mass opinion polling, neighborhood canvassing and a televised mass media forecasting the outcome based on real-time precinct level voting (Donaldson, 2007, Rorabaugh (2009)). De Sola Pool and McPhee would go on to be employed as consultants for the J. F. Kennedy presidential campaign in 1960 (Lepore, 2021).

The second change involves data. De Sola Pool and McPhee founded a company, Simulmatics, to apply these same novel methods of analyzing data to consumer behavior. As explained by historian Jill Lepore , the company would fail within 10 years, bankrupted in part and ironically from a lack of data (Lepore, 2021). There were simply not enough publicly available surveys of consumer behavior, and massive sales records of millions of consumers would not be available for decades. It was an attempt at a data science of politics before data science could exist. In Lepore's history of the company and the era, their failure presaged our world where algorithms and data patterns rule, constructing social media feeds, directing advertising, and mobilizing voters. Being literate in data — having basic competencies in the organization and analysis of data — is more important than ever.

From another perspective, some parts of contemporary political science look broadly similar to De Sola Pool and McPhee's work in 1959. Political scientists investigate research questions, construct data-driven measurements of voter characteristics, and polling is everywhere. For example, their 480 voter types would lack parsimony by today's standards, but the goal of categorizing and characterizing voters is a ubiquitous one in survey research. The Pew Research Center, in their series of reports, "Beyond Red vs. Blue", categorize voters into a typology of nine different segments on a left-to-right continuum (Pew Research Center, 2021).

Political scientists often make use of multiple data sources, accessing the data that De Sola Pool and McPhee lacked but sorely needed. The study of politics, even public opinion itself, is no longer solely focused on sample surveys resembling the sampling methods of the 1950s. Data is everywhere, so the challenge is making sense of the enormous scale of it, from campaign polls and financial statements, to social media activity, country level human rights reports, parliamentary debate transcripts, to global development indicators spanning decades. Organizing and interpreting it requires skill.

1.1 What we will do: data science + statistics

What we will learn is a combination of two closely related fields, statistics and data science in the study of politics. In political science, statistics often has often meant inferring likely population characteristics from sample surveys, much like those involved in the work of De Sola Pool and McPhee. The boundaries between the two are fuzzy, but 'data science' has largely emerged in response to the sheer volume of information we all have at our fingertips (Donoho, 2017). At its core, data consists of the measurement of facts. Data science is the process of applying systematic methods of analysis to these facts; practitioners, the data scientists, "are people who are interested in converting the data that is now abundant into actionable information that always seems to be scarce" (Baumer et al., 2017). As a result, the book is about more than sample statistics, because the starting point for data analysis is rarely ever a neat and tidy sample survey. The sea of data at the disposal of political science requires learning more than inferential statistics. Learning how to assemble and organize data for analysis is as much a critical skill, reflecting the same basic idea that De Sola Pool and McPhee engaged in back in 1959. However, unlike their efforts, we actually do have the data. And if anything, too much of it. So in this text, you will learn some fundamentals of data science (and statistics) in the study of politics.

In our study of political data analysis, we will learn about the concepts embedded within it, the required assumptions about data, and the interpretation of the results. There is very little math, certainly no more than is encountered in high school algebra. Because the textbook and course are ultimately about how we apply techniques of data analysis to learn about politics, the emphasis is on the application — how we use the techniques to learn about politics. In place of math, what is required is the mindset to approach how we will analyze data — through coding, which requires patience and persistence. Coding for data analysis requires attention to detail and breaking big problems down into smaller, sequential steps.

1.2 How we will do it: coding

Political scientists, not just computer scientists, often write code in the analysis of data. The code is collected into a step-by-step set of instructions within a text file, often known as a "script" file (Donoho, 2017). Writing the code does not require extensive knowledge of a computing language; a small set of coding tools can accomplish a great deal of analysis (Baumer et al., 2017).

In this text you will learn to write code in the statistical software application, named **R** (R Core Team, 2022), within a helper application Posit Workbench (also known as "RStudio"), which will be useful for organizing our work.[1] **R** is widely regarded as the *lingua franca* of data analysis; in political science, **R** is arguably the most commonly used language. In developing your skills as a data analyst, you will become part of this global community of **R** users.

Before getting started with R coding for data analysis, let us consider the bigger picture of how data analysis fits into political science research. Then we will briefly review some applications of data analysis with two examples, one from the study of global development and the other from US presidential elections.

1.3 Data analysis in the research process

Data analysis — in the context of political science — is situated within a particular part of a broader research process, an empirical (evidence-based) investigation about how some part of the political world works. It is worth briefly reviewing some key concepts in research design prior to getting introduced to data analysis coding in **R**. This introduction is necessarily brief; more comprehensive reviews of the subject are found in research design oriented textbooks.

The research process is depicted as a cyclical process in Figure 1.1. Typically, researchers begin with a research question, a 'how' or 'why' puzzle. Research questions relate to how the world of politics works. Questions can be either descriptive or explanatory. A descriptive research question, as is suggested by the phrase, is a question that describes a political phenomenon or concept of interest: How does the quality of democratic government support for democracy vary around the world? An explanatory question, however, queries the relationship between two concepts. Examples of such questions include the following: Why have party caucuses in the US Congress become more polarized?; How does social trust affect government effectiveness? The 'how' or 'why' part leads to the investigation of causal relationships or puzzles to explain.

From the research question follows a literature review, a search for prior research. From there, a theory emerges. A theory is a tentative answer to a research question. Then the researcher translates theoretical concepts into empirically observable quantities. This translation is often referred to as "operationalization", the process of deciding how to develop data-based (empirical)

[1] R is a statistical programming language first written and published openly under a General Public License in 1995 by Ross Ihaka and Robert Gentleman of the University of Auckland. As open source software, R is used by millions of people around the world.

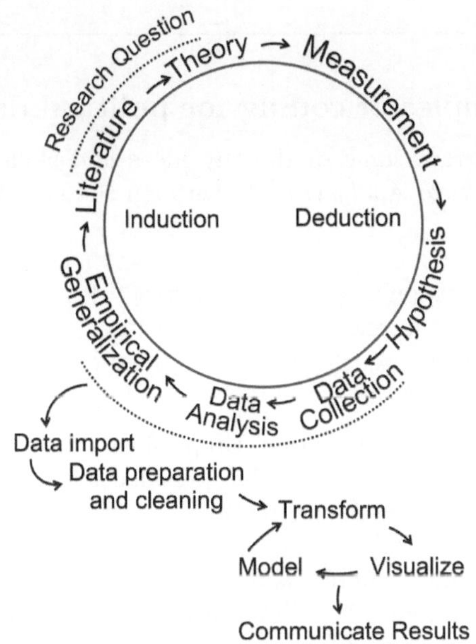

FIGURE 1.1 Data analysis as part of a broader cyclical research process. Adapted from Wallace (1971) and Wickham and Grolemund (2016).

measures of concepts. Next, hypotheses are derived from the theory; expressed in terms of the measured concepts, often referred to as "variables", hypotheses express an anticipation of how variation in one measured concept is related to another. A 'directional hypothesis' expresses how variation in an "independent" variable (the cause) relates to a particular direction of variation in the "dependent" variable (the effect).

As illustrated in Figure 1.1, data analysis in **R** plays a central role in the research process, beginning after data collection to describe a phenomenon or test a hypothesis. The steps outlined in the wheel, starting with "Data Import," present a data analysis cycle. "Data Import" involves reading external data into **R**, followed by "Data Preparation and Cleaning," where messy or unformatted data is organized into a tidy, analyzable format. During the "Transform" step, researchers manipulate variables — rescaling or creating new ones to prepare for analysis. Next, "Visualize" involves creating graphics to explore or present patterns. "Model" refers to identifying statistical relationships. Finally, "Communicate Results" emphasizes sharing insights from the analysis, completing the cycle of induction in which data driven insights contribute to the literature and inform future research questions. These steps integrate data analysis into the broader research process, linking empirical generalizations to theoretical contributions.

1.4 Two examples of coding for political data analysis

Two examples illustrate some of the key ideas and skills we will learn in *Introduction to Political Analysis in R*, through the use of coding in the **R** language.

1) managing rectangular datasets to extract descriptive measures of global economic development, life expectancy, and poverty, visualizing how the relationships between these country-level measures have changed over time.
2) explaining how the outcomes of US presidential elections are shaped by the so-called 'fundamentals': presidential approval, economic conditions, and incumbency status.

The presentation of these examples is mostly conceptual and illustrative, apart from a few lines of **R** code. We will learn the technical details — including the code — in subsequent chapters.

Measuring changes in global development

In 2007, Hans Rosling, a Swedish professor of public health at the Karolinksa Institute delivered a TED talk entitled, "The best stats you've ever seen" (Gapminder Foundation, 2024). The talk became famous for his visualization of country-level global development. In a series of yearly scatterplots showing the relationship between (human) life expectancy and fertility rates, Rosling visualized country- and regional-level changes in a dynamic graphic. Encoded within each of the scatterplots were the population (point size) and region (point color) of each country, showing over time how population, fertility, and life expectancy have changed in tandem, making uneven but steady progress (Rosling et al., 2018).

A sense of the dynamic visualizations and evidence Rosling displayed is found in the panels in Figure 1.2. Both figures display the same country characteristics, but for different years, 1975 and 2015. Because of the varying size of the points in the scatterplot, figures such as this are known as 'bubble charts'. Countries within the South Asia region are labeled. Comparing the movement of these countries from the upper to lower figure panels in the figure illustrates one of the points Rosling made. In 1975, there is wide variation in fertility rates, with much of the world population having a relatively low life expectancy compared to North America; in South Asia, all but Sri Lanka hold a life expectancy below 60 years old. Yet by 2015 improvements are such that life expectancy for much of the global population is above 60 years old.

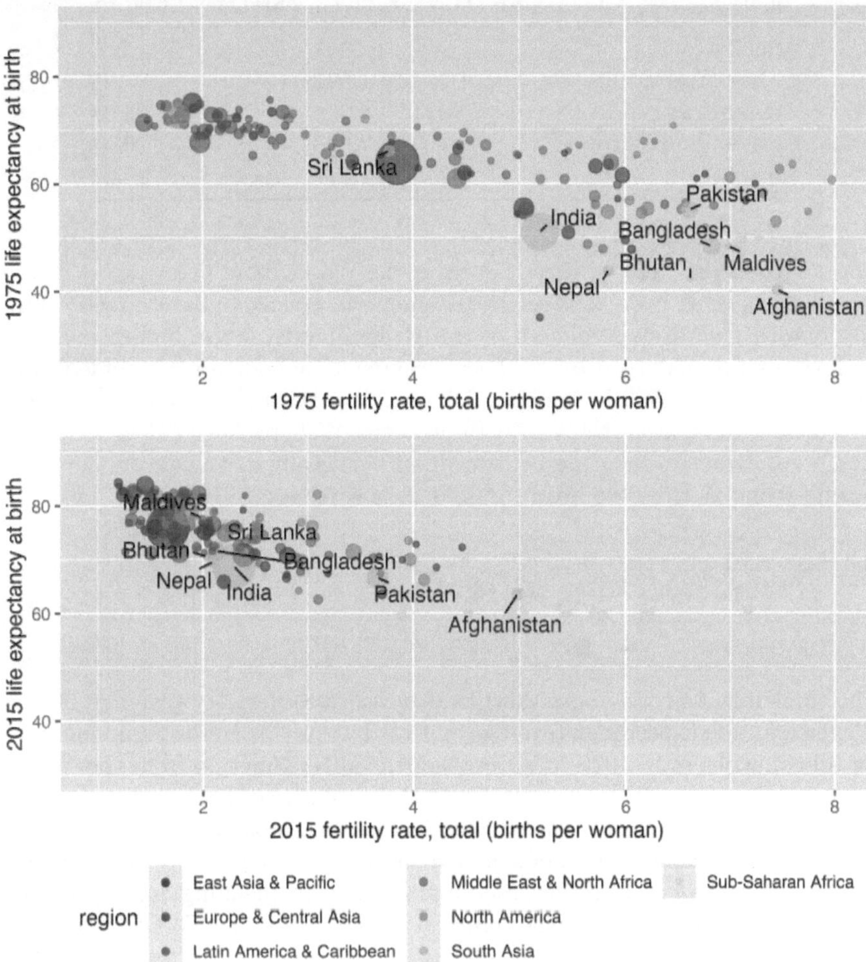

FIGURE 1.2 Scatterplots of life expectancy at birth (years) by fertility rate (births per woman) for 1975 (upper panel) and 2015 (lower panel), with countries colored by world region. Marker size indicates population size. In 1975, countries exhibited wide variation in both fertility and life expectancy, with limited regional clustering. By 2015, most countries experienced increases in life expectancy and declines in fertility, resulting in tighter regional clustering, except for Sub-Saharan Africa, which remains more dispersed in both measures.

TABLE 1.1 Excerpt of World Bank data for Figure 1.4. The second column lists fertility rate and the third life expectancy

country	SP.DYN.TFRT.IN	SP.DYN.LE00.IN
Andorra	NA	NA
United Arab Emirates	1.541	77.28
Afghanistan	4.976	63.38
Antigua and Barbuda	2.003	76.48
Albania	1.677	78.03

The source of the data for these figures is the World Bank (The World Bank, 2024). Through **R** code, it is relatively simple to download data directly into an organized dataset. Members of the **R** community write bits of code to perform useful analytical tasks. These fragments of code are assembled into 'packages' that can be downloaded for use. For example, a popular package for accessing data from the World Bank is the *WDI* package (Arel-Bundock, 2022). For example, the code below will automatically download the data for Rosling's figures from the World Bank's data servers.

```
wbdata1<-WDI(country="all",
        indicator=c("SP.DYN.TFRT.IN", "SP.DYN.LE00.IN", "SP.POP.TOTL"),
        start=1975, end=2015, extra=TRUE)
```

The three lines of code above download measures of fertility (average births per female), life expectancy (average expected at birth), and total population for all countries from 1975 to 2015 using the `WDI()` function from the **WDI** package. Functions are commands, and the parentheses contain the specific instructions of what to do. The dataset is stored in an object named wbdata1, which could be named anything, though shorter, meaningful names are easier to use. The <- symbol is the assignment operator, meaning "gets," and is used to assign the results of the function to the object. The WDI() function specifies arguments inside its parentheses, such as `country="all"` to include all countries, the indicator argument to select specific measures, and the `start=` and `end=` arguments to set the time range. When the code is run, **R** downloads the specified data and stores it in the *wbdata1* object for further analysis.

Table 1.1 displays an excerpt of the downloaded data for 2015, structured as a rectangular dataset. Each row represents a country, while the columns record features of the countries, life expectancy and fertility rates. Missing values, such as *NA* for Andorra, indicate data that is not available for that year. Overall, this example highlights code and the structure of datasets you will work with in **R** for learning skills in data organization, visualization, and analysis in later chapters.

Modeling U.S. presidential elections

Election forecasting models provide another example of the data analysis concepts and skills covered in this book. A "model" refers to a simplified representation of a political phenomenon, built using data and assumptions about how different aspects of the data are related. During U.S. presidential campaigns, forecasts of election outcomes frequently appear in the news, ranging from expert historical judgments to computational predictions based on data analysis.

These dynamic models build on static forecasts from the 1970s and 1980s, which focused on the so-called "fundamentals", contextual factors like economic conditions and presidential approval that shape elections beyond the influence of campaigns (Lewis-Beck, 2005). For example, Abramowitz's "Time for Change" model (' Abramowitz, 2016) predicts the incumbent party's share of the two-party vote based on three factors: (1) the incumbent president's approval rating in early summer, (2) second-quarter GDP growth, and (3) whether the incumbent party is running for a third term. Figure 1.3 illustrates this relationship using a bubble chart: the incumbent party's vote share appears on the Y-axis, GDP growth on the X-axis, and approval rating is encoded in the size of the points. The positive slope of the trend line shows how GDP growth is strongly associated with incumbent support. (Approval rating influences the size of the points but not the line itself.) The trend line, $y = 50.70 + 0.80X_{gdp}$, reveals the linear relationship. Here, the y-intercept (50.70) is the predicted incumbent vote share when GDP growth is zero, and the slope (0.80) indicates the expected change in vote share for each one-unit increase in GDP growth. This model and its forecasting potential will be explored further in later chapters.

By fitting the model $y = 50.70 + 0.80X_{gdp}$ to presidential elections from 1948 to 2020, we can substitute the GDP growth rate for each election year to generate predicted vote shares for the incumbent party. In some years, the predictions align closely with reality: for example, in 2012, the model predicted a 51.01 percent vote share for President Obama, nearly identical to the observed 51.10 percent, showing the economy's expected influence on his reelection. However, in 2008, the model predicted 51.09 percent for the incumbent Republican party ticket, yet the actual vote share was 45.70 percent, reflecting a significant economic under performance. These differences highlight how elections can sometimes align with a "fundamentals" based model while in other years, much less so.

This example illustrates purpose of data analysis: understanding historical patterns, incorporating data to measure the patterns, and making predictions or simply exploring relationships. While fitting a model to historical data reveals important relationships, using it to forecast an upcoming election, like 2024, introduces additional uncertainty. Substituting in projected GDP values can generate a prediction, but this process requires careful consideration of

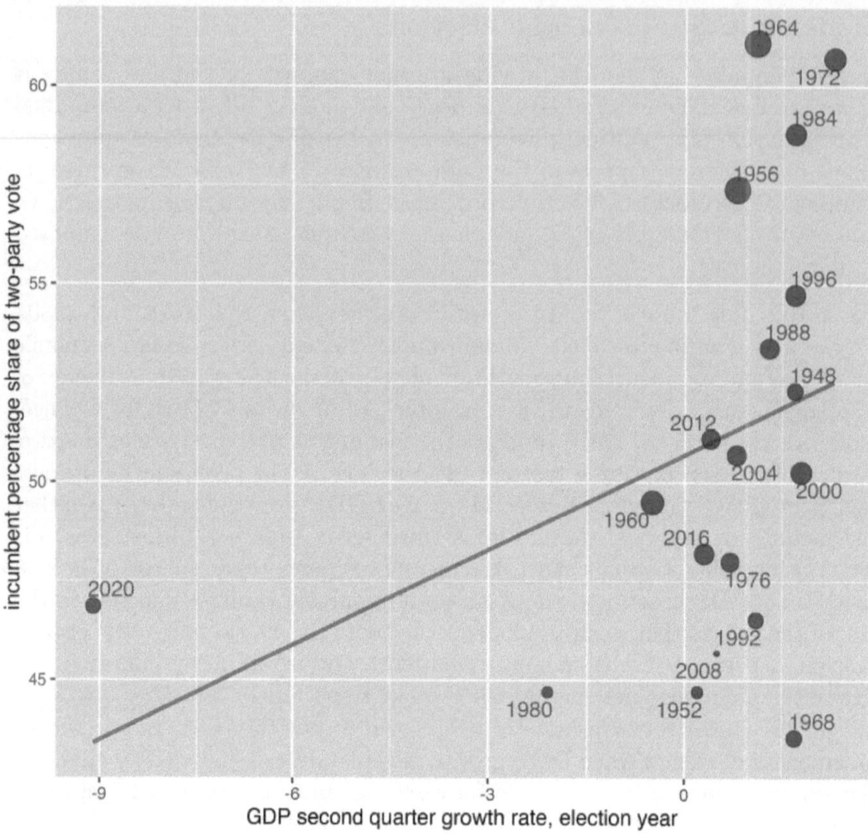

FIGURE 1.3 Scatterplot of incumbent party two-party popular vote share (in percent) by second quarter GDP growth rate (in percent) for U.S. presidential elections (1948–2020). Each point represents an election year, with point size reflecting the incumbent president's approval rating at mid-year (larger points indicate higher approval). The red trend line summarizes the positive linear relationship between GDP growth and vote share: incumbents tend to receive a higher share of the popular vote when economic growth is stronger.

the model's assumptions and limitations, which highlights the complexity of data analysis: it guides us toward insights, but also reminds us of the need for critical thinking about what the numbers mean. This theme — connecting data, models, and real-world interpretation — underscores the purpose of the book.

A roadmap to the chapters

The book organization is as follows. In Chapter 2, we will get **R** apps up and running. We will learn some of the fundamentals of using the **R** language for data analysis and useful features of Posit Workbench (also known as RStudio), an integrated development environment (IDE) for **R**. As an IDE it is an interface for working more efficiently with R code, importing datasets, saving code and output such as visualizations.

In Chapter 3 we will learn about the organization and manipulation of rectangular datasets. For example, we will learn about the organization of data into tabular (tables or rows and columns) form, how to organize data for observations collected across time and space – observations on one or many points in time, and how to merge multiple data sources. We will learn about different measurement levels, storage types in R, and common summary statistics.

In Chapter 4 we will learn about data visualization, common visualization types, and we will investigate visualizations for describing data or comparing relationships, such as histograms and box-whiskers plots, scatterplots, and 'bubble' chart variants. While this chapter focuses on the fundamentals, because data visualization is so commonly used to communicate ideas with data, subsequent chapters introduce new data visualization types based on these fundamentals.

In Chapter 5 we will learn about 'data wrangling', or how to extract meaningful information from relatively large datasets. We will learn to filter, sort, rearrange, summarize, and create new measures. We will focus our attention on changes in global development and the challenges of tidying data tables detailing election returns.

In Chapter 6 we will learn how to manage spatial or geographically defined data. We will learn various ways to create maps of election results tied to geographic areas, such as 'red — blue' election maps. We will also learn how to evaluate claims of 'gerrymandering' of U.S. Congressional districts.

In Chapter 7 our attention turns to two different methods of data analysis, scaling and clustering. We will study patterns of roll call voting in the U.S. Congress and the electorate's feelings toward American social and political groups.

In Chapter 8 we turn to the study of 'text as data'. We will first apply scaling and clustering to texts, including the study of disputed authorship of *The Federalist* papers. Next we will study ways of summarizing the thematic content of the text, including public comment on proposed federal regulations on Title IX civil rights policy affecting transgender students.

In Chapter 9 we will study public opinion through election sample surveys. We will learn fundamental skills in the analysis of survey data, from sample survey statistics to inferring population-level characteristics. Along the way we will consider how to account for the sample design of contemporary election surveys, and post-sampling adjustments made to observations to make the samples more representative. Our analysis will focus on the American National Election Studies (ANES) survey fielded in 2020.

Chapters 10 to 12 turn to the workhorse of the social sciences, linear modeling, much like the model of presidential elections reviewed in this chapter. We will learn the tools and common pitfalls to avoid in constructing models. Chapter 10 introduces the subject of simple linear and multiple regression models, while in Chapter 11 we consider complications to the models. In Chapter 12 we will construct models often expressed in probabilistic terms, binary logistic regression, for assessing questions such as how increasing educational attainment affects the probability of an individual supporting or opposing a particular policy issue.

Finally, the appendix presents a description of the datasets available to download from the author's website https://faculty.gvsu.edu/kilburnw/inpolr.html for use within the examples presented within the text.

Resources

FRED, St. Louis Federal Reserve, https://fred.stlouisfed.org/ an extensive collection of economic data.

Our World in Data: tracking global change in public health, economic, social, and political development, collected from many official sources https://ourworldindata.org/.

Party Facts: an extensive aggregation of data related to political party organizations around the world https://partyfacts.herokuapp.com/.

World Bank Global Development Indicators: https://databank.worldbank.org/source/world-development-indicators A repository of data on national level political, economic, and social characteristics.

R Graphs Gallery: https://r-graph-gallery.com/ A wide ranging gallery of data visualizations generated with **R** code.

Gapminder: https://www.gapminder.org/ Hans Rosling's work on global development data and visualization.

2

Getting Started with R for Data Analysis

Chapter 2 introduces **R** and the RStudio interface, working through an example of importing a data table and writing **R** code.

Learning objectives and waterials

By the end of Chapter 2, you should be able to (1) install **R** software, (2) identify key parts of the RStudio interface, (3) write **R** code interactively through the Console prompt, (4) import a dataset into RStudio, (5) open, edit, and save a text script file containing **R** code, (6) create a data visualization, (7) save and knit **R** code into an RMarkdown (.Rmd) document to create a self-contained Word or PDF document. The material uses the datafile *gdpfert.csv* and a text file of **R** code *chapter2.r* from https://faculty.gvsu.edu/kilburnw/inpolr.html

2.1 Getting started with R and RStudio

Our tool for learning data analysis is programming language **R** in the application RStudio. **R** is both a language and a specific application — an app — and it powers the interface RStudio, which provides many useful features for **R** coding. There are two main routes to get started: one is through the 'computing cloud', in which the **R** software is set up for you on a server, and you interact with it through a web browser. (You would know if this option applies to you, having been given the web address to access it.) More likely, however, you will go the second route, which is to install **R** on your personal computer.

The chapter begins with instructions for downloading and installing the software. If you are working in RStudio through a website (such as the Posit Workbench), skip these sections and go to the section "Orientation to the RStudio interface".

Downloading R

The base R software is available free of charge to download. The first step is to go to http://cran.r-project.org, the Comprehensive R Archive Network, where you will see a set of links for the operating system of your personal computer. Click the corresponding link, shown in Figure 2.1.

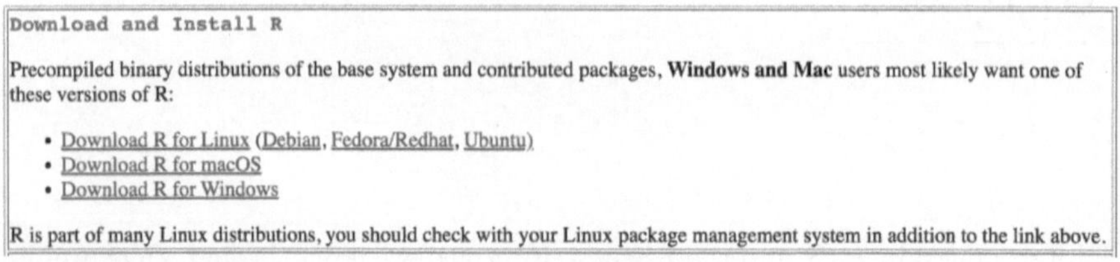

Download and Install R

Precompiled binary distributions of the base system and contributed packages, **Windows and Mac** users most likely want one of these versions of R:

- Download R for Linux (Debian, Fedora/Redhat, Ubuntu)
- Download R for macOS
- Download R for Windows

R is part of many Linux distributions, you should check with your Linux package management system in addition to the link above.

FIGURE 2.1 Comprehensive R Archive Network (CRAN) is the home of the **R** computing language. At the top of the CRAN home page, click the link for your computer operating system to download the base **R** app.

For MacOS, download the .pkg file for your version of the MacOS operating system; unless you are on an older Mac (check by going to the Apple menu, then "About This Mac"), you will want to download the .pkg file under the subheading "Latest release". The file will have a sequence of numbers that identifies the most recent version of R available for downloading, similar to "R-4.X.X.pkg". (As of this writing, it is version 4.3.1.). Open the .pkg file and install the R base software, following the default prompts to complete the installation. If you encounter trouble, check the MacOS FAQs[1].

For Windows, at the top of the page, find the line beginning with the "base" file, and click the link to "install R for the first time." From that page, you will see a link that appears similar to "Download R-4.X.X for Windows". (As of this writing it is version R-4.3.1.) Install the .exe file as you would other software on your Windows system. If you encounter trouble, check the Windows FAQs[2].

After downloading and installing the R app, move on to doing the same for RStudio. Generally, you will never need to open the R app directly. R will run in the background, while you interact with RStudio. The RStudio app provides a more polished and user-friendly way to do data analysis through writing R code. RStudio provides support for accessibility screen readers, for example.[3]

[1] https://cran.r-project.org/bin/macosx/RMacOSX-FAQ.html

[2] https://cran.r-project.org/bin/windows/base/rw-FAQ.html

[3] Check the link https://support.posit.co/hc/en-us/articles/360045612413-RStudio-Screen-Reader-Support for more information on screen reader support for RStudio.

Downloading RStudio

Next, download the RStudio IDE for R. The app is free and open source, developed by the data science company Posit, and available at https://posit.co/downloads/. Go to this page, then you will see a link to the RStudio desktop, as shown in Figure 2.2.

FIGURE 2.2 Click the Download RStudio icon at the Posit.co website to find the version of RStudio Desktop for your operating system.

Once you click that link, you will see two steps, downloading base R (which you have already done) and then RStudio. The second step will show a link to the RStudio software "Recommended for your system". Download and install it.

After you have it installed, you are now ready to get started with R coding. Remember that when you analyze data, you will start up and use RStudio rather than the base R software. Each time you begin any data analysis, you will open up RStudio, which will automatically run the base R software in the background.[4]

2.2 Orientation to the RStudio interface

To begin, start up the RStudio app you just installed. The Rstudio interface has many features; to keep it simple, we will use only the essential parts. Figure 2.3 displays the RStudio interface along with an arrow and annotation pointing out the essential parts. No matter whether you are on an RStudio server or your own computer installation, your view of RStudio will have these same components.[5]

(1) On the left-hand side, an arrow and annotation points out the

[4]Periodically, (perhaps once a semester), you should update your copy of R and RStudio, by reinstalling each one.

[5]If you are working on an RStudio website, the functionality of RStudio is the same, for the most part. The only major difference is getting files into and out of the cloud. For that, there is an 'upload' button above the files pane, and for downloading, an 'export' button.

Console pane. The Console is where **R** code is interpreted to follow instructions. **R** code is either entered directly at the > prompt or saved in a file and sent to the prompt to be interpreted.

(2) On the upper right-hand side, an arrow points to the Environment pane, , which says it is "empty", but later will be filled with data. The Environment pane displays data loaded into RStudio, and saved objects such as data visualizations.

(3) Next to the Environment pane is the History pane, which reports the history of commands you have run in RStudio.

(4) On the lower right-hand side is the Files pane that displays your machine's files and folders, starting with a default directory. Note that it shows all files and folders, not just those associated with RStudio.

(5) Next to the Files pane is the Plots pane, which displays data visualizations.

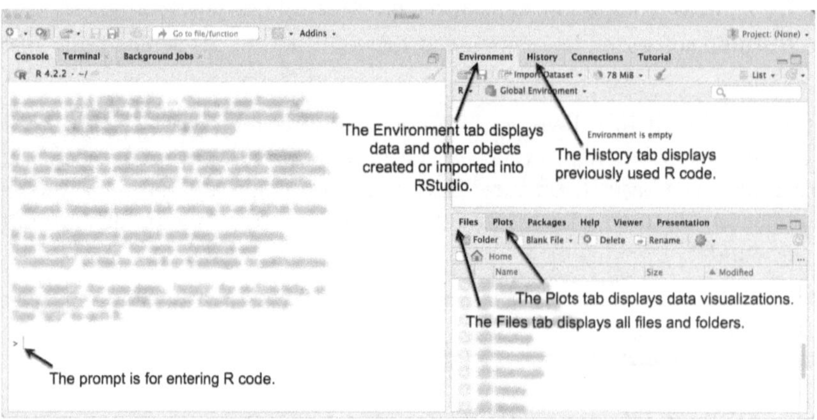

FIGURE 2.3 RStudio interface. The left pane displays the Console window, through which commands are entered at the > prompt. The upper right pane shows an Environment tab, which displays data loaded into RStudio. The History tab shows previously issued commands. On the lower right side, the Files tab displays all files, while the Plots tab displays data visualizations.

Enter R code at the command line prompt

Locate the RStudio console pane, on the left-hand side of your computer screen. (If unsure, compare your screen to Figure 2.3.) Click to the right of the > sign; the > is the R prompt, signaling that R is waiting for you to issue a command.

The most basic of all data analysis commands is to use **R** as a calculator , `+` for addition, `-` subtraction, `*` multiplication, and `/` division. Try out some simple arithmetic expressions at the prompt: enter `5 * 10`. Press the return key. Observe the result — `50` — prefaced by `[1]`. R usually identifies the output with a number in brackets. This feature – the identification of lines of output – will be useful later.

```
5*10
```

```
## [1] 50
```

Try typing `5 *` at the prompt, and then press return. You will see that now in place of the > prompt, there is a `+` symbol. The `+` symbol means that **R** is waiting for you to complete the command. Enter the number `10` at the `+` sign and notice that RStudio performs the arithmetic and returns the result as expected. When a command is recognized by **R** as incomplete, you will see a `+` sign in place of the `>` prompt. Pressing the Escape key ("esc") will interrupt an incomplete command and return you back to the `>` prompt.

2.3 Use functions for data analysis

In our analyses, we will generally not apply arithmetic operations directly to numbers, entered into the command prompt, as in `2+2`. Instead, we will use functions.

R functions are instructions or commands to perform on data. For example, an arithmetic mean, or average, is a function. It is a set of operations or steps that are performed on a set of numbers. In **R**, functions are identified by parentheses. The name of the function – a hint at what the function does – precedes a set of parentheses, `()`. The parentheses hold the arguments of the function, the input to the function.

An example of an **R** function is `seq()` , in which `seq` is short for "sequence". The sequence function automatically creates a sequence of numbers from a specified start and end point in increments of one. The start and end points (the arguments) are placed within the parentheses, separated by a comma. To create a sequence of integers from 0 to 20:

```
seq(0,20)
```

```
##  [1]  0  1  2  3  4  5  6  7  8  9 10 11 12 13 14 15 16
## [18] 17 18 19 20
```

An **R** function always includes open and closed `( )` parentheses. The function operates on the arguments contained within the parentheses. Try an argument and observe the result, such as `seq(50,100)`. In the `seq()` function, the required arguments are the beginning and end of the sequence.

Functions are nested

Functions can be nested within other functions. For example, the `mean()` function calculates an arithmetic mean. To calculate the mean of the sequence of integers from 0 to 20, the `seq(0,20)` function is placed within the `mean()` function:

```
mean(seq(0,20))
```

```
## [1] 10
```

There are multiple ways to use functions to achieve the same result. For example, instead of using `mean()` we could use the `sum()` function. The mean, 10, can be found with the `sum()` function, which will calculate a sum – adding up the elements as it is instructed, then dividing by 21 (the sequence starts from 0):

```
sum(seq(0,20))/21
```

```
## [1] 10
```

Functions are case sensitive

Case sensitivity applies to all parts of the R language. Notice the output when the function `Seq()` is entered:

```
sum(Seq(0,20))/20 # Notice the typo in the function Seq().
```

```
## Error in Seq(0, 20): could not find function "Seq"
```

Hashtags are for explanatory comments and output

In the example of R code above, an explanatory comment about the typo in the function was preceded by a hashtag, `#` . The hashtag in R refers to the beginning of a comment. In input code, everything to the right of a hashtag is ignored by R. In output, hashtags are used for formatting.

Finding a help file for a function

At the console >, type `?seq()`. The question mark before the name of the function results in the help file for the function appearing in the lower right-hand pane of RStudio. Scan through this help file entry to see that the `seq()` function accommodates a third object, which specifies the count sequence of steps. For example, to create a sequence of numbers from 0 to 50, by 10s:

```
seq(0,50,10)
```

```
## [1]  0 10 20 30 40 50
```

The help file for any **R** function can be accessed with the `?` prefix. The entries, however, are somewhat cryptic and often written with the purpose of providing a detailed, technical explanation of how the function works. So getting help is usually easier with the use of web searches and other resources, described at the end of this chapter.

Create objects to store data

In **R**, an object is a stored value or data container, such as *wbdata1* from Chapter 1. Objects are fundamental to R programming, serving as building blocks for storing data, analysis results, or visualizations. A common saying in R is "everything is an object," meaning nearly any element — data, functions, or results — can be stored and reused, provided it fits the purpose of the function in which it appears.

To create an object, use the assignment operator `<-`, which assigns everything on the right to the object name on the left. For example, in `a <- seq(1, 20)`, the sequence of numbers from 1 to 20 is assigned to the object *a*. Typing the object name afterward will print its contents:

```
a<-seq(1,20)

a
```

```
##  [1]  1  2  3  4  5  6  7  8  9 10 11 12 13 14 15 16 17
## [18] 18 19 20
```

Object names can be almost anything but must follow a few rules: they cannot begin with a number (e.g., *1a* is invalid, but *a1* is valid) or an underscore (e.g., *_dataobject* is invalid, but *data_object* is valid). Using underscores to separate words is called "snake case," which we will use as a standard throughout the book.

Objects can be passed into functions as arguments. For instance, `mean(a)` and `mean(seq(1, 20))` both return the same result, `10.5`, because `a` stores the sequence seq(1, 20):

```
mean(a)
```

```
## [1] 10.5
```

```
mean(seq(1,20))
```

```
## [1] 10.5
```

Objects can also interact with one another. For example:

```
a<-seq(1,10)

b<-1

c<-a + b

c
```

```
## [1]  2  3  4  5  6  7  8  9 10 11
```

Here, *a* is a vector of integers from 1 to 10, and *b* is a single number. Adding *b* to *a* creates a new vector, *c*, where each element of *a* is increased by 1, resulting in integers from 2 to 11. These examples illustrate three key concepts in *R*:

(1) Analyses are performed using functions, which modify or summarize data.
(2) Data are stored as objects that can be reused or manipulated.
(3) Functions typically require arguments—objects or instructions placed inside parentheses `( )` to perform specific tasks.

2.4 Packages extend the functionality of R

"Base **R**" refers to the core functionality of **R**, but its capabilities can be greatly expanded by using packages. Packages, like **WDI** from Chapter 1, are collections of functions, data, and documentation created by **R** users and shared via repositories like the Comprehensive R Archive Network (CRAN). As of this writing, CRAN hosts over 17,000 packages, ranging from general purpose tools to highly specialized resources for advanced statistics and specific fields of study.

Among the most widely used packages are those in the **tidyverse** ecosystem (Wickham, 2023b), which provides a cohesive framework for data science. Key packages in the **tidyverse** include **dplyr** for data manipulation (Wickham et al., 2023b), **ggplot2** for data visualization (Wickham et al., 2023a), and **readr** for importing and exporting data (Wickham et al., 2023c). For instance, the **readr** package includes functions to load external datasets into **R** as objects, ready for analysis.

Installing packages

To install a package, use the `install.packages()` function. For example, the tidyverse suite can be installed with:

```
install.packages("tidyverse")
```

If R cannot find the package, you may need to specify the CRAN repository with `repos=`: `install.packages("tidyverse", repos="http://cran.r-project.org")`. Package installation might prompt additional steps, such as handling dependencies (other packages required for the installation). If prompted with a statement "There is a binary version available but the source version is later…", typically, choosing "no" when asked to install from source will ensure a smoother installation process. Once installed, packages do not need to be reinstalled unless you update **R** to a new version. You can update all packages with `update.packages()`.

Attaching or loading a package

After installation, packages must be attached (loaded into memory) each time you start **R**. Use the `library()` function to do this. For example, to load the **tidyverse**, run:

```
library(tidyverse)
```

Two important steps to remember:

(1) Install each package once.

(2) Each time you open up RStudio after it has been closed, or starting a new **R** session, attach packages with `library()` to make the package available for use in your analysis.

2.5 A brief example from data import to report writing

In this section we will work through a simplified analysis workflow in **R**, covering data import, visualization, and report creation. We'll start by organizing files, setting a working directory, and writing **R** code in a script file to keep track of our work.

Organize folders for analysis files

First, create a folder for your work, such as `political_analysis`. From the text website, save the files, *gdpfert.csv* (data) and *chapter2.r* (script), into this folder. If working in RStudio through a web browser, upload the files into your workspace instead.

Set the working directory

The working directory is the folder where RStudio looks for files to import and saves exported data. Check your current working directory by running `getwd()`. To change it, use the function `setwd()` with the file path to your folder. Getting this path exactly correct is error prone; it is easier with RStudio to set the working directory through the top-level menus: "Session" > "Set Working Directory" > "Choose Directory...". Figure 2.4 shows this process. Alternatively, navigate to your folder in the Files pane and set it via the "More file commands" icon > "Set As Working Directory".

Use the Session menu in RStudio to set your working directory to the folder containing your chapter2.r script and gdpfert.csv dataset, so R can locate these files.

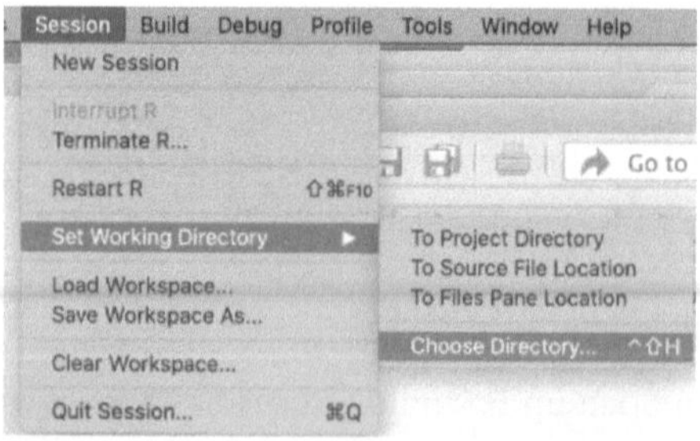

FIGURE 2.4 Set the working directory in RStudio using the Session menu, so **R** can locate your *chapter2.r* script and *gdpfert.csv* dataset.

Once set, the files in your working directory, such as *chapter2.r* and *gdpfert.csv*, will appear in the `Files` pane, as displayed in Figure 2.5.

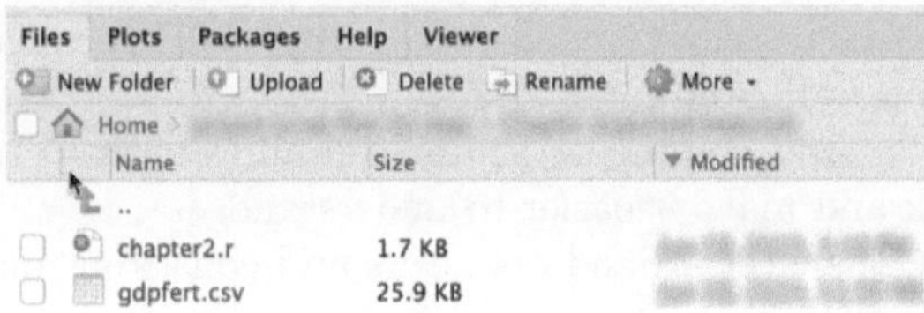

FIGURE 2.5 The Files pane in RStudio showing *chapter2.r* and *gdpfert.csv* after setting the working directory.

Writing and running code in a script file

A script file is a plain text file in which you write and save **R** code for future use. Open the *chapter2.r* script file by clicking it in the Files pane. This action opens the script in the top-left pane of RStudio (as in Figure 2.6).

In the script, highlight the code you want to run and click the `Run` button, or place the cursor on a line and press `Run` to execute it directly in the Console. For example, line 20 in the script identifies an external dataset, *gdpfert.csv*, imports it, and stores it in an R object named *gdpfert*, `gdpfert <- read_csv("gdpfert.csv")`.

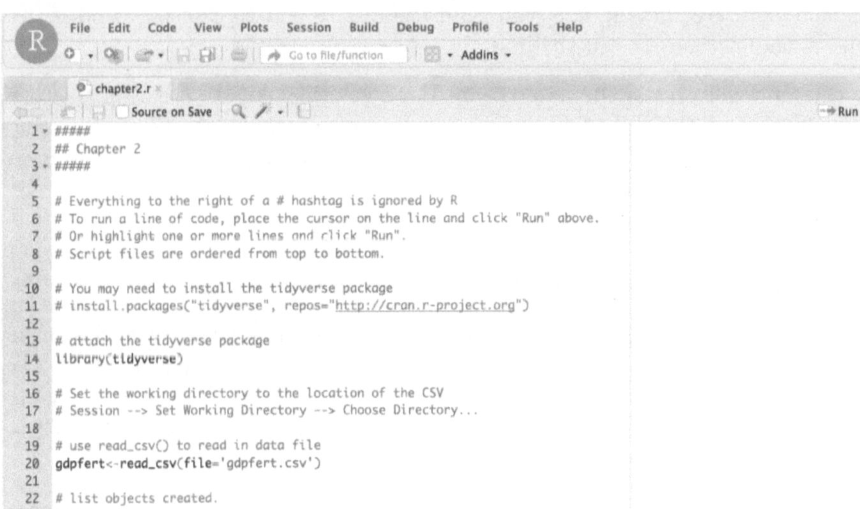

FIGURE 2.6 The *chapter2.r* script file after it has been opened in RStudio.

Key points about script files as as follows:

(1) Purpose: Script files preserve code step-by-step for future use or as a record of completed work.

(2) Structure: Keep scripts organized logically—load packages first, then import data, followed by analysis.
(3) Comments: Add comments (lines beginning with `#`) to document your work and make it easier to understand.
(4) Format: Use spaces around operators and consistent indentation to improve readability.

To create a new script file, go to "File" > "New File" > "R Script." Save it with a `.R` or `.r` extension to take advantage of RStudio's tools, such as predictive text and the `Run` button. And most important of all, remember that scripts are intended to be self-contained and read from top to bottom.

Running R code to import the gdpfert.csv dataset

If you have not already done so, click on line 20 in the *chapter2.r* script file. Highlight it – or just leave the mouse cursor on this line:

```
gdpfert<-read_csv(file='gdpfert.csv')
```

In the upper-right corner of the script file window (as in Figure 2.6), click the `Run` icon. By clicking `Run`, you send the line to the **R** Console window; **R** follows the instructions in the line.

In this line, `gdpfert` is the name of an **R** object that we are creating to store the contents of the data from `gdpfert.csv`, an external datafile in "comma-separated value" (CSV) format. This format is perhaps the most common storage type for sharing data, and more will be discussed about this format in subsequent chapters. For now, though, you could open up the file within a text editor or a spreadsheet editor, such as Microsoft Excel, to get a sense of what it means that values in the file are separated by commas.

Assigning the dataset to an object saves the dataset as a dataframe. Now that the dataset is stored as a dataframe object in **R**, the dataset is available for analysis. Nevertheless, if you encounter an error `does not exist in current working directory`, you either did not save the file to the correct folder or the working directory is set to a different location.

As we will see throughout the study of data analysis, this is the basic process of reading in data to be analyzed in **R**: it exists as an external file somewhere, and we read it into **R** with a function such as `read_csv()`. The dataset is read into **R** and assigned to an object in **R**, which the script file named `gdpfert`; we could have named it anything.

An alternative to running lines with a mouse click is to use shortcut keys. Hover the cursor over the 'Run' icon on the top right-hand side of the Console window. A tooltip will appear under the cursor, displaying the shortcut key to

Run a line of code. Run the line by either clicking `Run` or pressing the shortcut key — on a Windows machine, *Control* plus *Enter* or *Return* (both pressed at the same time), and on a Mac, *command* (the *command* key is to the left and right of the space bar) plus *return.*

Dataframes

A **dataframe** is the term in **R** for a rectangular dataset, a two-dimensional table with rows for units of observation (such as countries) and columns for measures or variables (such as GDP) that record different features of the rows. Dataframes are the most common way to store and interact with data in **R**. For example, externally to **R**, rectangular datasets are often stored as comma-separated value files; CSV files are organized in the same way as dataframes. (A CSV is a plain text file where for each observation, a comma separates different values; we will create a CSV file in Chapter 3.) When a CSV file — or any other external dataset — is imported into R for analysis, it is stored as a dataframe.

```
gdpfert<-read_csv(file="gdpfert.csv")
```

The CSV file *gdpfert.csv* focuses on a set of three measures from the World Bank Global Development Indicators. The first is a measure of the size of a country's economy, the Gross Domestic Product (GDP) per capita in 2010 U.S. dollars; GDP measures the sum total economic value of all goods and services produced within a country during a specified time period . As a per capita measure, this means that the total GDP for each country is divided by the human population of the country, which produces a roughly comparable measure of economic size, around the world.[6] The second measure is human fertility rate, an average number of births per woman within each country.[7] The third is average life expectancy at birth[8]. There are country and regional indicators, along with some additional information about income level of the country.

The function `read_csv()` from the *readr* package (in the *tidyverse*) will read and interpret the contents of a datafile in CSV format. But just reading the datafile with `read_csv()` isn't enough. We also want to store the results of `read_csv()` as a dataframe. That is what the assignment operator `<-` does: it assigns `read_csv(file="gdpfert.csv")` to `gdpfert`. With the assignment operator, the contents of the dataset are stored as a dataframe object named `gdpfert`.

[6]See the World Bank's notes on the measure here: https://data.worldbank.org/indicator/NY.GDP.PCAP.KD.

[7]For an explanation of how the fertility rate is calculated, see https://data.worldbank.org/indicator/SP.DYN.TFRT.IN.

[8]See https://data.worldbank.org/indicator/SP.DYN.LE00.IN.

As with any object stored within **R**, typing its name will print out its contents. Observe the results of typing `gdpfert` below:

```
gdpfert
```

By default, **R** prints out the first ten rows and ten columns of the dataframe. Above and below this printout are two comments. The first, `# A tibble: 211 x 10` identifies the dataframe as something called a *tibble*. A *tibble* is the name of a method for printing out the contents of a dataset. It takes the contents of a dataset as input and outputs a tidy summary of the data. A *tibble* is also the name of the package in the *tidyverse* from which this print method originates. Because it is useful for printing out tidy summaries of data, we will store our data as tibbles whenever this print method is useful.

Due to the space limitations of the printed page, only a few columns from the dataset can be printed without spilling over the left and right margins of the paper. Typing `gdpfert` will print out the 10-by-10 snippet of the dataset, similar to the more limited view of the data printed below with the `glimpse()` function:

```
glimpse(gdpfert)
```

```
## Rows: 211
## Columns: 10
## $ country        <chr> "Andorra", "United Arab Emirat~
## $ year           <dbl> 2010, 2010, 2010, 2010, 2010, ~
## $ NY.GDP.PCAP.KD <dbl> 39639.4, 34341.9, 569.9, 13017~
## $ SP.DYN.TFRT.IN <dbl> 1.270, 1.868, 5.746, 2.130, 1.~
## $ region         <chr> "Europe & Central Asia (all in~
## $ longitude      <dbl> 1.522, 54.370, 69.176, -61.846~
## $ latitude       <dbl> 42.508, 24.476, 34.523, 17.117~
## $ income         <chr> "High income: nonOECD", "High ~
## $ gdp            <dbl> 39639.4, 34341.9, 569.9, 13017~
## $ fert           <dbl> 1.270, 1.868, 5.746, 2.130, 1.~
```

RStudio features for examining datasets

Beyond printing the dataset through the console, there are features of RStudio for examining datasets. Figure 2.7 displays the Environment pane of RStudio, which after importing a dataset provides some useful information about it. The Environment panel shows datasets available in the **R** Workspace, datasets like `gdpfert` that have been imported. The panel shows that the dataset consists of 214 observations with 13 variables. If you imagine the dataset as a rectangular matrix, the observations — individual countries – are rows, while variables — measurements of country characteristics — are the columns.

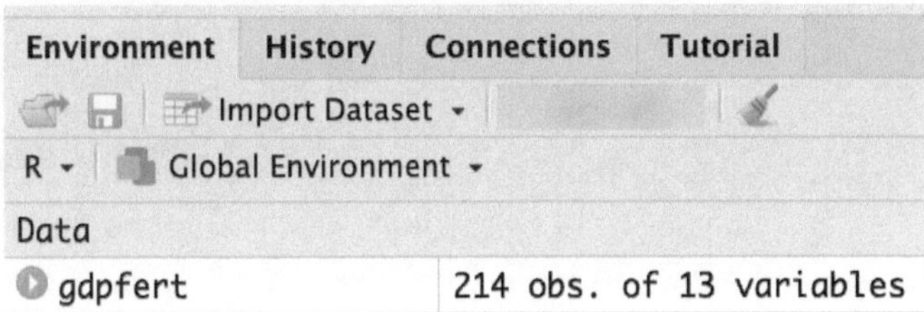

FIGURE 2.7 After importing *gdpfert.csv* the Environment pane displays the dataset with 214 observations and 13 variables, confirming a successful import.

The 13 variables of the `gdpfert` dataset are the following selected measures from the World Bank's Global Development Indicators database. We can see this dataset by clicking on the matrix icon within the Environment tab, shown in Figure 2.7. (The icon looks like a tiny spreadsheet.) Clicking on this icon reveals a spreadsheet-like view of the data. Scrolling up and down, or left and right within this spreadsheet reveals the observations on countries and variables. To close this tab, click on the `x` in the tab. (Incidentally, the function to view the data is `View(gdpfert)`.)

To analyze the data, however, we need to return to the script file. Let us illustrate a few uses of analysis commands within the script file. Try running lines 20 to 30 on your own and observe the results.

To illustrate one analysis of the data, we will construct the scatterplot of GDP by fertility. Run the lines in the script file that correspond to a different figure. For example, run lines 42-43; highlight both lines together and click `Run`. Try the other figures.

A simple scatterplot

We will learn about creating data visualizations in Chapter 4. But for now, let's see one example of how a function from the *ggplot2* package (part of the *tidyverse*) will quickly plot the relationship between two measures, such as GDP per capita and the fertility rate by country within the `gdpfert` dataset.

A function to draw quick graphics is named `qplot()`. Depending on the type of data stored, the function will determine a graphical form. The arguments within the `qplot()` function require a column of data to plot on `x=`, another on `y=`, and finally the name of the dataset to find the columns, `data=`. Figure 2.8 displays the result:

```
qplot(x=NY.GDP.PCAP.KD, y= SP.DYN.TFRT.IN, data=gdpfert)
```

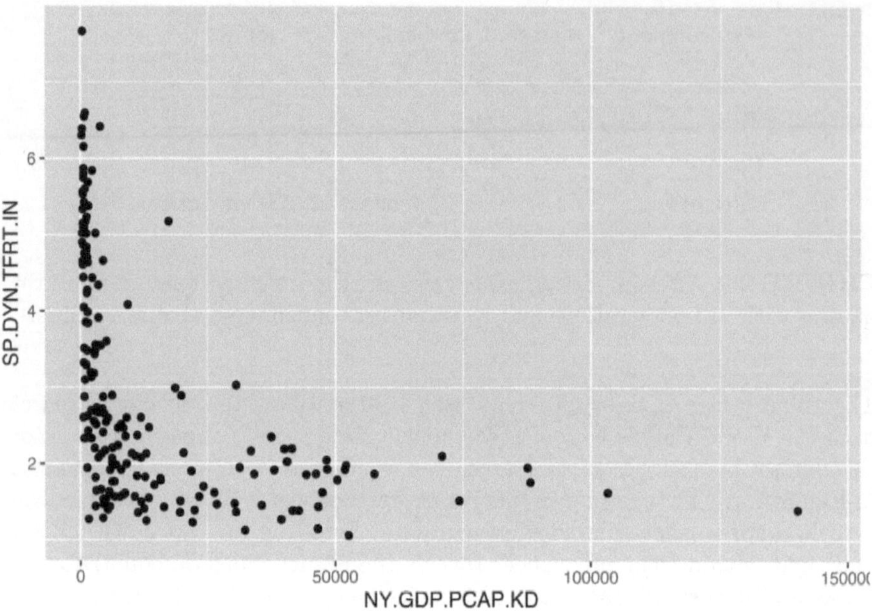

FIGURE 2.8 Output of `qplot()` showing GDP per capita and fertility rate for countries in the gdpfert.csv dataset.

While not a presentation-ready graphic, it is a useful starting point. We will work on refining visualizations throughout subsequent chapters. But for now, the `qplot()` function is a useful demonstration of a function that draws a graphic.

Saving data in an R Workspace

Once we have imported or otherwise created a dataset, we will want to save it. Dataframes, and any other **R** object created during an **R** work session, are saved within a "Workspace". So the Workspace contains more than a single dataset; it can contain many. **R** graphics can be stored as objects, and since those are objects, those can be saved in an **R** workspace as well. Once a dataframe (and any other **R** objects) are saved as a Workspace, the Workspace can be opened later. From the drop-down menus, `Session` then `Save Workspace As` will save a Workspace containing all current objects in the **R** session.

2.6 Authoring a data analysis report: RMarkdown

RMarkdown is a tool for data analysis and report writing in **R**. It allows you to interweave narrative explanations, data visualizations, and code all within a self-contained document. In an RMarkdown file, code is included with the document; when the document is "knit" together, the **R** code is interpreted and the results such as tables and graphs appear alongside the code.

We will create an RMarkdown report based on the *chapter2.r* script file; the report will contain the code in *chapter2.r* along with output.

To do so, go to the "File" menu, then choose "New File" and then "R Markdown..." as in Figure 2.9. After clicking "R Markdown..." a popup menu will appear as in Figure 2.10. Fill in the title and name. And click "OK".[9] Unless you are working in RStudio via a web browser, after you click through Figure 2.9, you will be prompted to install some additional packages. Click `OK` to do so. Agree to install the packages. Next, once the additional packages are installed, R will prompt you to name the RMarkdown document, as in Figure 2.10.

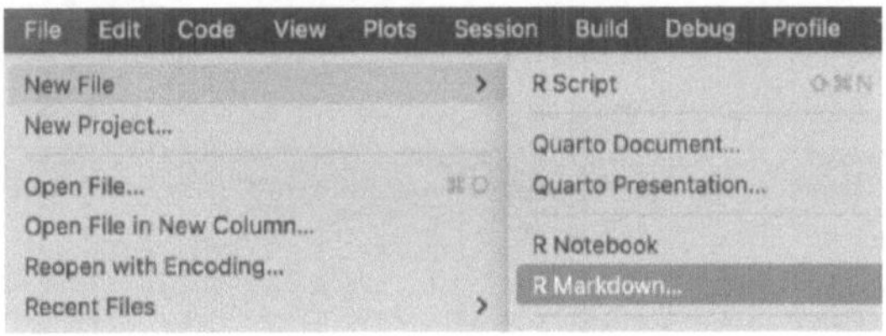

FIGURE 2.9 File menu in RStudio to select R Markdown and begin creating a new report.

Once you click OK, RStudio will display a new tab for the RMarkdown document you just created. The file will appear as in Figure 2.11.

The RMarkdown file contains example **R** code, headings, and brief narrative explanations. The **R** code appears within code chunks, the sections of the document that begin and end with three backticks (the key next to the number 1 on most keyboards). To observe the purpose of an RMarkdown file, how it weaves together code and interpretation into a self-contained report, click the **down triangle** button labeled "Knit" next to a ball of yarn. Choose "Knit to Word" and observe what happens. After a few seconds, you should see a Word document appear in your working directory, revealed within the Files pane.

[9]Note that on the lower left side of the popup menu a button for creating a "Blank Document"; when you get used to creating an RMarkdown document, you may find it easier to create a blank one and including the minimal elements or an RMarkdown document, explained further below.

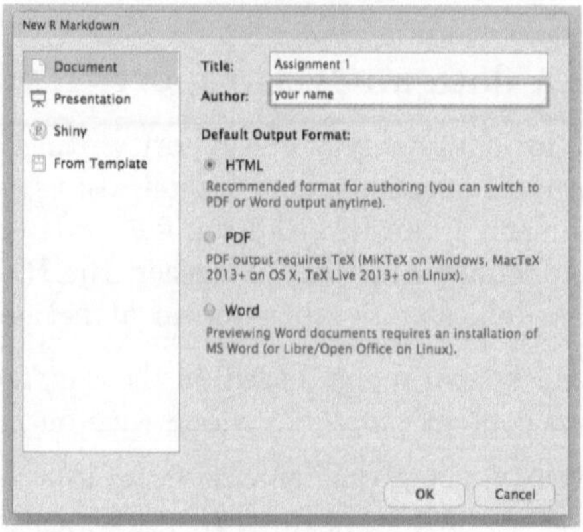

FIGURE 2.10 Dialog box for creating a new RMarkdown document. Enter a title and your name as the author.

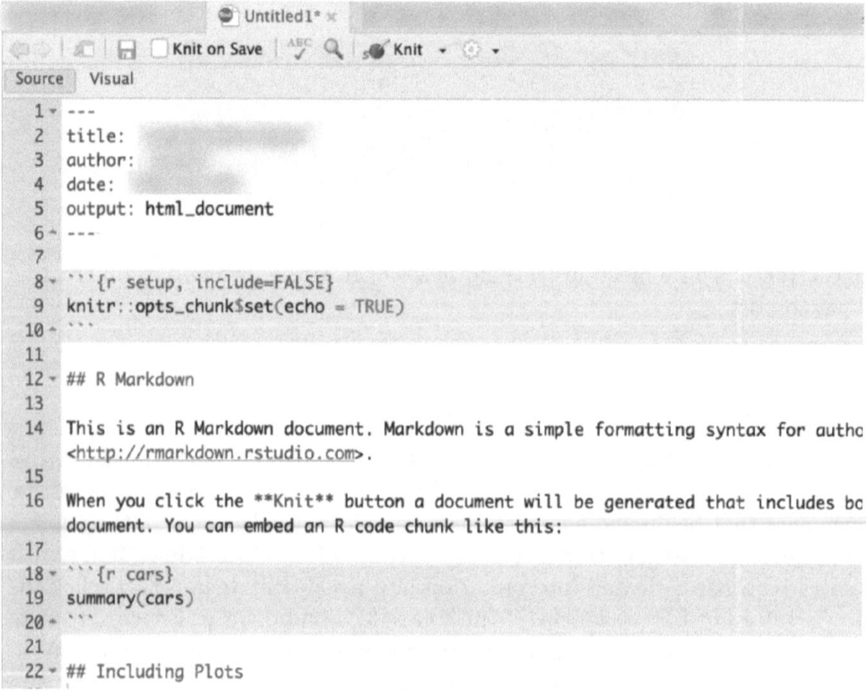

FIGURE 2.11 Default RMarkdown template, including a header, setup code chunk, narrative text, and an example **R** code chunk.

Open up the Word document to see the contents; pay attention to how the comments and **R** code are translated from the RMarkdown document to the Word document.

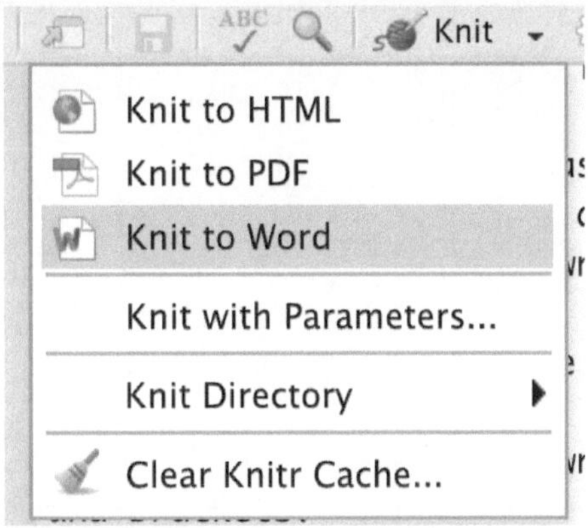

FIGURE 2.12 Click the triangle next to the Knit button in RStudio to open the output format menu. Select Knit to Word to generate a Word document, which will appear in your working directory.

We now want to edit the RMarkdown document to interpret the code from the script file *chapter2.r*. We need to delete the lines of the RMarkdown file from lines 11 to the end. (Keep lines 1 through 10 as in Figure 2.11.) In place of the deleted sections, we will add

(1) an example of narrative text describing results of the **R** code; and
(2) an **R** code chunk, denoted by triple backticks (“‘) and a bracketed `{r}` that tells RStudio to begin interpreting **R** code and present the resulting output. The code chunk ends with triple backticks (“‘). Note the backtick key is usually to the left of the number 1 on most computer keyboards.

Figure 2.13 displays the RMarkdown file with a blank **R** code chunk, beginning on line 13 and ending at line 16. Try copy-pasting all of the **R** code from the script file into the **R** code chunk at line 14 in Figure 2.13. Any interpretive comments could be placed before or after the code chunk.

Then click to “Knit to Word” and observe how the results of the Word document change. You should now see the familiar summary statistics and scatterplots from the **R** code in *chapter2.r* printed neatly in the Word document.[10] When RMarkdown is running code, it works in a “fresh” **R** session. RMarkdown only

[10]Within this template, changing the `output:` header line to `output: word_document` will ensure that when clicking the “Knit” yarn icon, the document will automatically knit into a Word document.

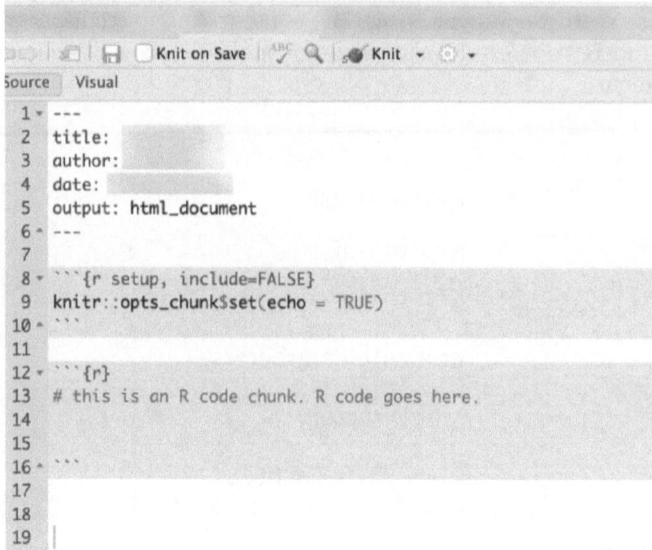

FIGURE 2.13 RMarkdown file prepared for use with **R** code from the *chapter2.r* script. Replace the placeholder text inside the **R** code chunk with your own analysis code. Any content within a code chunk (delimited by triple backticks) is treated as **R** code.

has access to the code, packages, and data objects that are referenced within the **R** code chunks. And the code is read from top to bottom. So it is essential that the code is in the correct order and all relevant packages are included first. When knitting together code, RMarkdown will search the folder in which it is located to find any datafiles.

Saving R work

Finally, you'll want to get into the habit of saving your work in order to return to it later. Doing so in RStudio is easiest with an **R** Project (.RProj) file; each time you save the project from the "File" menu, RStudio will automatically prompt you to save the Workspace and script file. If you are not working within an **R** Project file, make sure you do the following:

(1) Save all edited **R** code in a script file (Go to "File" then "Save").
(2) Save the **R** Workspace and then reopen it later to continue working on it. The **R** Workspace is saved via the "Session" menu, choose "Session" then "Save Workspace...". The default file extension for an **R** Workspace is `.RData`. Double-clicking on a saved **R** Workspace will open up RStudio; then use the "File" menu to open up your last script file. Third, save any changes to an RMarkdown file.

2.7 Getting help

You will, at some point, need help resolving error messages or figuring out how to resolve a problem when you feel stuck. There are three different approaches for doing so.

(1) Consult the help menu in RStudio within "Help" tab on the lower right pane of RStudio. Search for topics or specific functions such as by entering `?summary()` for help with the `summary()` function. Once you have attached a package to your R session, you can pull up help files for package-specific functions, such as `write_csv()` from the **tidyverse** by typing `?write_csv()`. If you click "R Help" under the "Help" top menu in RStudio, it will pull up a directory of **R** Resources. There are a few very useful "Cheat Sheets" that summarize common features; start with the RStudio cheat sheet. Overall most of the "R Resources" are either for longer term help or more advanced niche **R** tools.

(2) Two of those resources, however, stand out: "R on StackOverflow" and "RStudio Community Forum". These two fora are for posting questions, searching past discussion threads, and hopefully finding answers from other users. Follow the links to visit each forum. Once there, search for an error message or relevant discussion thread. A related resource is social media; /rstats is popular on Reddit.com, for example. A common help strategy is to web search on the error message; doing so with Google will most likely point you toward StackOverflow or the RStudio forum. And of course there are many **R** help pages all over the web. Searching simple statements such as "How to create a scatterplot in R" will bring up multiple web pages.

3. This search strategy leads to the third major approach, though one fraught with ethical problems: ask an artificial intelligence (AI) chatbot for help. It "knows" **R** in the sense that it can diagnose syntax errors, tell you the name of the **R** function you cannot immediately recall but can describe, and even write **R** code from scratch given your instructions. Nevertheless, it has important limitations. It often makes mistakes. It may suggest outdated code or code that combines packages or functions in ways that do not work. Or it provides working code that is an oddly roundabout solution to the problem you identified in the prompt. It usually does, however, find basic syntax errors in code. Give it your code and explain the error message or what you are trying to do. It will likely identify the source of the error or at least give you some places to start looking.

It would be a poor learning strategy, however, to attempt to rely on an AI to write **R** code for you as a starting point, and doing so is unfeasible without some prior **R** knowledge to make sense of the results. Most importantly, you should be aware of how your use of it may comply with course policies and an Honor Code. Talk to your instructor about guidelines for the ethical use of AI chatbots.

Key points

(1) RStudio is the interface for coding with **R**.

(2) Startup the RStudio app to do any data analysis; the **R** app runs in the background.

(3) **R** code is entered at the Console window `>` prompt.

(4) Data analysis proceeds through the use of functions, such as `seq()`, and the assignment operator `<-` to create data objects.

(5) A script file is a plain text file with an *.R* or *.r* suffix; the file contains R code and explanatory comments that begin with a hashtag, `#`. Script files organize **R** code.

(6) **R** project files (*.RProj*) are helpful for organizing work, though not essential.

(7) RMarkdown files (*.Rmd*) are used to generate self-contained reports, displaying code alongside any tables and graphics.

Resources

Education portal for RStudio https://education.rstudio.com/learn/

The official website for the Tidyverse https://www.tidyverse.org/

RStudio Cheat Sheet https://github.com/rstudio/cheatsheets/raw/main/rstudio-ide.pdf

RMarkdown Cheat Sheet https://github.com/rstudio/cheatsheets/raw/main/rmarkdown-2.0.pdf

2.8 Exercises

(1) Create a sequence of integers with `seq()` from 1 to 10, and store it as the object `B`. Type the name of the object and observe the results.

(2) Following the previous exercise, create a second object and call it `C`. Store the following numbers: from 0 to 50, every fifth number, such as 0, 5, 10, etc. Use the `seq()` function, but add a third argument to it, the number by which to skip or space the sequence.

(3) Create an object `D` that is the product of `C` times `B`. Inspect the contents of `D`. Given that `B` and `C` are different lengths, how would you explain the results of `D`?

(4) The `mean()` function calculates a mean on an object. Store a sequence of integers from 0 to 40 in an object, call it `testobject`, and calculate the mean with the `mean()` function. Do the calculation two ways: by evaluating the function on the object name, after it is created, and then directly on the function generating the sequence of integers.

(5) Open the *chapter2.r* script file. Edit the file to include only the following code, line by line: (a) attach the **tidyverse** package, (b) import the *gdpfert.csv* dataset, and (c) use the `qqplot()` function to create a scatterplot of Life expectancy by GDP. Paste the three lines (as an **R** 'code chunk'), then knit the file to a Word document.

(6) The **datasets** package is pre-installed and attached as part of the base R system, containing various datasets. Enter `library(help = "datasets")` at the Console prompt and a tab will open up in RStudio explaining the contents. One of the US States datasets is `state.x77`, an excerpt from the 1977 Statistical Abstract of the United States, with observations on 50 States for the following variables: "Population, Income, Illiteracy, Life Exp, Murder, HS Grad, Frost, and Area". To view the dataset, simply type `state.x77` at the prompt. Use the `summary()` function to calculate and interpret some summary statistics on Life Expectancy and the Illiteracy rates across the US. Try using `qplot()` – from **ggplot2** in `library(tidyverse)` – to construct a scatterplot of Life Expectancy by Income. Interpret what you have observed in a paragraph; include the code and discussion in an RMarkdown document, knitted to Word or a PDF.

3

Importing, Cleaning, and Describing Data

Chapter 3 introduces the organization of rectangular datasets in **R**, fundamental operations on datasets, along with measurements of, and relationships between, columns in datasets.

Learning objectives and chapter resources

By the end of this chapter you should be able to: (1) organize and differentiate forms of tabular, tidy data tables; (2) use summary and descriptive statistics for central tendency (mean, median, mode) and variability (range, IQR, standard deviation, MAD) to describe data tables given specific levels of measurement; (3) import, export, and merge data tables; (4) construct common measurement scale transformations. The material in the chapter requires the following packages: **tidyverse** (Wickham, 2023b) and **WDI** (Arel-Bundock, 2022), which should already be installed. The material uses the following datasets *gdpfert.csv*, *hdi_2021.csv*, *vdem_data.csv*, and *countries_tsxs.csv* from https://faculty.gvsu.edu/kilburnw/inpolr.html.

Introduction

In Chapter 3, we will learn about the organization and description of data in a form usually preferred for analysis: a rectangular matrix of rows and columns. This form, with observations on (horizontal) rows and the characteristics of the observations on (vertical) columns, is a standard form in which researchers analyze data. The term "Data" does not refer exclusively to information organized into this format, but typically the analysis of data will involve its representation in this form. Beyond organizing data, we will learn about common summary statistics and measurement scales.

TABLE 3.1 National GDP per capita and fertility rate, in 2010, downloaded from the World Bank

country	NY.GDP.PCAP.KD	SP.DYN.TFRT.IN
Andorra	34667.3	1.270
United Arab Emirates	32534.8	1.819
Afghanistan	526.1	5.977
Antigua and Barbuda	13986.8	1.988
Albania	3577.1	1.660
Armenia	2958.8	1.722
Angola	2994.4	6.194
Argentina	13551.3	2.346
American Samoa	12256.3	NA
Austria	43334.5	1.440

3.1 Rectangular data structure

The process of collecting and organizing data is often the most time consuming in any research project. A common rule of thumb is the '80/20 rule'. You should expect to spend about 80 percent of your time cleaning and organizing data to get it into a clean, usable format, and only about 20 percent actually analyzing it (Dasu and Johnson, 2003). So, learning how to organize and clean data into an amenable format for analysis is an essential skill.

Table 3.1 displays a tidy (or clean) data table from the World Bank in this structured form. The units of observation are countries, in rows (read horizontally), while the characteristics of each country (gross domestic product per capita, labelled *NY.GDP.PCAP.KD* and average births per woman, labelled *SP.DYN.TFRT.IN*). In this example, if the plan were to collect data on economies and fertility rates around the world, a necessary step would be to organize the observations into this tabular form. The rows organize the units of observations, countries, while the columns organize the features, often referred to as variables regardless of how much the features of each country actually vary. Such a dataset, with only a single entry for each cell of the table, would be regarded as a tidy, clean dataset or data table. In the first row of a clean data table, there may or may not be a header row, the first row of the table containing the names of each of the columns. If the table consists of nothing else besides a header row and the entries for each unit of observation (no additional footnotes or additional text explanations below the table), the dataset is also machine readable, meaning it can be imported into **R** for analysis.

In sum a simple clean, tidy, rectangular datafile has the following structure:

(1) Rows (across the horizontal) that record observations on units of analysis. And the data file may also include an additional row — appearing in the position of the first row — containing names of the features.

(2) Columns (down the vertical) that record features of the observations — also referred to as 'variables' — measurements of observation characteristics.

(3) And nothing else – no additional notes or explanations of either the rows or columns, appearing anywhere else in the file.

Your goal in analyzing data about politics is nearly always to get the data into tidy form.

Within **R**, a data table typically exists as a *dataframe* in an **R** Workspace. There are different ways to get data tables into an **R** Workspace. The most common is to import an existing clean data table into R, by applying an **R** function to an external dataset. Below we will review doing this via the `read_csv()` function in the **tidyverse** suite of packages.

3.2 Importing and exporting data tables

The CSV file

A Comma-Separated Value (CSV) is perhaps the most common way of storing a rectangular, clean dataset for future use. Rows organize observations. Columns organize variables (with or without a header row containing variable names). Across each row, for each observation the variables are separated by commas, hence the "Comma" in the name for this data storage format. Files in this format usually carry the "*.csv*" or "*.CSV*" file extension, such as the datafile *gdpfert.csv*. Files of this type are simply plain text files with alpha numeric records of observations and variables.

A CSV datafile can be created with a simple text editor, such as WordPad or Notepad (on Windows) or TextEdit (on a Mac). Any app capable of creating a plain text file (such as a file with the extension *.txt*) can be used to view, edit, or create a CSV file. Try opening the *gdpfert.csv* file in a text editor. Adding observations, or even creating datasets, can be accomplished with this method. The key is to make sure that the CSV is clean and tidy — only rows and observations, with a header row containing column names. Each row must have the same number of columns. Of course, some observations might have a missing value for a particular feature, when the measure of that feature is missing for the observation. In those cases, a typical solution is to either use a period symbol (.) or `NA` for missing values. The data analysis software will either guess, or will need to be instructed, what exactly the missing value indicator is.

Importing, editing, and exporting CSV files

In RStudio, a CSV file appearing in the Files pane can be opened by clicking on it. If the file has the extension *.csv*, from within the Files pane, click on it and choose `view file`. Then it can be edited exactly like in the example discussed above. Try opening the file *gdpfert.csv* in RStudio. Note that a file can be organized as a CSV file but have `.txt` as the file extension.

In Chapter 1, we used the `read_csv()` function (part of the **tidyverse** set of packages), without drawing much attention to it. The function reads the datafile from CSV format and saves it as the object we choose. And it will show us the storage type of the measures that it reads. Try it with *gdpfert.csv*; set the working directory to the location of the file. Then run the `read_csv()` function:

```
library(tidyverse)
```

```
gdpfert<-read_csv(file='gdpfert.csv')
```

The output from the function shows the "Column specification", the comma delimiter separating observations as well as the storage type of the columns. Three columns were read as text characters "chr", and seven were imported as numeric "dbl".

To see the contents of the dataset, enter the name of the data object `gdpfert`.

```
gdpfert
```

```
## # A tibble: 214 x 13
##    iso2c country     year NY.GD~1 SP.DY~2 SP.DY~3 iso3c
##    <chr> <chr>      <dbl>   <dbl>   <dbl>   <dbl> <chr>
##  1 AD    Andorra     2019  39414.  NA        NA   AND
##  2 AE    United Ar~  2019  43785.    1.33    79.7 ARE
##  3 AF    Afghanist~  2019    584.    4.87    63.6 AFG
##  4 AG    Antigua a~  2019  17697.    1.47    78.7 ATG
##  5 AL    Albania     2019   4543.    1.41    79.3 ALB
##  6 AM    Armenia     2019   4562.    1.58    75.4 ARM
##  7 AO    Angola      2019   2570.    5.44    62.4 AGO
```

```
## 8 AR    Argentina    2019  12716.     1.99     77.3 ARG
## 9 AS    American ~   2019  13288.   NA         NA   ASM
## 10 AT   Austria      2019  46647.     1.46     81.9 AUT
## # ... with 204 more rows, 6 more variables:
## #   region <chr>, capital <chr>, longitude <dbl>,
## #   latitude <dbl>, income <chr>, lending <chr>, and
## #   abbreviated variable names 1: NY.GDP.PCAP.KD,
## #   2: SP.DYN.TFRT.IN, 3: SP.DYN.LE00.IN
```

The result is a neatly formatted excerpt of the dataset commented as `# A tibble: 214 x 13`. The `tibble` formatting is the result of using the `read_csv()` function from the *tidyverse*, explained in Chapter 5.

To write out the `gdpfert` dataset to an external file, we use `write_csv()`, specifying the data table (or dataset object) to write out as a CSV file, and then the name of the CSV file we would like to have saved. For example, the line below would write out `gdpfert` as a CSV file, saving it as `gdpfertdata.csv`":

```
write_csv(gdpfert, "gdpfertdata.csv")
```

The datafile *gdpfertdata.csv* appears as a file in the current working directory.

Other datafile formats

Of course, beyond comma-separated, there are many other potential delimiters — a 'tab' key delimited (*.tvs*) file is common. Another popular file format is JSON (*.json*), short for Java Script Object Notation . It is a file format commonly used by coders building web applications, and is increasingly used to store data accessible to the public. While JSON-formatted data can be organized into multiple nested sets of observations, for a simple pair of CSV rows, JSON organizes observations into key-value pairs.

Opened in a text file editor, a JSON formatted snippet of the *gdpfertdata.csv* datafile would appear like this:

```
[
  {
    "country": "United States",
    "year": 2020,
    "gdp_per_capita": 63543,
    "fertility_rate": 1.73
  },
  {
    "country": "Canada",
    "year": 2020,
```

```
    "gdp_per_capita": 46292,
    "fertility_rate": 1.50
  }
]
```

Within the brackets, braces `{ }` separate observations that are neatly labeled by the corresponding column header. In Chapter 6 we will work with JSON-formatted map data — datafiles recording the boundary lines of US Congressional districts, for example. The package **jsonlite** (Ooms, 2014) will read in JSON data. Once the file is read into **R** as a dataframe, regardless of the format of the existing dataset, you would apply analysis functions to it like any other dataframe.[1]

Summary measures of data objects

Regardless of the file format, once imported as a data object, basic summary statistics are accessed with the `summary()` function. We can access summary statistics for the entire `gdpfert` dataset by entering `summary(gdpfert)`. With large datasets, it's better to identify a specific variable by name. Individual columns of data can be identified with the `$` symbol:

```
summary(gdpfert$SP.DYN.TFRT.IN)
```

```
##    Min. 1st Qu.  Median    Mean 3rd Qu.    Max. 
##   0.918   1.590   2.106   2.585   3.329   6.961 
##    NA's 
##       7
```

We will review the importance of these summary statistics further below. For now, just note that the default output is to report — from left to right — the minimum value, 25th percentile (`1st Qu.`), median or 50th percentile, arithmetic mean, 75th percentile (`3rd Qu.`), and maximum value. The summary means, for example, at the minimum one country averages just .918 births per woman, while the average country level rate is 2.585 births per woman. The entry `NA's` refers to seven countries that are missing fertility rates from the dataset.

[1]Datafiles are often made available in the proprietary file formats of commercial software publishers such as SAS or Stata. There are various **R** packages designed to import data from these formats into an **R** Workspace. The **foreign** (R-core, 2023) package will read in commercial formatted data and store it as a dataframe, while the **haven** (Wickham and Miller, 2023) package will store it as a dataframe with the print formatting of a *tibble*. Both packages interact differently with various data storage types found in commercial software titles. For a package specifically designed for reading in worksheets in Microsoft Excel format (rather than CSV), see the **readxl** (Wickham and Bryan, 2023) package.

Other functions useful for summarizing the contents of dataframes include `glimpse(gdpfert)`:

```
glimpse(gdpfert)
```

```
## Rows: 214
## Columns: 13
## $ iso2c          <chr> "AD", "AE", "AF", "AG", "AL", ~
## $ country        <chr> "Andorra", "United Arab Emirat~
## $ year           <dbl> 2019, 2019, 2019, 2019, 2019, ~
## $ NY.GDP.PCAP.KD <dbl> 39413.6, 43785.4, 584.4, 17697~
## $ SP.DYN.TFRT.IN <dbl> NA, 1.334, 4.870, 1.468, 1.414~
## $ SP.DYN.LE00.IN <dbl> NA, 79.73, 63.56, 78.69, 79.28~
## $ iso3c          <chr> "AND", "ARE", "AFG", "ATG", "A~
## $ region         <chr> "Europe & Central Asia", "Midd~
## $ capital        <chr> "Andorra la Vella", "Abu Dhabi~
## $ longitude      <dbl> 1.522, 54.370, 69.176, -61.846~
## $ latitude       <dbl> 42.508, 24.476, 34.523, 17.117~
## $ income         <chr> "High income", "High income", ~
## $ lending        <chr> "Not classified", "Not classif~
```

And to see a snippet of the top and bottom rows of the table, the functions `head()` and `tail()` will print these rows. For example, `head(gdpfer)` will print the top several rows of the table:

```
head(gdpfert)
```

```
## # A tibble: 6 x 13
##   iso2c country      year NY.GD~1 SP.DY~2 SP.DY~3 iso3c
##   <chr> <chr>       <dbl>   <dbl>   <dbl>   <dbl> <chr>
## 1 AD    Andorra      2019  39414.  NA          NA AND
## 2 AE    United Ara~  2019  43785.   1.33     79.7 ARE
## 3 AF    Afghanistan  2019    584.   4.87     63.6 AFG
## 4 AG    Antigua an~  2019  17697.   1.47     78.7 ATG
## 5 AL    Albania      2019   4543.   1.41     79.3 ALB
## 6 AM    Armenia      2019   4562.   1.58     75.4 ARM
## # ... with 6 more variables: region <chr>,
## #   capital <chr>, longitude <dbl>, latitude <dbl>,
## #   income <chr>, lending <chr>, and abbreviated
## #   variable names 1: NY.GDP.PCAP.KD,
## #   2: SP.DYN.TFRT.IN, 3: SP.DYN.LE00.IN
```

Of course, the data table can be inspected visually in spreadsheet form with the `View()` function, which with `View(gdpfert)` would bring up the spreadsheet view of the data. The use of this function is equivalent to clicking on the spreadsheet icon next to the data in the Environment pane of RStudio.

3.3 Data organization: cross-sectional and time series

Data tables record variation, which can occur in two different directions, cross sectionally (across observations) and time series (across time), or a combination of both. For example, the data tables generated from **gdpfert.csv** are cross-sectional — each row of the table consists of an observation on countries, with each column recording different characteristics of the countries each at one point in time.

The file *countries_tsxs.csv*, which varies both across countries and over time, is a cross-sectional time series dataset.

To illustrate the organization of data varying across the two dimensions, use the `read_csv()` function to read in the datafile to the data object *countries_tsxs*.

```
countries_tsxs<-read_csv(file="countries_tsxs.csv")
```

The imported data object, *countries_tsxs* contains observations of various development characteristics on countries around the world, from 2000 to 2019. The data is both cross sectional and time series — on any one unit of analysis (a country) there are observations on development characteristics over the 20 year period. The first five rows and seven columns of the dataset display the variation across Afghanistan on two development indicators. This portion of the data table is selected with subscripts, bracketed rows and columns `[r,c]` to select on a dataframe. The rows for entries one through five are selected with `1:5` and one through seven with `1:7`:

```
countries_tsxs[1:5, 1:7]
```

```
## # A tibble: 5 x 7
##   country     region  iso2c iso3c  year gdp_c~1 gdp_g~2
##   <chr>       <chr>   <chr> <chr> <dbl>   <dbl>   <dbl>
## 1 Afghanistan South ~ AF    AFG    2003 8.22e 9    8.83
## 2 Afghanistan South ~ AF    AFG    2014 1.97e10    2.72
## 3 Afghanistan South ~ AF    AFG    2007 1.11e10   13.8
## 4 Afghanistan South ~ AF    AFG    2013 1.92e10    5.60
## 5 Afghanistan South ~ AF    AFG    2017 2.10e10    2.65
## # ... with abbreviated variable names 1: gdp_constant,
## #   2: gdp_growth
```

A time series data table contains measurements over time on a single unit of observation. Quite simply, an example of a time series component of *countries_tsxs* could be illustrated by simply subsetting it on a single development indicator on a single country, such as the observations on `gdp_growth` and `gdp_pc_constant` on Afghanistan. Using row and column subscripts to select only columns recording the year and those two indicators `c(5, 7, 8, 9)` on the first 20 rows `1:20` of the dataset:

```
countries_tsxs[1:20, c(5, 7, 8, 9)]
```

```
## # A tibble: 20 x 4
##     year gdp_growth gdp_constant_int_pp gdp_pc_constant
##    <dbl>      <dbl>               <dbl>           <dbl>
##  1  2003      8.83         29265058689.            363.
##  2  2014      2.72         70158264545.            603.
##  3  2007     13.8          39589169630.            429.
##  4  2013      5.60         68297470349.            608.
##  5  2017      2.65         74711922906.            589.
##  6  2010     14.4          57116893447.            569.
##  7  2016      2.26         72785293848.            590.
##  8  2004      1.41         29678901146.            354.
##  9  2012     12.8          64675178731.            596.
## 10  2001     NA                     NA              NA
## 11  2000     NA                     NA              NA
## 12  2011      0.426        57360414055.            551.
## 13  2002     NA            26890054382.            360.
## 14  2005     11.2          33011757108.            380.
## 15  2019      3.91         78557606649.            584.
## 16  2008      3.92         41143038133.            437.
## 17  2006      5.36         34780330056.            384.
## 18  2018      1.19         75600418109.            579.
## 19  2015      1.45         71176481723.            592.
## 20  2009     21.4          49943751387.            512.
```

Implicit in the organization of the data is that it refers to a single country, Afghanistan, which is not recorded explicitly anywhere within it. Adding any additional country observations would introduce cross-sectional variation, turning it back to being a cross-sectional time series dataset.

3.4 Data measurement levels and storage types

The measures, or columns of a data table can be categorized into different groups: qualitative and quantitative. These terms do not have a precise definition. Typically, the term "qualitative measure" refers to a variable that records a set of categories without an inherent numerical measure. For example, party identification "Are you a Democrat, Republican, Independent, or what?" is a qualitative measure. Another example is a survey question asking, "How many of the people running the government are corrupt?", where responses are recorded on a scale including categories such as "All", "Most", "About half", "A few", and "None". A quantitative measure typically refers to broadly a meaningful numerical score. A measure of GDP per capita measured in US dollars, a percentage or a count of votes cast, would be quantitative measures. Quantitative measures are often further described as discrete (a count of votes — integers) or continuous (a measure of GDP per capita — in theory any real number). The characteristics of these measures can be further refined into four levels of measurement, which matter for determining appropriate summary statistics applied to each one.

Measurement as categorical, ordinal, interval or ratio

A fundamental measurement distinction is between nominal, ordinal, interval, or ratio. In a data table, any variable can be described by one of these terms, either due to the inherent qualities of the measure or assumptions about it.

Nominal measures are variables with distinct categories without an underlying order. For example, in a survey, a measure of religious identity ("Buddhist, Christian, Hindu, Jewish, or Muslim?") is nominal. A nominal measure is qualitative, and can be summarized with a mode (the most frequent category). While it would not make sense to ask what the average religious identity is, it would make sense to tally up the most common identity.

Ordinal measures are ordered categories reflecting an underlying rank quantity. For example, a Likert scale in survey research in which respondents indicate their level of agreement on a five-point scale ranging from "Strongly Disagree" to "Strongly Agree" is an ordinal scale. An ordinal measure could be interpreted as qualitative or quantitative. There is no inherent numerical quality to differentiate "Strongly Disagree" from "Disagree", only a rank order of stronger disagreement. An ordinal measure can be summarized by a median or mode. Yet as explained further below, ordinal measures are often treated as quantitative, interval measures in which the categories as assigned numerical scores, such as a 1 for "Strongly Disagree" and 2 for "Strongly Agree" and so on. Only with this added assumption of equal difference (from 2 to 1) between categories could an ordinal scale convert into an interval measure.

Interval and ratio measures are inherently quantitative. With the assumption of equal intervals between measured points, a survey question asking individuals how strongly they identify as Democrats or Republicans on a seven-point scale ranging from 1 for "Strong Democrat" through "Independent" to 7 for "Strong Republican" would be an interval scale. A ratio measure is a meaningful zero indicating the absence of a quantity. For example, a measure of percentage change in Gross Domestic Product (GDP) includes a meaningful zero point and is ratio in scale. Vote counts are ratio. The term "ratio" references the fact that because there is a meaningful zero, quantities can be compared in ratio form – such as one candidate receiving twice as many votes as another. For interval and ratio measures, summary statistics of a mode, median, and a mean are appropriate.

Storage types in R: numeric, factor, character

In **R** levels of measurement roughly correspond to different storage types in dataframes. The three primary types of storage you will encounter are "numeric", "character", and "factor". Each type has its own unique characteristics and uses.

The function `str()` will reveal the storage type of a each column in a dataframe, such as `str(gdpfert)`. When reading in an external CSV file with `read_csv()`, you will generally find that the storage types are either numeric or character. For example, in the `str(gdpfert)` output the three-letter code `chr` refers to character storage for the variables *iso2c* and *country*.

A character data type stores text strings, such as the names of countries. Counterintuitively, character storage types can also include numbers. Typically character storage types are reserved for text-heavy data without an inherent order or any quantitative measurement quality. Nonetheless, it is often better to have text-heavy data stored as a "factor" storage type when text values are repeated frequently. Factor storage types are explained after numeric types:

Most of the columns in *gdpfert* are of course numeric and store numbers. Numbers can be either integers or decimals. What is important to recognize is that for any data analysis, numeric types are required. Even if a column of numbers appears as if it is numeric, for various reasons it could be stored as a character set. For example, the functions to convert columns of data between numeric and character are `as.numeric()` and `as.character()`, although these should be used with caution as the results may not always be as intended.

To illustrate the differences, we could create a (nonsensical) character version of GDP per capita:

```
gdpfert$gdp_character<-as.character(gdpfert$NY.GDP.PCAP.KD)
```

It appears at the end of the dataset. (Try `names(gdpfert)`.) We can compare the storage types and data entries to observe how differently the character and numeric versions are treated, viewing the first four observations on these two variables:

```
str(gdpfert[1:4, c(4, 14)])
```

```
## tibble [4 x 2] (S3: tbl_df/tbl/data.frame)
## $ NY.GDP.PCAP.KD: num [1:4] 39414 43785 584 17697
## $ gdp_character : chr [1:4] "39413.6344732996" "43785.4172694435"
```

```
head(gdpfert[1:4, c(4, 14)])
```

```
## # A tibble: 4 x 2
##   NY.GDP.PCAP.KD gdp_character
##            <dbl> <chr>
## 1         39414. 39413.6344732996
## 2         43785. 43785.4172694435
## 3           584. 584.386882896106
## 4         17697. 17697.2311549703
```

The numeric version by default does not print out decimal places, athough these values are stored within the variable. The character version may appear to be more precise with five decimal places represented in the dataframe snippet. But applying the `summary()` function displays a problem:

```
summary(gdpfert$NY.GDP.PCAP.KD)
```

```
##    Min. 1st Qu.  Median    Mean 3rd Qu.    Max.
##     270    2431    6430   16840   20140  191194
##    NA's
##      12
```

```
summary(gdpfert$gdp_character)
```

```
##    Length     Class      Mode
##       214 character character
```

As a character storage type, **R** is unable to calculate any summary statistics as expected. So interval or ratio scaled measures should be stored as numeric types. Nominal and ordinal measures could be stored as characters or a third type of storage, “factor”.

A factor is a hybrid of character and numeric, combining numeric scores and verbal labels. A factor has numeric scores associated with verbal labels. For example, country names may be stored numerically within a dataset, but **R** indexes the countries to a set of verbal labels associated with each country score. Factors are preferred over characters for ordinal data because the label still facilitates data analysis. For example, you can easily calculate the frequency of each category or use factors in **R** functions that require categorical (not continuous) input.

We create a new variable *region_factor* with the function `as.factor()`:

```
gdpfert$region_factor <- as.factor(gdpfert$region)
```

Then observe the storage type:

```
str(gdpfert$region_factor)
```

```
##  Factor w/ 7 levels "East Asia & Pacific",..: 2 4 6 3 2 2 7 3 1 2 ...
```

Notice in the storage type in `str()` there are seven `levels` or labels of the factor that are associated with numeric placeholders from 1 through 7.

```
summary(gdpfert$region)
```

```
##    Length     Class      Mode 
##       214 character character
```

```
summary(gdpfert$region_factor)
```

```
##        East Asia & Pacific      Europe & Central Asia 
##                         36                         56 
##  Latin America & Caribbean Middle East & North Africa 
##                         42                         21 
##              North America                 South Asia 
##                          3                          8 
##         Sub-Saharan Africa 
##                         48
```

The summary function as applied to a factor allows a tabulation of frequencies.

3.5 Constructing and merging data tables

To practice running **R** functions and applying some basic principles of organizing data tables, we will download and construct a data table from the World Bank. Then we will merge it with country level data on quality of democracy. We will use functions from the **tidyverse** set of packages and download World Bank data with the **WDI** package, which was introduced in Chapter 2 but will need to be installed if not already (`install.packages('WDI', repos='http://cran.us.r-project.org')`.

The **tidyverse** package is already attached from earlier parts of Chapter 3. We will attach the **WDI** package:

```
library(WDI)
```

We are constructing the data table from scratch, downloading data directly from the World Bank. We first need to find the indicators (variables) that interest us.

We will first search for data using the `WDIsearch()` function. We search on a specific key term, which depending on the function argument could be the World Bank's descriptive label for the indicator, the alphanumeric abbreviation such as *NY.GDP.PCAP.KD* for GDP per capita. The descriptive label for the World Bank's data is called the "name", while the "indicator" is the World Bank's term for the alphanumeric identifier of the measure.The default is to search the verbal label. We will search for any measures mentioning fertility:

```
WDIsearch(string="fertility")
```

```
##             indicator
## 17316 SP.DYN.TFRT.IN
## 17317 SP.DYN.TFRT.Q1
## 17318 SP.DYN.TFRT.Q2
##                                                             name
## 17316                        Fertility rate, total (births per woman)
## 17317  Total fertility rate (TFR) (births per woman): Q1 (lowest)
## 17318           Total fertility rate (TFR) (births per woman): Q2
```

In this text the output is abbreviated to save space. In your console the result is a list of all the indicators that include "fertility" in the name. Each separate row is an indicator, a separate measure. There are various fertility-related metrics available; some appear identical in subject matter, perhaps only differing by source data. Indicators `17316` and `17315` both record a total fertility rate. We will download data after searching for a few other measures; the `WDIsearch()` function searches the database, but does not download observations on countries.

Try searching for GDP per capita, `WDIsearch(string="GDP per capita")`. Then try searching for "life expectancy" (`string="life expectancy"`) and "income". Note that for general measures or subject matter at the core of the World Bank's business, such as GDP, the result can be several hundred indicators.

We will download data on GDP per capita and fertility. The function `WDI()` downloads the data. The required arguments are a list of countries (`"all"` for all countries) and a list of indicators, as well as a start year and end year. For a cross-sectional dataset, the `start=` and `end=` arguments are the same. In this case, we will download the data for the year 2010 and save it as the data object wbdata1. The data are downloaded and stored through the assignment operator `<-` as data object `wbdata1`.

```
wbdata1<-WDI(country="all", indicator=c("NY.GDP.PCAP.KD",
          "SP.DYN.TFRT.IN"), start=2010, end=2010, extra=TRUE)
```

Inspect the first few lines of data by entering `head(wbdata1)`. The dataset includes regional aggregate statistics. For a dataset intended to comprise of country-level observations, these regional aggregates should be excluded. We use the `subset()` function to remove aggregate totals Arab world, East Asia, etc. We subset the data by the condition `region!="Aggregates"` as the subsetting criterion. The symbol `!=` means "not equal to". So we instruct RStudio to subset the data, keeping all rows of the data that are **not** `Aggregates`.

```
wbdata1<-subset(wbdata1, region!="Aggregates")
```

The subsetted data is overwritten on the existing data object `wbdata1`. Try typing the name of the object to observe that it now no longer includes the regional aggregates.

We can create a neat presentation table with a function from the **knitr** package (Xie, 2023b) that powers RMarkdown, `kable()` (as in 'table' but with a k). The main argument to `kable()` is the name of a dataframe. While the function will produce table output in the Console window, the main purpose of it is to generate tables in RMarkdown that are knitted to PDF or Word format. Try running `kable(wbdata1)` and a large unwieldy table will scroll through the Console.

If, for example, we wished to print a table consisting of the country name, GDP and fertility rate, we could subscript the dataframe to include just these

TABLE 3.2 Table of National GDP per capita and fertility rate in 2010, created with the kable() function.

	country	NY.GDP.PCAP.KD	SP.DYN.TFRT.IN
1	Afghanistan	542.9	6.099
4	Albania	3577.1	1.656
5	Algeria	4456.6	2.843
6	American Samoa	12446.3	NA
7	Andorra	36277.3	1.270
8	Angola	3114.7	6.194
9	Antigua and Barbuda	16132.6	1.785
11	Argentina	13387.2	2.346
12	Armenia	2796.1	1.501
13	Aruba	27324.6	1.941

three columns, which occur in column positions 1, 7, and 8: `kable(wbdata1[, c(1,7,8)])`. And to limit the number of observations, we will select just the first ten rows: `kable(wbdata1[1:10, c(1,7,8)])`.

Adding `caption=` yields Table 3.2:

```
kable(wbdata1[1:10, c(1,7,8)], caption="Table of National GDP per capita
    and fertility rate in 2010, created with the kable() function.")
```

Note that when using the function in an RMarkdown document, you will need to call up the function from the **knitr** package with `knitr::`, as in `knitr::kable(wbdata1[1:10, c(1,7,8)])`, or attach the **knitr** package with `library(knitr)` prior to using the function. If knitting directly to PDF instead of Word, the argument `booktabs=TRUE` adds additional formatting lines.

A simple way to make a quick scatterplot is with `qplot()`, a function from the **tidyverse** suite of packages. Enter the name of the X and Y variables, and identify the dataset. The result is Figure 3.1:

```
qplot(NY.GDP.PCAP.KD, SP.DYN.TFRT.IN,  data=wbdata1)
```

The function picks a plot type given the measurement qualities of the input data; in this case it produces a scatterplot since the two measures are stored as numeric. We have constructed a tidy data table and created a visualization based on two of the measures. Next we will combine data from multiple data tables.

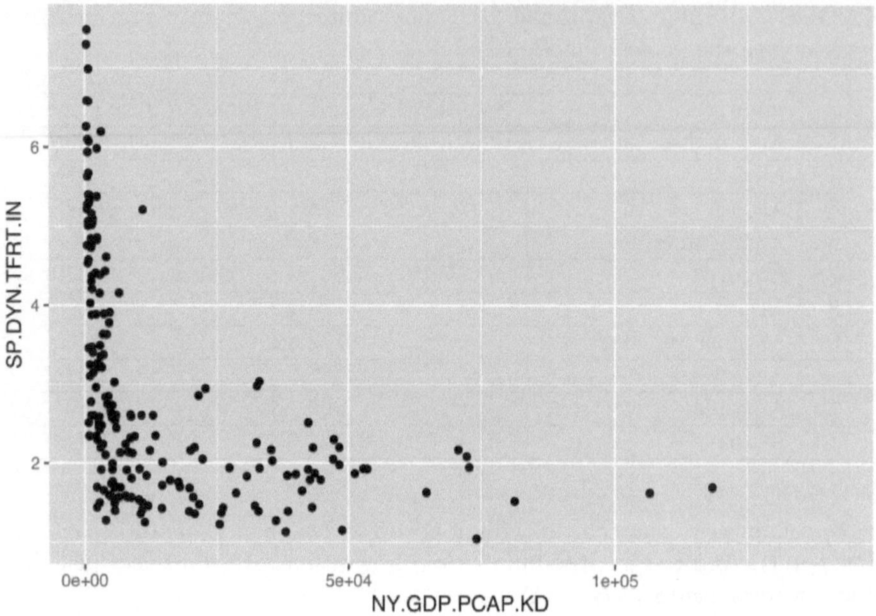

FIGURE 3.1 Scatterplot of country-level fertility rate (births per woman) by GDP per capita (constant 2010 USD). GDP per capita is right-skewed, with most countries clustered at lower values and a few outliers at high levels. This skew compresses the majority of observations along the X-axis, making it difficult to discern the relationship between GDP and fertility.

Merging data tables

Researchers often need to combine data from multiple sources, for example where different data tables hold different indicators or measurements on units of interest. Many different research programs collect international-level data, releasing extensive tables containing multiple political, economic, and social conditions of countries around the world. Yet despite the comprehensiveness of different sources, it may be necessary to merge together the sources.

For example, in a country-level dataset one source contains various measures of public health, while another contains measures of the quality of democracy, and the challenge is to combine the two data tables into one. If the units of observation in each data table match exactly and are few in number, a merge is feasible with a simple spreadsheet. Even when merging one to one, where each row from one table matches up exactly with a row from another table, cut-and-paste manipulation of a spreadsheet is error prone. And once the units differ slightly from table to table, or cover multiple years, such a merge quickly becomes unwieldy and requires the use of a merge function in code.

Merges are performed pairwise — two tables are compared at a time. And before beginning, start the merge with the end product in mind. For example, will the merged dataset be created as a new data table, one that contains only rows in common between the two tables? Or will the rows of the merged dataset contain all rows of one table, combined with any matching rows from another? Or perhaps are there many rows from one table to match with single rows from another table, a so-called "many to one" match?

After thinking ahead to the type of merge, the next step is to identify how the tables will be merged. It requires a "key" field (a column or set of columns in the data table) that uniquely identifies each row — each unit of observation. For example, in merging two data tables collected at the country level, the key field could be the name of each country, but if the data tables included multiple years of observations, then the key field would be the country name and year. (In the language of databases, the key field in the main table is the "primary" key, while in a second table – to merge on to the main table – the key is referred to as a "foreign" key. See Wickham and Grolemund (2016) for a further explanation.)

As an example, we will work with a small subset of the Varieties of Democracy (Coppedge et al., 2022b) core dataset (Coppedge et al., 2022a), two indicators, measures of polyarchy and liberal democracy. And we will merge these country-level indicators with a data table containing the United Nations Development Programme's Human Development Indicators (HDI) index (2022)[2].

We will construct a data table containing only the rows (observations) that match within each source table; we will see that this type of merge reduces the total number of countries included within the data, to the subset of countries found within both tables. Then we will compare this table to a merge that preserves the country rows from either table.

An excerpt from the Varieties of Democracy data is the file *vdem_data.csv*. So we first read in the excerpted data with the `read_csv()` function from the *tidyverse* package:

```
vdem_data<-read_csv(file="vdem_data.csv")
```

And we can see it consists of 179 rows and four columns:

```
vdem_data
```

```
## # A tibble: 179 x 4
##    country_name country_text_id v2x_libdem v2x_polyar~1
##    <chr>        <chr>                <dbl>        <dbl>
```

[2]See the Human Development Reports data center https://hdr.undp.org/data-center/human-development-index#/indicies/HDI.

```
## 1 Afghanistan  AFG                 0.021      0.16
## 2 Albania      ALB                 0.403      0.478
## 3 Algeria      DZA                 0.145      0.284
## 4 Angola       AGO                 0.19       0.351
## 5 Argentina    ARG                 0.658      0.819
## 6 Armenia      ARM                 0.556      0.739
## 7 Australia    AUS                 0.808      0.854
## 8 Austria      AUT                 0.749      0.825
## 9 Azerbaijan   AZE                 0.068      0.194
## 10 Bahrain     BHR                 0.052      0.122
## # ... with 169 more rows, and abbreviated variable
## #   name 1: v2x_polyarchy
```

We are merging it with the HDI, in the comma-separated value file *hdi_2021.csv*. Read in the HDI data to see what might potentially be a matching key field for `vdem_data`:

```
hdi_2021<-read_csv(file="hdi_2021.csv")
```

Then the HDI file consists of 191 columns and six rows:

```
hdi_2021
```

```
## # A tibble: 191 x 6
##    Country              hdi life_~1 expec~2 mean_~3 gni_pc
##    <chr>              <dbl>   <dbl>   <dbl>   <dbl>  <dbl>
##  1 Afghanistan        0.478    62      10.3     3     1824
##  2 Albania            0.796    76.5    14.4    11.3  14131
##  3 Algeria            0.745    76.4    14.6     8.1  10800
##  4 Andorra            0.858    80.4    13.3    10.6  51167
##  5 Angola             0.586    61.6    12.2     5.4   5466
##  6 Antigua and Ba~    0.788    78.5    14.2     9.3  16792
##  7 Argentina          0.842    75.4    17.9    11.1  20925
##  8 Armenia            0.759    72      13.1    11.3  13158
##  9 Australia          0.951    84.5    21.1    12.7  49238
## 10 Austria            0.916    81.6    16      12.3  53619
## # ... with 181 more rows, and abbreviated variable
## #   names 1: life_expect, 2: expect_school,
## #   3: mean_school
```

Combining rows of two tables

The process of merging begins with identifying the key fields. In the *hdi_2021* dataframe, the only possibility is "Country". The *vdem_data* frame includes both a three character abbreviation and "country_name". The starting point for merging the files is to find matches (and the observations that do not match) in comparing "Country" to "country_name". For the functions to find the matching exact rows within the key fields, the country names have to be exact matches in capitalization, spacing, and use of any additional characters such as parentheses or slash marks. So, for example, in an exact match "Vietnam" would not be identified as matching "Viet Nam".

Before merging the files together, it is often useful to first inspect which rows (countries) are not matched.

This detection works in two directions:

1) from the VDEM data to find countries not present in the HDI data; and
2) from the HDI data to find countries not present in the VDEM data.

To find the rows that do not match, we use the `anti_join()` function from the `tidyverse` package. Like other join functions from the package, the `anti_join()` function requires three arguments, two dataframes and the key field, `anti_join(DATAFRAME1,DATAFRAME2, KEY_FIELD)`. The function will identify all rows in *DATAFRAME1* that are missing from *DATAFRAME2*. For example, to identify the rows from *DATAFRAME1* that do not match, we enter the names of the datasets to compare, then a `by=c("KEY_FIELD1" = "KEY_FIELD2")` to identify the names of the key fields that should be compared for matches. If the names of the key fields were the same across the two datasets, the identification of the key field would be `by="KEY_FIELD"`. In this example, the `anti_join()` function specifies the two dataframes, `vdem_data`, `hdi_2021`, followed by the key fields corresponding to the dataframes in order, `by=c("country_name" = "Country"))`.

Let's see how many mismatches are identified:

```
anti_join(vdem_data, hdi_2021, by=c("country_name" = "Country"))
```

```
## # A tibble: 8 x 4
##   country_name          country_text_id v2x_lib~1 v2x_p~2
##   <chr>                 <chr>               <dbl>   <dbl>
## 1 Kosovo                XKX                 0.421   0.601
## 2 North Korea           PRK                 0.015   0.085
## 3 Palestine/Gaza        PSG                 0.06    0.134
## 4 Palestine/West Bank   PSE                 0.143   0.253
## 5 Somalia               SOM                 0.093   0.163
## 6 Somaliland            SML                 0.259   0.413
## 7 Taiwan                TWN                 0.699   0.813
## 8 Zanzibar              ZZB                 0.198   0.263
## # ... with abbreviated variable names 1: v2x_libdem,
```

```
## #   2: v2x_polyarchy
```

The tables shows the eight rows — eight countries — of the VDEM dataset that do not match any rows in the HDI dataset. To find the reverse, the countries in the HDI dataset that do not appear in the VDEM dataset reverse the order of the datasets and key fields : `anti_join(hdi_2021, vdem_data, by=c("Country"="country_name"))`. Try running this reversed `anti_join()` function to observe how many countries are not matched in *vdem_data* from *hdi_2021*.

From the `anti_join()` function, it's clear that the intersection of the two datasets, the matching country observations, will be fewer than either single dataset. To find the intersection of the two datasets, we use the `inner_join()` function. The function is symmetric with respect to which observations (rows) are selected, so in that sense it does not matter which dataset is specified first in the function:

```
inner_join(hdi_2021, vdem_data, by=c("Country"="country_name"))
```

```
## # A tibble: 171 x 9
##    Country   hdi life_~1 expec~2 mean_~3 gni_pc count~4
##    <chr>   <dbl>   <dbl>   <dbl>   <dbl>  <dbl> <chr>
##  1 Afghan~ 0.478    62      10.3     3     1824 AFG
##  2 Albania 0.796    76.5    14.4    11.3  14131 ALB
##  3 Algeria 0.745    76.4    14.6     8.1  10800 DZA
##  4 Angola  0.586    61.6    12.2     5.4   5466 AGO
##  5 Argent~ 0.842    75.4    17.9    11.1  20925 ARG
##  6 Armenia 0.759    72      13.1    11.3  13158 ARM
##  7 Austra~ 0.951    84.5    21.1    12.7  49238 AUS
##  8 Austria 0.916    81.6    16      12.3  53619 AUT
##  9 Azerba~ 0.745    69.4    13.5    10.5  14257 AZE
## 10 Bahrain 0.875    78.8    16.3    11    39497 BHR
## # ... with 161 more rows, 2 more variables:
## #   v2x_libdem <dbl>, v2x_polyarchy <dbl>, and
## #   abbreviated variable names 1: life_expect,
## #   2: expect_school, 3: mean_school,
## #   4: country_text_id
```

The results show the 171 matching rows from each dataset, with columns from *hdi_2021* first followed by *vdem_data*. The functions for these joins find the respective rows, but do not automatically save the results into a new dataframe. The new dataframe must be specified with an object name and the assignment operator. To name the merged data *hdi_vdem*:

```
hdi_vdem<-inner_join(hdi_2021, vdem_data,
                     by=c("Country"="country_name"))
```

The order of the columns in the new dataset appears in the order specified in the function, *hdi_2021*, followed by *vdem_data*, including the column "country_text_id" from *vdem_data*, but only the key field from *hdi_2021*.

Merging columns on an existing table

While `inner_join()` finds the intersecting observations, other functions add any matching observations onto an existing dataset — to preserve all row observations within the existing table.

A `left_join()` of `hdi_2021, vdem_data`, starts at the data on the left-hand side, *hdi_2021*:

```
left_join(hdi_2021, vdem_data, by=c("Country"="country_name"))
```

The function adds countries from *vdem_data* to *hdi_2021* — keeping all rows of *hdi_2021*.

Notice that the row entries for the two democracy indicators of *vdem_data* include a few missing values, `<NA>` entries, for the countries in the *HDI* data not included in it.

This `left_join()` would be useful to merge any additional datasets onto the *hdi_2021* dataset. For example,

```
hdi_2021<-left_join(hdi_2021, vdem_data, by=c("Country"="country_name"))
```

would replace the existing *hdi_2021* with the `left_join()` version. A complementary function, `right_join()` would keep observations in `vdem_data` and add rows to it from `hdi_2021`.

For most purposes, these inner, left and right joins are sufficient. For a more detailed discussion of different types of data merges, see Wickham and Grolemund (2016). Note that for data tables in which the units of observation in rows within two tables match each other exactly in their row order, the function `cbind()` appends one dataset onto another by adding columns; hence "c" is for "column". (To append rows to a table, in which the order of the columns match exactly, the row bind, or `rbind()` will add rows.)

3.6 Measurement characteristics: center and spread

Given a clean data table, summary statistics are the starting point for a data analysis. While fundamental, the statistics provide insight into a typical ('average' value) or by how much the measures vary. With the two measures from the assembled World Bank dataframe *wbdata1*, such statistics could help answer questions such as, "What is the typical fertility rate globally or in regions?" and "How much does fertility vary within those regions?" To be clear, we will define these measures and review functions for calculating each one.

Mean, median, and mode

We have already seen the mean from `summary()` applied to a numeric variable. The arithmetic mean (the "average") is the most commonly used measure of central tendency. Given a quantitative numeric measure, interval or ratio in scale (such as a country's life expectancy), represented by x, on a total of n different units of observation: $x_1, x_2, x_3, ..., x_n$, the mean is calculated as

$$\bar{x} = \frac{x_1 + x_2 + x_3 + ... + x_n}{n} = \frac{1}{n}\sum_{i=1}^{n} x_i$$

where $\bar{x}$ is the mean, n is the number of values or units of observation, and x_i is the ith observation value.

Beyond the `summary()` function, the `mean()` function will calculate a mean, but it requires an argument you will see repeatedly when working with election survey data, `na.rm=TRUE`, which ensures that missing values are dropped from the calculation. `mean(wbdata1$NY.GDP.PCAP.KD, na.rm = TRUE)`. (The default in `summary()` is to drop missing observations.)

The median is another measure of central tendency, but one that is less sensitive to highly unusual observations such as high GDP countries, the observations often referred to as "outliers". Like the mean, in **R** the `summary()` function reports the median, as does `median()` with `na.rm=TRUE()` as needed. The median accounts for the middle position within a set of (n) observations that are ordered from lowest to highest (or the reverse), separating the lower half of the data from the upper half. The median is also the 50th percentile. Conceptually, the median is simply the middle value if the number of observations is odd in total number, and if even, the median is the average of the two middle values. The median is a positional measure of central tendency; it only accounts for the order of the observations, not the values of the observations like the mean. A graph displaying the mean and median of GDP per capita is shown in Figure 3.2.

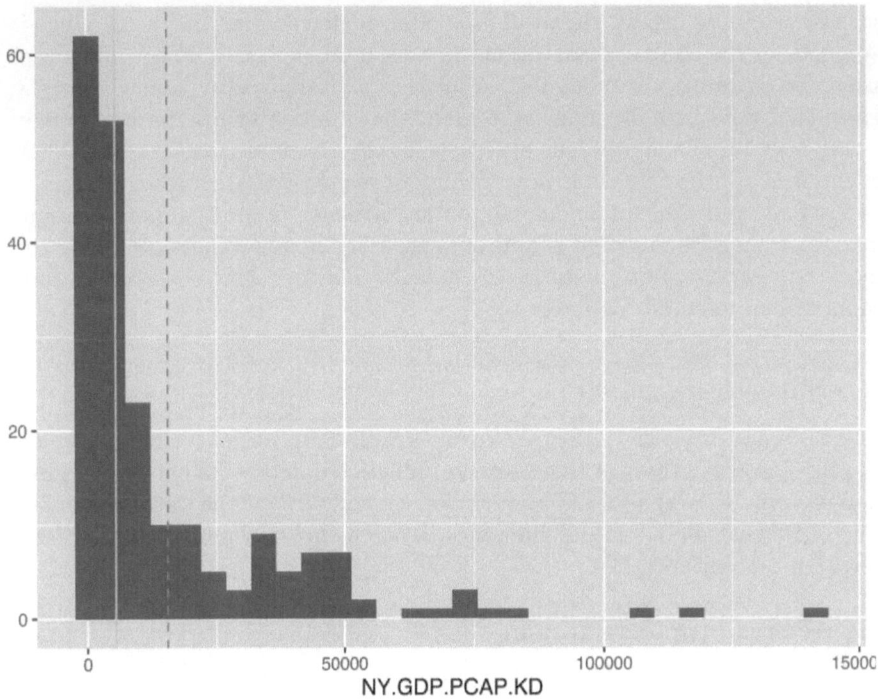

FIGURE 3.2 Histogram of country-level GDP per capita (constant 2010 USD) for 2010. The dashed vertical line marks the mean (14,947 USD), while the solid vertical line marks the median (5,206 USD). The large gap between the mean and median reflects the right-skewed distribution of GDP per capita, where a small number of wealthy countries pull the mean far above the median.

The mean and median are graphed as lines intersecting the X-axis by adding an additional layer separated by a + symbol, drawn with `geom_vline()`. Most countries fall below the mean level of 15,538 USD, but the skewed distribution of GDP per capita — with some countries having much larger economies than the rest of the world — is pulling higher than the median. The median level 5,417 USD is less than half the mean, showing that half of the countries have a GDP per capita at or less than this value.

One last measure of central tendency is the **mode**. The mode is simply the most frequent observational value. For a continuous, ratio-level measure such as GDP per capita, the modal value is not very illuminating. But for a nominal (versus ordinal or interval/ratio) measure, it is the only appropriate measure of central tendency.

Different measures of central tendency are appropriate for different levels of measurement. For a continuous, interval or ratio level, measure, the mean, median, and mode are all valid summary statistics. For highly skewed measures, such as GDP per capita, the median is often preferred over the mean, since it accounts for the order of observations, not the numeric values. In the absence of skew, the mean may be preferable. A mode is unlikely to provide much insight, given that the range of values is so great that a mode will be proportionally a tiny fraction of the different observations. For ordinal measures, a median or mode is appropriate. It is not uncommon to see ordinal measures treated as interval, and a mean calculated. In the absence of an assumption about interval scaling, the median is appropriate. With nominal measures, only the mode is an appropriate measure of central tendency, since the observations cannot be ordered on the measure.

Interquartile range

The interquartile range (IQR) encompasses both data centrality and dispersion. It, too, is reported through the `summary()` function applied to numeric measures. It represents the range within which the central 50% of the data values fall. The IQR is calculated as the difference between the 75th percentile (Q3) and the 25th percentile (Q1).

```
summary(wbdata1$NY.GDP.PCAP.KD)
```

```
##    Min. 1st Qu.  Median    Mean 3rd Qu.    Max.
##     290    2116    5417   15538   20641  141815
##    NA's
##       9
```

Percentiles (and the nearly identical "quantile") divide a numeric variable into groups (counts of observations) of equal size, providing insight into the relative position of a value compared to the rest of the variable. The `summary()` function divides a variable into quartiles, marked by the 25th, 50th, and 75th percentiles.

The `quantile()` function will calculate quantiles with an input argument of specific values to create. By default, it calculates quartiles. For example, to find three equal-sized groups of observations: `quantile(wbdata1$NY.GDP.PCAP.KD, c(0,1/3,2/3, 1), na.rm=TRUE)`. In the argument `c(0,1/3,2/3, 1)` the second entry `1/3` is evaluated as close to .3333. Or specifying ten equal groups (deciles) would require the argument `c(0,.1,.2,.3,.4,.5,.6,.7,.8,.9,1)`. Thus the IQR is a calculation built from quartiles. The IQR encompasses both data centrality and dispersion. It is reported through the summary() function applied to numeric measures and represents the range within which the central 50 percent of the data values fall. The IQR is calculated as the difference between the 75th percentile (Q3) and the 25th percentile (Q1). A quantile of .50 corresponds to the 50th percentile, the only difference being the scale. The 50th percentile for GDP tells us that at the value of 5,417 USD marks the midpoint in the distribution of GDP per capita around the world. The

0.25 quantile of 2,116 USD (Q1 or `1st Qu.`) marks the point below which 25 percent of the data fall, while the .75 quantile 20,641 USD (Q3 or `3rd Qu.`) marks the point below which 75 percent of the data fall. This division of the data into four groups (quartiles) of equal size — the 25th, 50th, and 75th quantile — yields the IQR, the difference between Q3 and Q1.

IQR = (20641 - 2116) = 18525

There is an `IQR()` function in **R**, but of course it can be easily gleaned from the `summary()`. The IQR provides a measure of the spread of the middle half of the data, one which is less sensitive to extreme values (outliers) than other measures of data spread reviewed further below. Thus the IQR is especially useful in understanding the variability in skewed distributions or datasets with outliers. In the case of GDP, the IQR range of 18,528 USD shows us that fully half of the countries fall within the range of the IQR of 2,116 to 20,641. Comparing the width to the minimum and maximum shows how skewed the distribution is, with typical countries falling relatively much lower on the scale compared to the maximum.

Measuring spread

While average values are important to investigate, another characteristic is the spread – how dispersed or spread out from the average is a measure? For example, two countries could have similar average or arithmetic mean income levels, but strongly differ by the degree of income inequality, a difference that would be measured in spread or dispersion.

Range is the simplest measure, the difference between the largest and smallest value on a set of measurements. The range is useful for comparing overall variation across a measure, and of course is easily observable from the `summary()` function. The `Min.` and `Max.` output correspond to the 0th percentile and 100th percentile, respectively. The range of 141,815-290 USD, or 141,525 USD, relative to the median value of 5,417 USD reveals the highly skewed distribution of GDP per capita.

The median absolute deviation (MAD) is a measure of disperson relative to the median. The measure is constructed from 'absolute' deviations of each observation from the median, the absolute difference between each observation and the median of the distribution. Then the median — of the absolute deviations — summarizes the spread. Formulaically, the MAD is as follows:

$$\text{MAD} = \text{median}(|x_i - \text{median}(x)|)$$

where x_i is an observation on a variable and $median(x)$ represents the median of the variable x. Then $\text{median}(|x_i - \text{median}(x)|$ calculates the absolute deviation of each observation from the median. Like the median itself, as a measure of dispersion the MAD is less sensitive to outliers.

The function for calculating a MAD is the `MAD()` function:

```
mad(wbdata1$NY.GDP.PCAP.KD, constant=1, na.rm=TRUE)
```

```
## [1] 4555
```

Note that by default, the `mad()` function in **R** 'scales up' the MAD to make it comparable to the standard deviation, a dispersion measure discussed below. Adding the argument `constant = 1` calculates the raw MAD as in the formula. The argument `na.rm=TRUE` removes missing values from the calculation.

The standard deviation

By far, the two most common and important measures of dispersion are the standard deviation and its related quantity, the variance. Both measures capture spread relative to the mean. Let's break down conceptually the "deviation" and the "standard". The "deviation" refers to the difference between an observation, x_i and the distribution mean $\bar{x}$. Some observations will be above the mean and others below it, of course, and by definition of the mean the sum of the deviations $\sum_{i=1}^{n}(x_i - \bar{x})$ must equal zero. So to sum up the total deviations, the standard deviation measures the squares of the deviations, $\sum_{i=1}^{n}(x_i - \bar{x})^2$. This quantity is also known as a sum of squares, which we will return to in later chapters. The 'squared' units of this metric is known as the variance, discussed below. By taking the square root $\sqrt{\sum_{i=1}^{n}(x_i - \bar{x})^2}$ the quantity is converted back into its original unit of measurement. Finally, by averaging out this total over the total number of observations, n, the standard deviation measures an average distance of each observation from the mean, a measure of population-level standard deviation. For a sample of n observations, the divisor in the sample standard deviation is $n - 1$. The sample standard deviation sd is:

$$sd = \sqrt{\frac{\sum_{i=1}^{n}(x_i - \bar{x})^2}{n - 1}}$$

The sample standard deviation and variance are divided by n-1 instead of n (as an average) to correct for bias in the use of sample data to estimate population-level standard deviation and variance. The adjustment is meant to account for the fact that sample data tend to underestimate these quantities. (Consult a statistics textbook for a further explanation.) The base **R** `sd()` function calculates a standard deviation and assumes the data at hand are sample data.

```
sd(wbdata1$NY.GDP.PCAP.KD, na.rm=TRUE)
```

```
## [1] 22297
```

The result, 22,297 USD indicates the average amount by which an individual country differs from the mean level of GDP per capita.

The variance is simply the squared standard deviation — or in the formula above, the sum of the standard deviations divided by $n - 1$.

$$var = \frac{\sum_{i=1}^{n}(x_i - \bar{x})^2}{n - 1}$$

In base *R*, calculated directly by the `var()` function or squaring the standard deviation `sd(wbdata1$NY.GDP.PCAP.KD, na.rm=TRUE)^2`, the answers are identical.

```
var(wbdata1$NY.GDP.PCAP.KD, na.rm=TRUE)
```

```
## [1] 497169686
```

Thus in squared GDP per capita units, the variance measures the dispersion as 497,169,686 "squared US dollar GDP per capita".

The empirical rule for standard deviations

For distributions that are approximately bell-shaped, unimodal, and symmetric, such as Figure 3.3, the standard deviation has a particular property that is useful in understanding data.

For a symmetric, bell shaped distribution, from the mean:

- Approximately 68 percent of the observations fall within one standard deviation, between $\bar{x} - s$ and $\bar{x} + s$.
- Approximately 95 percent of the observations fall within two standard deviations, between $\bar{x} - 2s$ and $\bar{x} + 2s$.
- Approximately 99.7 percent (or nearly all) of the observations fall within three standard deviations, between $\bar{x} - 3s$ and $\bar{x} + 3s$.

Figure 3.3 displays the bell-shaped, symmetrical 'normal' distribution, which is a theoretical construct – an idea based on infinite sampling. In the real world, while not perfectly normal, data distributions can be approximate enough that the empirical rule provides a close enough reference.

The `dnorm()` function will draw the contour of a 'normal' curve given an input mean, standard deviation, and a set of X values over which to draw on Y the curve, representing the density of the area under different portions of the curve.

For example, the curve in Figure 3.3 is drawn from first setting the values on X with `seq()`: `Xvalues <- seq(-4, 4, length = 1000)`. The large number

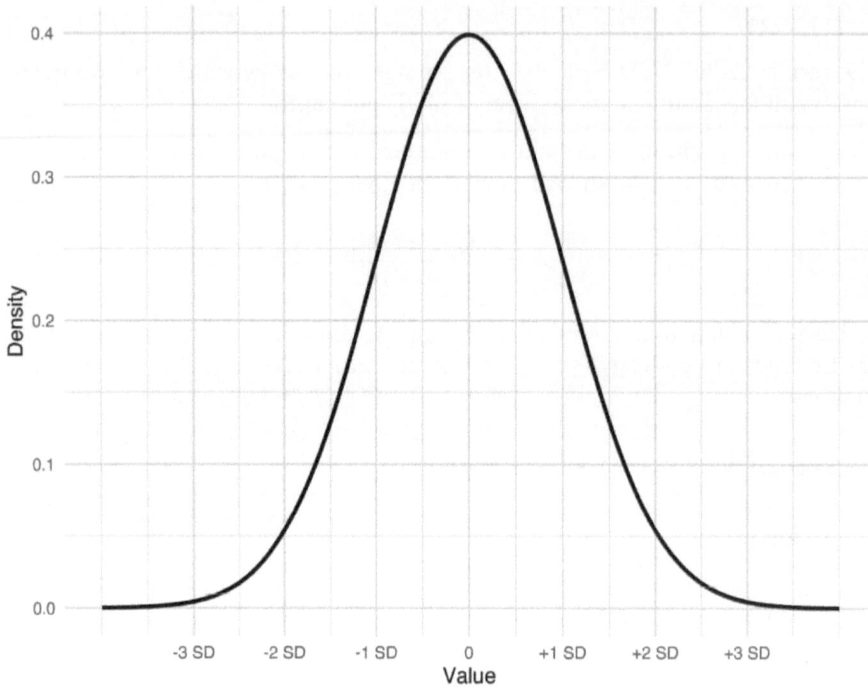

FIGURE 3.3 Bell-shaped or 'normal' curve displaying the mean plus or minus one, two, and three standard deviations. This curve, with mean 0 and a standard deviation of 1 is often referred to as a 'standard normal' curve.

of points (`1000`) on X ranging from `-4` to `4` ensures a smoothly drawn curve. The function `dnorm()` draws the curve over the X values, by default with a mean of 0 and a standard deviation of 1,`dnorm(Xvalues)`. (The default can be changed by substituting in values for the mean and standard deviation; for example, a mean of 2.5 and a standard deviation of .7 would be `dnorm(2.5, .7)`.) Storing the result from `dnorm(0)` as `Yvalues`, then the curve can be drawn with `qplot()`.

```
Xvalues <- seq(-4, 4, length = 1000)
Yvalues <- dnorm(Xvalues)
qplot(Xvalues, Yvalues)
```

These three lines will draw the same curve as in Figure 3.3, without the labels.

3.7 Measuring relationships between variables

Covariance and Pearson's correlation coefficients are fundamental measures to describe the relationship between two continuous variables. Covariance quantifies how much two variables change together. If the covariance is positive, the variables tend to increase or decrease together; if negative, one variable increases while the other decreases.

The formula for the covariance between two variables X and Y is

$$cov(X,Y) = \sum_{i=1}^{n}(x_i - \bar{x})(y_i - \bar{y})\frac{1}{n}$$

The covariance captures the average product of the deviations of X and Y from their respective means, $\bar{X}$ and $\bar{Y}$, whether large values of X tend to correspond to large values of Y, and similarly for small values.

The `cov()` function calculates covariance. For example, in the merged *hdi_vdem* dataset, the covariance between the quality of liberal democracy and gross national income (GNI) per capita is:

```
cov(hdi_vdem$v2x_libdem, hdi_vdem$gni_pc)
```

```
## [1] 2750
```

The result is 2750, which is not in interpretable units, but does show that the two variables tend to move in the same direction. Because covariance is measured in the combined units of the two variables, it is not readily intuitive to interpret across different measurement scales. From the covariance, Pearson's r correlation is a standardized, more intuitive metric.

The formula for a Pearson's r correlation is essentially the covariance of the two variables divided by the product of their standard deviations.

$$r = \frac{\sum_{i=1}^{n}(x_i - \bar{x})(y_i - \bar{y})}{\sqrt{\sum_{i=1}^{n}(x_i - \bar{x})(y_i - \bar{y})}}$$

Pearson's correlation coefficient standardizes the covariance by dividing it by the product of the standard deviations of the two variables. The correlation is a standardized measure of the strength of linear association, a value that ranges from -1 to 1, where 1 indicates a perfect positive linear relationship, -1 indicates a perfect negative linear relationship, and 0 suggests no linear relationship. The `cor()` function calculates the correlation:

```
cor(hdi_vdem$v2x_libdem, hdi_vdem$gni_pc)
```

```
## [1] 0.5275
```

The result, .5275, suggests only a moderate correlation between democracy and income.

3.8 Measurement scaling: z, linear, and log

Beyond percentiles, another measure of an observation's position on a distribution is the z score. The z score is also commonly used to facilitate the comparison of distributions on different measurement scales. To measure position or to facilitate comparisons on different scales, the z score combines information about an observation's distance from the mean and the distribution's standard deviation – to measure position in standard deviation units. To illustrate the use of a z score, consider the following example with World Bank data.

In 2020 the USA's life expectancy was 77.3 years. To make sense of this life expectancy (is it relatively long? short?), we would need to know at least the mean life expectancy around the world. That figure is 72.3 years. So we know the USA is 5 years above the mean, but is this higher-than-average life expectancy unusual or not? Of course, we could visually skim the data to get a sense of it, but for any level of precision, we need to know the standard deviation of life expectancy as well — how spread out was life expectancy around the world in 2020? The meaning of the USA's life expectancy changes considerably if we knew the standard deviation of life expectancy was, for example, 5 years versus 10 years. For a standard deviation of 5 years, the USA's life expectancy would be $5/5 = 1$ standard deviation above the mean. Yet, for a standard deviation of 10, the USA's life expectancy would be $5/10$ only .5 standard deviations above the mean — much closer to the mean.

Supporting this intuition is the calculation of a z score. In this example of life expectancy in 2020, the standard deviation is 7.6 years. The USA's life expectancy is $77.3 - 72.3 = 5.0$ years above the mean. The USA's distance from the mean is $5/7.6 = .66$ standard deviations above the mean; this measure of the USA's life expectancy relative to the rest of the world, .66, is the USA's z score:

$$z = \frac{observation - mean}{standard\ deviation} = \frac{77.3 - 72.3}{7.6} = .6579$$

More generally, the formula for a z score is

$$z = \frac{x_i - \bar{x}}{s}$$

In this formula, x_i is an observation from a distribution, $\bar{x}$ is the distribution mean, while s is the distribution standard deviation. The z score measures the distance of an observation from the mean, per standard deviation — for any observation, the number of standard deviations from the mean. Thus in these units, any variable converted to a z distribution will have a mean of 0 and a standard deviation of 1. In converting to z scores, a z distribution now has a mean of 0 and a standard deviation of 1.

One additional example shows why the z score is essential for comparing position across different distributions. For example, how does the USA's better than average (or z= .66) life expectancy compare to the USA's position in the world on fertility rates, given a fertility rate of 1.64 average births per woman in 2020?

To answer we construct a z score, given the mean of the corresponding z score for the USA is $(1.64 - 2.55)/1.29 = -0.70$ standard deviations below the mean. So, in addition to being below the mean fertility rate, the USA's fertility rate is more unusual than its life expectancy. Z scores provide insight into measures collected over time. For example, while we can observe that the USA's life expectancy grew from 69.8 years in 1960 to 77.3 years in 2020, to know how the USA's 2020 standing in the world compares to 1960 requires the calculation of a z score: $(69.8 - 54.0)/11.78 = 1.34$ standard deviations above the mean, illustrating how the rest of the world caught up to the USA.

In base R, the function `scale()` will construct a z transformation of a variable. For example:

```
wbdata1$zgdp<-scale(wbdata1$NY.GDP.PCAP.KD)

summary(wbdata1$zgdp)
```

```
##          V1
##  Min.   :-0.684
##  1st Qu.:-0.602
##  Median :-0.454
##  Mean   : 0.000
##  3rd Qu.: 0.229
##  Max.   : 5.663
##  NA's   :9
```

The positive skew is revealed in the comparison of quartiles. The 1st quartile is a z score of -.596, and a median at -.453 showing that fully 50 percent of the countries are nearly half a standard deviation below the mean. We can use tools from Chapter 5 to sort the data by the z scores and print a table.

```
wbdata1<-as_tibble(wbdata1)

wbdata1 %>%
  arrange(desc(zgdp)) %>%
  select(country, zgdp, NY.GDP.PCAP.KD) %>%
  print(n=15)
```

```
## # A tibble: 215 x 3
##    country          zgdp[,1] NY.GDP.PCAP.KD
##    <chr>               <dbl>          <dbl>
##  1 Monaco               5.66        141815.
##  2 Bermuda              4.61        118383.
##  3 Luxembourg           4.08        106544.
##  4 Switzerland          2.95         81315.
##  5 Cayman Islands       2.87         79483.
##  6 Macao SAR, China     2.63         74159.
##  7 Norway               2.57         72804.
##  8 Qatar                2.55         72334.
##  9 Isle of Man          2.48         70748.
## 10 Channel Islands      2.21         64727.
## 11 Australia            1.71         53579.
## 12 United States        1.67         52821.
## 13 Denmark              1.60         51209.
## 14 San Marino           1.53         49673.
## 15 Singapore            1.49         48752.
## # ... with 200 more rows
```

The code is syntax from packages in the **tidyverse**. In the first line, the `as_tibble()` function simply assigns the *tibble* print format. In what follows, each line is chained together by a `%>%` symbol. In English, the code would correspond to saying, "take the *wbdata1* dataframe, and then arrange it by descending order of *zgdp* and then keep only the variables *country*, *zgdp*, and *NY.GDP.PCAP.KD*, and then print the first 15 observations of the dataset".

The table clearly displays how 'unusual' or how much of an outlier is any particular observation. Take for example Monaco, while a per capita GDP of 132,094 USD is clearly large, the fact that Monaco is over 5 standard deviations above the mean is what makes it so much of an outlier.

Further exploring the empirical rule and z distributions

Regardless of the original scaling of the distribution, no matter its mean and standard deviation, when transformed to a standardized z score, the z-transformed distribution will always have a mean of 0 and a standard deviation of 1. An observation with a z score of 0 means that the observation is exactly at the distribution mean, and a z score of -1 (or +1) means that the observation is exactly one standard deviation away from the distribution mean. A z score communicates not just information about the individual observation, but also the mean and the standard deviation.

The function `pnorm()` will calculate the area to the left of a z score, the "cumulative distribution function" (CDF), with the z score as the sole argument to the function. In probabilistic terms, it returns the probability that a standard normal random variable (mean = 0, standard deviation = 1) is less than or equal to that value. For example, for a z score of 1.34:

```
pnorm(1.34)
```

```
## [1] 0.9099
```

The result, 0.9099, is the probability of a standard normal random variable being less than or equal to 1.34 is about .91 or 91%. In the language of area under the z curve, we could say that about 91% of the values in a standard normal distribution are less than or equal to 1.34. Also the z score of 1.34 is about at the 91st percentile of the standard normal distribution.

Subtracting the function from 1 provides the upper proportion:

```
1-pnorm(1.34)
```

```
## [1] 0.09012
```

Here, .0901, or about 9% of observations would fall above a z score of 1.34 in a standard normal distribution.

Measurement scaling: linear versus log

In each graphic and table presented thus far, all measurements have been presented in a linear scale. For example, life expectancy and fertility rates are both measured in numbers corresponding to how we would intuitively choose to count such concepts — births incrementing in countable averages, from 6.099 births per woman in Afghanistan, to 1.656 in Albania, to 2.843 in Algeria. This measurement of the concept of a fertility rate is placed on a linear scale — imagine a number line – in which the distance between any two fertility rates is equal to the numerical difference between the values they represent. For example, all along a number line representing the range of fertility rates, all equal intervals along the line correspond to equal differences in the measured fertility rates.

A logarithmic scale differs from a linear scale. In a logarithmic scale the distance between two observations increases exponentially as the values they represent increase — the distance is not constant as in a linear scale. The key to understanding the use of logarithmic scales is that equal intervals on a logarithmic scale represent equal ratios rather than equal differences.

For example, consider the difference between a value of 10 and 100 on a linear scale; the difference is 90 units. However, on a logarithmic scale, specifically using a base-10 logarithm (explained further below), the difference is represented by the ratio 10:100. This ratio simplifies to 1:10, meaning that each step on the scale represents a tenfold increase. In other words, moving from 10 to 100 on a logarithmic scale covers the same distance as moving from 1 to 10, because both represent an increase by a factor of 10. Why would researchers use a logarithmic scale? One reason, with respect to the GDP per capita measure in which Monaco was more than 5 standard deviations above the mean, is that it allows for a more 'compact' representation of highly skewed data, making it easier to visualize and compare values that differ greatly in size.

Log scaling, or logarithms, starts from different bases. A 'logarithm' or 'log' of a country's GDP is the number to which we need to exponentiate (raise) the base to recover that GDP. For example, let's compare the GDP per capita of Afghanistan (543 USD) with that of Monaco (13,2094 USD) and Sweden (48,370 USD). The two most used logs are a base-10 log (also known as a 'common log') and a base-e log (also known as a 'natural log').

Let's start with Afghanistan. Given a base of 10, the log of 543 USD is the number to which we need to raise 10 in order to recover 543. In base **R** logs can be calculated easily with the `log()` function. The functon arguments are the number to apply the logarithm, followed by the base:

```
log(543, base=10)
```

```
## [1] 2.735
```

Here, $log_{10}(543) = 2.735$. To observe that this quantity is, in fact, the number to which we raise 10 could be verified as $10^{2.735}$ or in **R** `10^2.735` which is approximately 543, differing only due to rounding of the logarithm. (The 543 USD can be recovered exactly in **R** with $10^{log(543,10)}$). Try calculating the base-10 logarithms for Monaco and Sweden.

Compiled together into a data table shows the differences in scaling. a dataframe from the calculations (using the `as.dataframe()` function) to illustrate the differences in scaling:

```
country <- c("Afghanistan", "Sweden", "Monaco")
GDP <- c(543,48370,  132094)
logGDP <- c(2.735,4.685,5.121 )

data.frame(country, GDP, logGDP)
```

```
##       country    GDP logGDP
## 1 Afghanistan    543  2.735
## 2      Sweden  48370  4.685
## 3      Monaco 132094  5.121
```

Compare the increase on the base-10 logarithm scale to the original unit scale. On any base-10 scale, a unit increase corresponds to a tenfold (or multiple of 10) increase in the linear scale. In the base-10 log scale, the $log_{10}(10) = 1$ and $10^1 = 10$ while $log_{10}(100) = 2$ and $10^2 = 100$. For example, the almost two-point increase in the *logGDP* variable from Afghanistan to Sweden $(4.685 - 2.735 = 1.94)$ corresponds to a nearly $10 * 10$ or 100x increase in the original GDP scale. Multiplying Afghanistan's GDP by 100, $543 * 100$ yields 54,300 USD, a few thousand more than Sweden's 48,730 USD. This change in the scale from linear to log illustrates how a log transformation 'pulls inward' the outlier large scores, reducing huge differences.

While a base of 10 is commonly used in political science because of the ease of interpreting unit changes, another common base to use is Euler's number e (2.718282) as a `natural logarithm`. So the natural log of 543 is the number to which we need to raise the constant e to recover 543. The natural logarithm, often abbreviated as $ln()$ to distinguish it from other logarithms is the default for `log()`:

```
log(543)
```

```
## [1] 6.297
```

Again, the value of 6.297 is the number to which we raise the base e. Each unit increase on a natural logarithm scale corresponds to the multiplicative or 'e-fold' increase by a factor of about 2.718. Natural logarithms are often used in subjects such as finance and demography, often because of particular mathematical properties of a natural logarithm. For our purposes in this text, the applications of logarithmic scaling make relatively little difference as to the base. Sticking to a common base-10 logarithm is a safe choice.

To illustrate the scaling difference for a highly skewed measure such as GDP, Figure 3.4 displays the difference between the scaled and unscaled measure of GDP. The visualization is a histogram, displaying values of GDP grouped into bins, or equally spaced intervals of GDP, with the number of countries falling in-to each bin plotted on the Y-axis. The extreme 'right' or positive skew is evident in the top portion of the figure. On the base-10 logarithmic scale, however, the extreme values are 'pulled inward'.

One consequence of transforming the scale of GDP is that in comparing it to other measures with which we should expect to see a clear linear dependency, the log transformation makes this dependency clearer. This purpose in converting a skewed measure such as GDP to a log scale is known as a

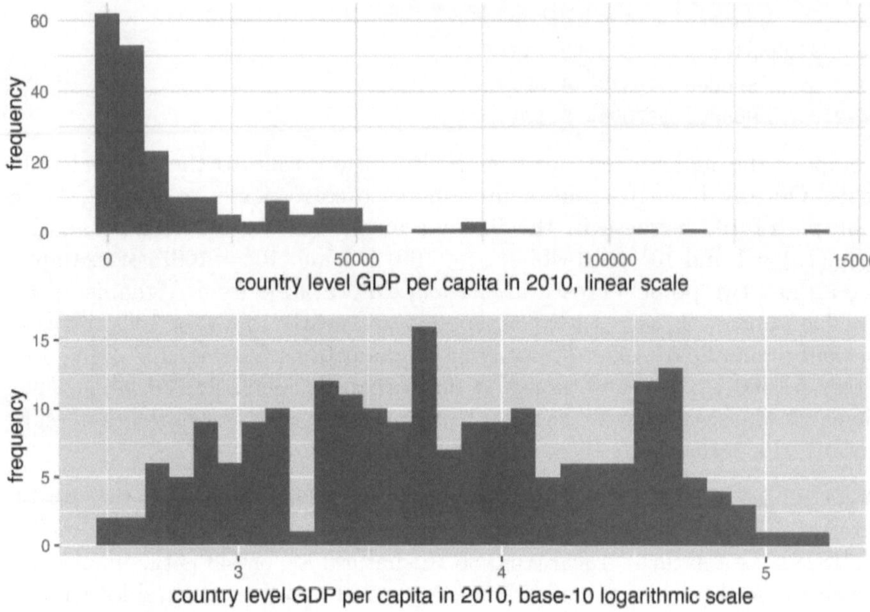

FIGURE 3.4 Histograms of country-level GDP per capita (constant 2010 USD), displayed on two different scales. The upper panel uses a linear scale, showing a highly right-skewed distribution with most countries clustered at lower GDP levels. The lower panel applies a base-10 logarithmic scale to GDP per capita, which compresses the higher values and spreads the lower values more evenly, reducing skewness.

'transformation to linearity', which appears in Figure 3.5. The lower panel displays a clear linear trend between fertility and GDP, although the bunching up of countries at low levels of fertility between 10 and 11 on the log GDP scale suggests that perhaps fertility, too, is a skewed measure that would benefit from a transformation.

The answer to the question of which scale — linear or logarithmic — is the 'correct' one is that they both are. Both are appropriate measurement scales depending on the analytical purpose. If the purpose is to emphasize the skewed nature of linear counting units, then the linear scale is appropriate. If the skewed distribution is problematic, then the logarithmic scale is appropriate, particularly in displaying the linear relationship with fertility rates in the bottom half of Figure 3.5.

In this chapter we reviewed key concepts in the organization and management of data tables. We reviewed data measurement levels and storage types, applying different summary statistics. We reviewed data transformations including the construction of z scores and log transformations. We have already constructed

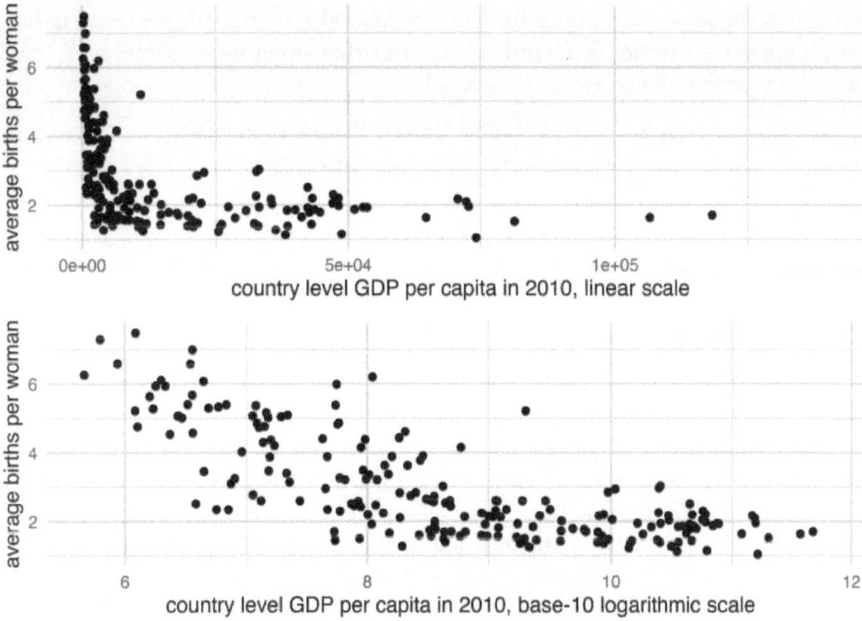

FIGURE 3.5 Scatterplots of fertility rate (births per woman) by GDP per capita (constant 2010 USD) for 2010, displayed on two scales. The upper panel uses a linear scale for GDP, where the relationship between GDP and fertility is obscured due to the right-skewed distribution of GDP. The lower panel applies a base-10 logarithmic scale to GDP per capita, revealing a clearer negative linear relationship: as GDP increases, fertility rates tend to decline.

histograms and scatterplots, two common data visualization types. In Chapter 4 we will learn about principles in visualizing data and additional **R** packages for doing so.

Resources

The Varieties of Democracy project https://v-dem.net/ and the Quality of Governance project https://www.gu.se/en/quality-government collect a wide range of politics related country level measures.

The Data Futures Exchange of the United Nations (UN) Development Programme collects data on UN related programs https://data.undp.org/. https://www.data.gov/.

Data.gov is the open data portal for the U.S. federal government. It is searchable and organized by topic. It includes links to other open data portals for U.S. States and governments around the world https://www.data.gov/open-gov/.

https://data.census.gov/cedsci/ The United States Census Bureau provides access to data tables to download and import into RStudio as CSV files.

3.9 Exercises

Answer the questions below within an RMarkdown document. Include code and relevant output as supporting evidence to your answers.

(1) Create a scatterplot of country-level fertility compared to log scaled GDP, as in Figure 3.5. Then compare it to another scatterplot converting fertility rate to a log scale. Describe how the observed trends in the scatterplots change from one to the other.

(2) Import the CSV file containing observations on the quality of democracy around the world. In a paragraph, describe the status of polyarchy — pay attention to the overall shape of the measure around the world, the presence of outliers, the center, and spread.

(3) Download data from the World Bank via the **WDI** package: find three indicators that interest you, such as particular measure of an economy, resources, or population. Create a dataset that includes these measures. Remove the regional aggregates from the data to create a clean data table. Summarize the center and variability of the measures and create at least one visualization with `qplot()`.

(4) For each of the measures in the dataset from Question 3, construct and compare z scores across each to answer the question of how the USA compares to other countries around the world on each of the different measures.

(5) Import the cross-sectional international dataset. Among different World Bank income groups, what is the average and variation in fertility rate among countries within each set of incomes? And within each group, how much does the fertility rate vary? Identify and report appropriate summary statistics.

Grouped summary statistics can be calculated through the `tapply()` function — "table apply". It requires three arguments: (1) a variable upon which to calculate summary statistics, (2) a grouping variable, and then (3) the name of the particular summary statistic function, such as `mean()`, `sd()`, or `range()`. For example, to calculate `mean` fertility rates across regions: `tapply(wbdata1$fertility_rate, wbdata1$region, mean, na.rm = TRUE)`. Another method is to chain together commands within **tidyverse** functions, a method covered in Chapter 5. The `group_by()` function identifies groups, and the `summarize()` function creates summaries. The result is a table of summary statistics such as:

```
wbdata1 %>%
group_by(region) %>%
  summarize(mean_level = mean(SP.DYN.TFRT.IN,na.rm=TRUE),
            std_dev = sd(SP.DYN.TFRT.IN,na.rm=TRUE))
```

(6) Find the two CSV datafiles containing measures the HDI and Vdem data, and create a merged dataset. Construct scatterplots to investigate whether stronger democracies tend to have higher HDI scores, and higher incomes. Use the `qplot()` function.

(7) Import one of the cross-sectional CSV datasets containing a measure of life expectancy. Use the `subset()` function to subset the data to only those countries with life expectancy over 70 years of age. From the example in the chapter, you will need to identify the life expectancy variable and use `> 70` and the name of the life expectancy variable in place of `region!="Aggregates"`.

4

Data Visualization

Chapter 4 introduces the tools for creating charts (data graphics) in **R**.

Learning objectives and chapter resources

By the end of this chapter, you should be able to (1) identify the required elements of a data visualization with the **ggplot2** package; (2) construct and interpret distributional graphics in the form of histograms, box whiskers, and density plots; (3) construct scatterplots and dynamic bubble plots; (4) construct a time series line plot. The material in this chapter requires the following packages: **ggplot2** (Wickham and Grolemund, 2016) part of the **tidyverse** (Wickham, 2023b, Wickham et al. (2019)), which should already be installed, and **ggrepel** (Slowikowski, 2023) and **plotly** (Sievert, 2020) which will need to be installed separately. The material uses the following datasets *gdpfert.csv*, *86_house_dw.csv*, *countries.csv*, *countries_tsxs.csv*, and *poll_ft.csv* from https://faculty.gvsu.edu/kilburnw/inpolr.html.

Introduction

The term "data visualization" is everywhere. Hundreds of books published on the subject, thousands of websites, and social media communities showcase data visualizations. (For examples, see the Reddit community r/dataisbeautiful, with over 15 million members). In the COVID-19 pandemic era, for many of us reviewing graphical displays of case counts and deaths became an integral part of getting caught up on the day's news. For students of political science, such visualizations are perhaps most commonly found in election results displayed across a geographic area through maps, or diagrams that show seat gains and losses for a political party within a legislature. The purpose of such visualizations is to enable greater insight into the data than would be accessible from an inspection of a data table alone.

A data visualization is information in graphical form, a picture that communicates a researcher's findings or insights. Data tables are effective for organizing data, or visually looking up specific information, but a visualization could be constructed to highlight intriguing features of a dataset, to prompt further investigation. For example, recall the table from Chapter 3 of countries, GDP, and fertility rates. Without a scatterplot — or prior knowledge of the two measures — how easily could we have communicated evidence to others of how the two measures are related? A good data visualization should help explore patterns and relationships within data, to facilitate the process of communicating analysis results, telling the story of the data (Gelman et al., 2002; Gelman and Unwin, 2013). For our purpose in analyzing data, a good visualization should enhance understanding; it should not be an ornament, something constructed as a decoration or mainly for graphical design purposes such as an infographic.

This chapter provides an orientation to some of the fundamental tools for constructing data visualizations in **R**. The field of data visualization is vast; in Chapter 4 we consider univariate visualizations for continuous measures, followed by bivariate and multivariate visualizations for continuous measures, to learn data visualization functions. Most of the visualizations within this chapter are static image graphics, although we review the construction of one dynamic graphic type. At the end of the chapter, there are links to learn additional code for an endless variety of different data visualization types, both static and dynamic.

4.1 Key concepts for data visualization in R

By far, the most common set of tools for visualizing data in **R** come from the **tidyverse**, the graphics package **ggplot2**. The subject of data visualization is exceptionally complex, and there is no single unified theory of how the analysis of data via visualizations should be conducted (Unwin, 2018). There is, however, greater agreement about the appropriate theory to describe what graphics are, what their components are and how the parts are put together into a coherent whole and how to correctly interpret graphics (Cairo, 2019). A set of standards, a grammar, for describing graphics by identifying their components (Wilkinson, 1999), is represented in the concepts used to build graphics in **ggplot2**. We will use these terms to build our own with the **ggplot2** graphics package from the **tidyverse**. Fortunately, this theory, and to the extent that there are some generally agreed upon ideas for how data visualizations should appear, are built into the construction of **ggplot2** functions. So while it is possible to get a graphic wrong with **ggplot2**, it is much easier by default to get it right.

In constructing visualizations from scratch, a starting point is to first consider the purpose of the visualization and the scale of measurement of the underlying variables. For example, if the purpose is to display quantities such as percentages that vary across groups, a bar chart would be an appropriate choice. If the underlying measure is continuous, and the purpose is to describe the shape of a distribution, then a histogram (a bar chart for continuous measures) would be appropriate. Of course, there are many different alternative visualization forms to consider, but a helpful starting point is to first identify the intended quantity or quality to be visualized, and the measurement type of the corresponding variables. Simply put, some visualization types are appropriate for categorical (nominal or ordinal) measures and others for continuous (interval or ratio) scaled measures. Another consideration in constructing charts is whether the graphic is either exploratory or explanatory. An exploratory graphic is intended to examine features or relationships embedded within the data, without necessarily having in mind a specific feature or relationship to highlight. It may be a graphic in rough, unfinished form. An explanatory graphic highlights important features or comparisons within the data, known previously to the researcher. The visualization, in a polished form, is intended to draw in the reader's attention to particular aspects of the data.

To introduce the code syntax for creating visualizations, we start with exploratory graphics for visualizing continuous distributions — histograms, box-whiskers plots, and density plots. Next, we turn to variations on bar plots for graphing quantities across groups. Finally, we examine variations on scatterplot based tools for comparing at least paired (measured on an X and Y) observations. The chapter concludes with an overview of some tools for modifying and adorning visualizations for explanatory purposes.

Visualizing continuous measures

For concepts measured on a continuous, interval or ratio, scale, typically we are interested in understanding a measure's distribution, how scores vary across units of observation, how scores vary within and across groups, or how one distribution relates to another. Typically when investigating a continuous measure, we start with four characteristics: shape (S), the presence of unusual or outlier (O) observations, the location of the center (C) such as a mean or median, spread (S) such as the range of scores or standard deviation, often abbreviated as "SOCS". The visualizations for examining these features of data are useful for learning how to construct visualizations within the **ggplot2** syntax. Perhaps the most common visualization type is the histogram, a sort of bar graph for continuous measures. We will review this visualization type, followed by others, box-whisker and density plots before moving on to types of scatter plots and line plots.

4.2 Histogram

A histogram represent the distribution of a continuous measure by displaying the frequency or count of observations within specified intervals, or "bins," along a continuous scale. Similar to a bar graph, a histogram represents the measure through variation in bar height. Higher bars correspond to more frequent values, which could be scaled by frequency or a relative frequency, such as a proportion (percentage) of the total observations. But unlike a categorical measure for a bar graph, because the measure is continuous, the bars are drawn from equally spaced groups, or bins, of values. The bar widths are centered over a midpoint in the bin, so the bars sit adjacent (or touching) each other, unless there are gaps in the measure. We have already seen an example of a histogram in the form of a histogram of GDP per capita in the *gdpfert.csv* dataset.

To demonstrate, import this dataset.

```
gdpfert<-read.csv(file='data/ch3 data/gdpfert.csv')
```

While the visualization functions are from **ggplot2**, we will do some data preparation using other packages, so we will load the entire suite of **tidyverse** packages:

```
library(tidyverse)
```

We could attach simply the **ggplot2** package as well (`library(ggplot2)`), but doing so would not make available the data preparation tools.

Consider a histogram of GDP per capita; the code is below. Figure 4.1 displays the histogram.

```
ggplot(data=gdpfert) +
  geom_histogram(aes(NY.GDP.PCAP.KD),
                 fill="gray", color="black", bins=25) +
  labs(x="GDP per capita (in constant 2000 US Dollars)",
       y="count")
```

Figure 4.1 displays on the X-axis GDP per capita (in 2010 US dollars) ranging from about 214 to 145,000 USD. The GDP values are grouped into 25 bins, and the height of each bar is drawn according to the number of countries within the bin. About 63 percent of the countries fall within the first two bins. Eight of the 25 bins are empty, with a frequency of 0; these bars are still spaced out on the histogram, because of the even spacing of the 25 bins across the range

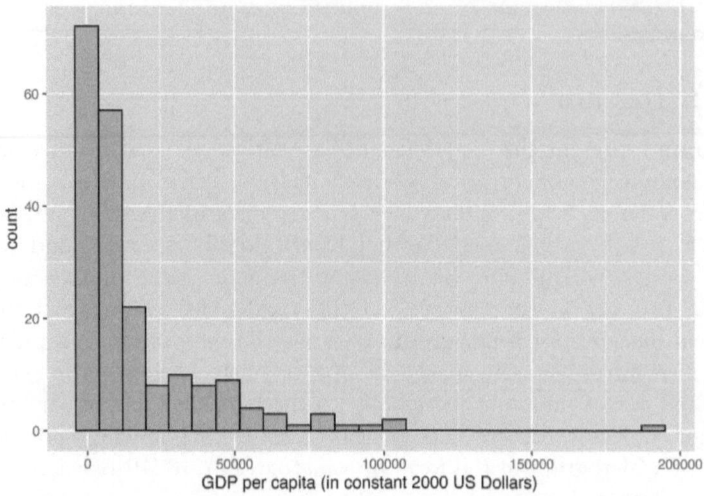

FIGURE 4.1 Histogram of country-level GDP per capita (in constant 2000 USD), created with the **ggplot2** package. This figure displays the default **ggplot2** style, including light gray background grids, black borders for bars, and axis labels.

of observations. Table 4.1 displays the data table converted by **ggplot2** to create the histogram.

The tallest bin is centered around zero and spans a distance of about 8000 USD, from below to above zero. The negative values do not make substantive sense, and there are no observations below 214 USD. But to draw the intervals equally spaced along the variable, and plot bars centered at the middle of each bin (25 total), this is necessary to do. The counts occur along the width of each bin through the scale, with several bins containing 0 observations, until reaching the highly skewed observations for the extraordinarily large GDP per capita countries.

There are various formulas devised to determine the number of bins or bin widths, given statistics such as the total number of observations and the inter-quartile range. However, your judgment as a researcher is preferable. Choose the number of bins based on your own judgment given the features of the data at hand (Unwin, 2018). There are no fixed rules for how many bins should be plotted in a histogram, but there are guidelines: select the number of bins that draws attention to features of the data that are of interest to the researcher. Perhaps these features could be the shape of the distribution, unusually frequent observations (such as modes), or outliers. Too few bins tend to obscure these differences. While too many will draw attention away from these features, while emphasizing noise or characteristics that are less relevant to your interests as a researcher. An example in Figure 4.2 shows three

TABLE 4.1 GDP bins, values, and counts for the histogram of GDP per capita

count	midpoint	bin min	bin max
72	0	-3978	3978
57	7955	3978	11933
22	15910	11933	19888
8	23865	19888	27843
10	31821	27843	35798
8	39776	35798	43753
9	47731	43753	51708
4	55686	51708	59664
3	63641	59664	67619
1	71596	67619	75574
3	79551	75574	83529
1	87507	83529	91484
1	95462	91484	99439
2	103417	99439	107395
0	111372	107395	115350
0	119327	115350	123305
0	127282	123305	131260
0	135238	131260	139215
0	143193	139215	147170
0	151148	147170	155125
0	159103	155125	163081
0	167058	163081	171036
0	175013	171036	178991
0	182968	178991	186946
1	190924	186946	194901

histograms of country level human fertility rate (average births per woman), with differences in the number of bins, from 10, 20, to 40 bins.

The first histogram on the far left panel of figure 4.2, with 10 bars hints a slightly bimodal distribution of scores, while not drawing attention to it as sharply as the histogram with 20 bars. For the purpose of displaying the bimodal distribution — fertility rates of less than three are common, but there is also a distinct group of countries with a relatively high fertility rate between four and six — the histogram with 20 bars is probably preferable, while with 40 bars the histogram is difficult to read with narrow multiple bars. Before breaking down the components of the **ggplot2** histogram, consider the most basic parts of the histogram, present in any graphic.

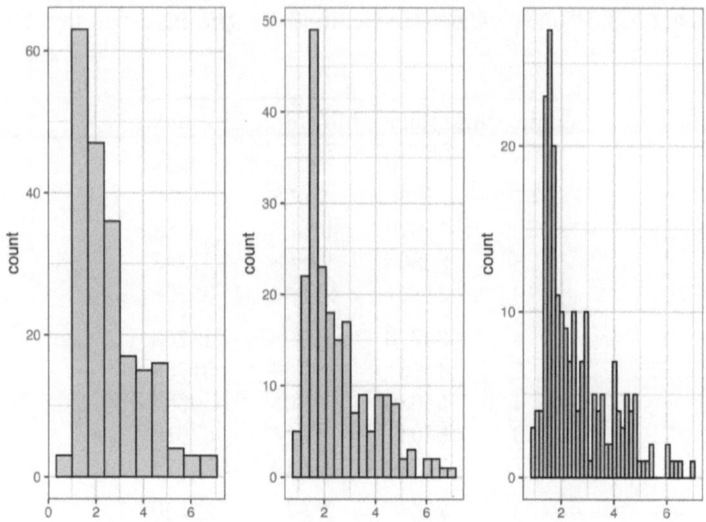

FIGURE 4.2 Histograms of country-level fertility rate, created with **ggplot2**, with 10, 20, and 40 bins (left to right). Increasing the number of bins reveals more detail in the distribution but introduces more irregular patterns.

Elements of a ggplot2 visualization

The name "ggplot" references an influential book, *The Grammar of Graphics* (Wilkinson, 1999) that presents a set of rules, or a grammar, for conceptualizing graphics. The "gg" refers to this "grammar of graphics", a translation of the grammar into a system for creating visualizations in **R** (Wickham, 2010). The histogram in Figure 4.1 displays parts of this grammar. We will review other parts with a review of the code for creating it.

There are four fundamental elements of a data visualization: (1) data, (2) coordinate system, (3) aesthetics, and (4) geometry.

(1) The histogram contains data. This part is obvious; without data, there is nothing to graph. The histogram references a data table containing a column of continuous GDP values by country, organized on rows. Perhaps a less obvious way of thinking about this first grammatical element is that it is about the organization of the data: in what form will it be represented in the graph, how the data cells are organized into tabular form in the data, and whether any measures within the data will be transformed or presented as is.

(2) The graph contains a coordinate system. The bars are presented on a Cartesian — right angle X and Y — set of coordinate axes. This set of X and Y axes is the most familiar way to structure a graphic

in a coordinate system.

(3) The graph contains aesthetics. An aesthetic is anything visible on a graph. We can break these aesthetics down into specific aesthetic items, some of which we tend to take for granted. One aesthetic consists of what we map onto the X (horizontal) and Y (vertical) axes. In Figure 4.1 we have mapped GDP per capita values. There are other aesthetics, such as the gray-colored background of the graphic, and the bounding box for the plot area. The size and choice of font for the axis labels are another aesthetic.

(4) The graph expresses the aesthetics through a geometry. The aesthetics are expressed through a particular geometric shape. Just like the mapping of a measure onto X and Y, the geometric shape is an aesthetic. In Figure 4.1 this shape consists of vertical bars.

The ggplot2 syntax

To learn how to construct a ggplot2 graphic, we will locate each of these elements of a graph in the syntax for the histogram in Figure 4.1, reproduced below:

```
ggplot(data=gdpfert) +
  geom_histogram(aes(NY.GDP.PCAP.KD),
                 fill="gray87", color="black", bins=25) +
  labs(x="GDP per capita (in constant 2000 US Dollars)",
       y="count")
```

The code for **ggplot2** graphics are organized by functions, one per line, which are layered together. The style of writing visualization code in **ggplot2** is to organize the layers line by line — combined across multiple lines.

The data, an implied coordinate system, the aesthetics, and the geometry are within the first two lines of the graphic. The `ggplot()` function identifies the data table in the argument `data=gdpfert`. A `+` sign is placed at the end of a line to signal that another will follow. The plus sign is interpreted as "and then". But we will also learn about another "and then" operator, `%>%`, which is used to link together multiple lines of code. The two symbols do basically the same thing; the ggplot2 package predates the invention of the `%>%` syntax. When working on graphics, remember to use `+` to add together layers of graphical elements in **ggplot2**. The next line specifies the remaining three of coordinate system, aesthetics, and geometry:

`geom_histogram(aes(NY.GDP.PCAP.KD)`. We have a histogram, so the obvious geometry on the graph is a set of bars; this geometry is specified with the function `geom_histogram()`.

Another function appears within the `geom_col()` function, the `aes()` function. The "aes" is an abbreviation for "aesthetic"; this function specifies the aesthetic mapping of the graph, the aesthetic that links measures within the dataset to the components of a coordinate system. Implicitly the "aes" argument means `aes(x=NY.GDP.PCAP.KD)` and Y for plotting the counts of countries in bins.

Within the `geom_histogram()` function, the number of bins is set by `bins=`, the default is 30. And there are additional, optional adornments: the bar fill color is set by the color name `fill="gray87"` and `color="black"` for the bar fill and outline.

As a basic template, a **ggplot2** graphic would look like the following, with the three essential components substituted for the uppercase letters:

```
ggplot(DATA) +
GEOM_FUNCTION(aes(MAPPING))
```

To generalize from this example to a template, it would require the name of a dataset for DATA, a geometry for GEOM_FUNCTION, which implies a coordinate system, and a corresponding aesthetic "mapping" on to the coordinate system in aes() for the MAPPING. The data can be any dataframe stored within the R environment. For the geometry, there are many different geometries; `geom_point()` is the geometry for a scatterplot, for example. For the aesthetic "mapping", think of it as the 'what' within the data, the aspect of it that should be graphed and to any axes such as X or Y. Within this function, you identify specific variables from a dataset.

There is one additional, optional element to the graphic, an argument for `labs()` to alter the default labels, specifying what appears on the X- and Y-axes. While left off the graphic, a title can be specified within `labs()` as `labs(title="title here")` next to the labels for `x=" "` and `y= " "`).

A "stat" is another graphical element. It is a feature of the data calculated for presentation in the graphic. Rather than being recorded directly as a variable in the dataset, it is calculated in the `aes()` function. For a histogram, a stat is useful for displaying the Y-axis in percentages, rather than frequency counts. The calculation of each histogram's percentage scaling is to multiply the width of each bar by the bar's "density" (the bar's proportional height, compared to one). The expression is simply `y=stat(width*density)`, which then instructs **ggplot2** to calculate the Y-axis in these terms based on the content of the `stat()` element. Try substituting this line for the histogram and observe the results: `geom_histogram(aes(NY.GDP.PCAP.KD, y = stat(width*density)*100), bins=20)`.

4.3 Box-Whiskers

The box-whiskers plot focuses attention on a set of five statistics, or five quantities, to summarize a distribution: (1) minimum value; (2) 25th percentile score; (3) 50th percentile, or median; (4) 75th percentile; and (5) maximum value.

The median, or 50th percentile rank, is the focal point of the box-whiskers plot. The box-whiskers plot is a visualization in which the median is marked, while the 25th and 75th percentiles (as quartiles — groups of four) are used to mark the upper and lower limits of a box. The "whiskers" extend to the farthest out points that are not outliers. Outliers are defined as any point further than the distance of 1.5 times the difference between the 25th and 75th percentiles (the interquartile range (IQR)); this definition of an outlier is specific to this graphic; it is not a general definition of an outlier.

Figure 4.3 displays these calculations, and the resulting box-whiskers plot for feeling thermometer scores for the late US Senator John McCain, measured during the fall 2004 presidential election campaign. The data are from the 2004 American National Election Studies survey, a nationally representative sample survey of US citizens. The 'feeling thermometer' survey question asks respondents to rate how warmly or coolly they feel toward, for example, Senator McCain.

The box-whiskers plot for McCain shows a skewed distribution; the public's feelings toward the Senator in 2004 were relatively warm. The median, or average respondent, felt relatively warm toward McCain, at 60 degrees, with the IQR centered fairly uniformly around that median. The presence of outliers at the minimum end of the distribution, but not the upper, reveals that relatively few respondents have cool feelings toward McCain, less than 20 degrees, and show that the distribution is left (or negative) skewed.

While the example of a feeling toward McCain shows a single box-whiskers graph, within *ggplot2* box-whiskers are intended to be visualized in a comparative fashion. The aesthetic of a *ggplot2* box-whiskers requires two variables: one is a continuous measure for graphing the boxplot, and the second is a categorical (factor) measure across which *ggplot2* will draw the box-whiskers. As many box-whiskers will be drawn as there are levels of the categorical measure.

We will read in an additional dataset, ideological "NOMINATE" scores for the U.S. House of Representatives. We take up these types of scores, based on roll call votes, in Chapter 7. The name refers to the technical details of how the scores are estimated. The key point to keep in mind is that the scale ranges from -1 (most left) to 1 (most right) on an ideological scale reflecting

Tabulation of feelings for John McCain, 2004 ANES

Degree	0	1	5	6	15	25	30	40	41	50	
Frequency	14	1	1	2	17	1	27	49	1	259	
Degree	55	60	65	70	75	77	80	85	90	95	100
Frequency	1	199	4	182	11	2	6	117	7	2	49

Summary statistics

Min.	25th percentile	Median	Mean	75th percentile	Max.
0.0	50.0	60.0	61.1	70	100.0

Inter-Quartile Range (IQR): 70-50=20.
Outliers: 1.5 * IQR: 30
High outliers: none
Low outliers: observations at and below 15 degrees

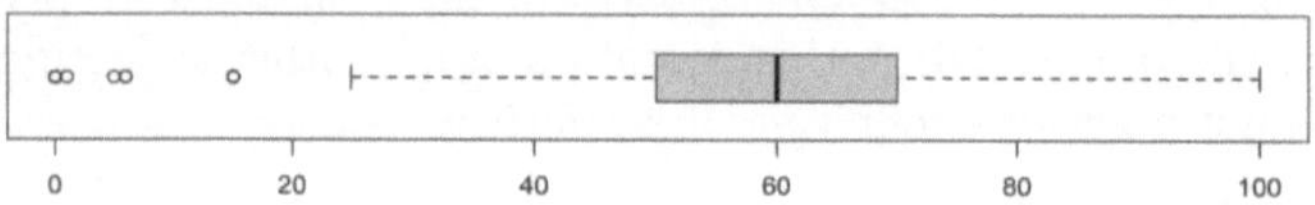

FIGURE 4.3 Tabular frequencies and summary statistics for feeling thermometer ratings of US Senator John McCain (2004 ANES), followed by a box-whiskers plot that visually summarizes the distribution. The frequency table and summary quartiles help interpret key features of the plot, such as the median, interquartile range, and outliers.

left to right economic conflict, the extent to which the federal government should intervene to reduce inequalities resulting from market forces (Poole and Rosenthal, 2006). Each representative is assigned a score, based on their roll call vote history during a session of Congress, much like the ratings interest groups give legislators. This specific dataset is for voting members of the U.S. House of Representatives during the 86th U.S. Congress (1959 - 1961), coinciding with election of John F. Kennedy to the White House.

```
dw_86<-read_csv("86_house_dw.csv")
```

Figure 4.4 displays the ideology scores by party caucus of the legislator in a box-whiskers graphic. Switching a histogram with a box-whiskers graphic in this instance requires only a change in the data and the geometry, with one addition to the aesthetic: `geom_boxplot(aes(x = party, y = nominate_dim1))`. The `geom_boxplot()` draws the boxplot but requires both a factor variable on the X-axis and the continuous measure on the Y-axis. The boxplot is drawn across groups identified on the X-axis.

The geometries are arranged vertically, with the party caucus membership arranged on the X-axis, and the ideology score plotted on the Y-axis. The

Democratic caucus is negative (or 'left-') skewed toward the liberal end of the scale. With a few outlying liberals, in the Democratic caucus of this Congress, there are far more moderates than liberals. The upper whisker extends to the 25th percentile of the Republican caucus. And combined with the corresponding overlap from the Republican lower whisker, the graph displays the same overlap of caucus members from the histogram, moderates from both party. It is interesting that the Republican caucus has more liberal and more conservative outliers, while the presence of outliers is one-sided among Democrats. While compared to the histogram, the box-whiskers does not display the count of representatives falling into this moderate middle; it does display concisely the ideological center (median) of each caucus, and how closely the caucus members are spread around the median. The Republican caucus is narrowly distributed around the caucus median compared to the Democratic caucus. In the exercises at the end of the chapter, you are asked to compare this plot to a histogram divided between party caucuses.

```
## # A tibble: 444 x 4
##   state_abbrev bioname                  nominate~1 party
##   <chr>        <chr>                         <dbl> <fct>
## 1 AL           ANDREWS, George William      -0.03  Demo~
## 2 AL           BOYKIN, Frank William        -0.105 Demo~
## 3 AL           ELLIOTT, Carl Atwood         -0.381 Demo~
## 4 AL           GRANT, George McInvale       -0.117 Demo~
## 5 AL           HUDDLESTON, George Jr.       -0.115 Demo~
## # ... with 439 more rows, and abbreviated variable
## #   name 1: nominate_dim1
```

```
ggplot(dw_86) +
     geom_boxplot(aes(x = party, y = nominate_dim1)) +
     labs(title = "Ideologies of Representatives to the
          U.S. House, \n 86th Congress (1959-1961)",
          x = "Party",
          y = "NOMINATE Score (left (-1) to right (1))") +
     theme_minimal()
```

To illustrate a useful feature for titling graphics, in the code, the title in the `labs()` function also adds the trick of a line break, `\n` to split the title into two lines. In addition to specifying a title, there are `labs()` options for a `subtitle=()` and `caption=()`.

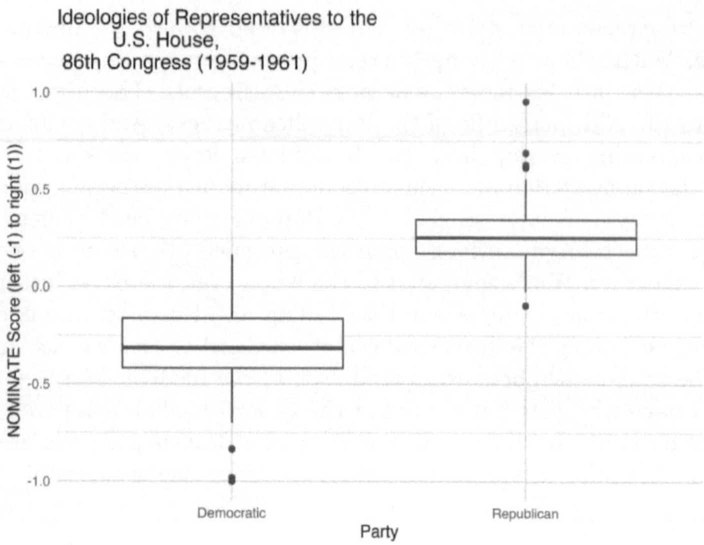

FIGURE 4.4 Box-whisker plot of legislator ideology scores by party caucus for members of the U.S. House, 86th Congress (1959–1961), created with **ggplot2**. The plot shows the distribution of ideology scores for Democrats and Republicans, with the box indicating the interquartile range, the horizontal line marking the median, and the whiskers and dots identifying outliers.

4.4 Density

A density plot is another common visualization for continuous measures. It is essentially a smoothed histogram, as if a curve is drawn over the height of the bars to represent the distribution of the data. The 'density' refers to the Y-axis scaling, where the axis is scaled so that the total area under the density curve is equal to one, making the curve a probability density. This same idea can be applied to histograms: when a histogram is scaled to show density rather than frequency, the total area of all bars also equals one.

Two density plots appear in Figures 4.5 and 4.6, displaying the densities for NOMINATE scores in the 86th Congress. The geometry of the density plot is simply `geom_density()` requiring only one aesthetic, the continuous measure to visualize as a density: `geom_density(aes(nominate_dim1))`. Figure 4.5 displays the scores for the House as a whole, while in Figure 4.6 the scores are grouped by party caucus. The shape of the densities resemble the histograms. Density plots are often used to compare distribution shapes, since the lines are overlaid and easily comparable. In this case, the density plots provide a sense of the degree of ideological overlap between the party caucuses.

```
ggplot(dw_86) +
  geom_density(aes(nominate_dim1)) +
  labs(title="Ideologies of Representatives to the U.S. House,
              \n 86th Congress (1959-1961)",
       y="NOMINATE score left (-1) to right (1)",
       x="party caucus")
```

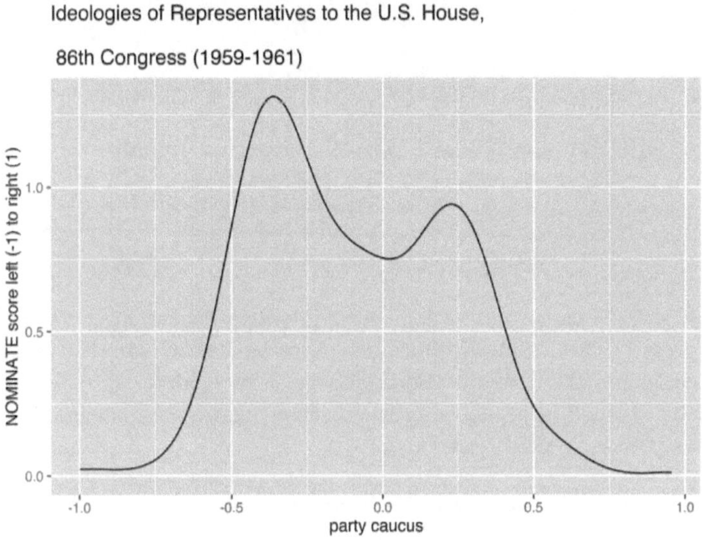

FIGURE 4.5 Density plot of legislator ideology scores in the U.S. House, 86th Congress (1959–1961), created with **ggplot2**. This plot provides a smoothed estimate of the distribution of ideology scores, illustrating the relative concentration of legislators along the left-right ideological spectrum.

```
ggplot(dw_86) +
  geom_density(aes(nominate_dim1, color=party, linetype=party)) +
  scale_color_manual(values=c("blue", "red")) +
  scale_linetype_manual(values=c("solid", "dashed")) +
  labs(title="Ideologies of Representatives to the U.S. House,
              \n 86th Congress (1959-1961)",
       y="density",
       x="NOMINATE score left (-1) to right (1)")
```

In Figure 4.6, within `geom_density()` the `aes()` argument maps the color and line type to party. Then two additional functions specify the colors and the line types.

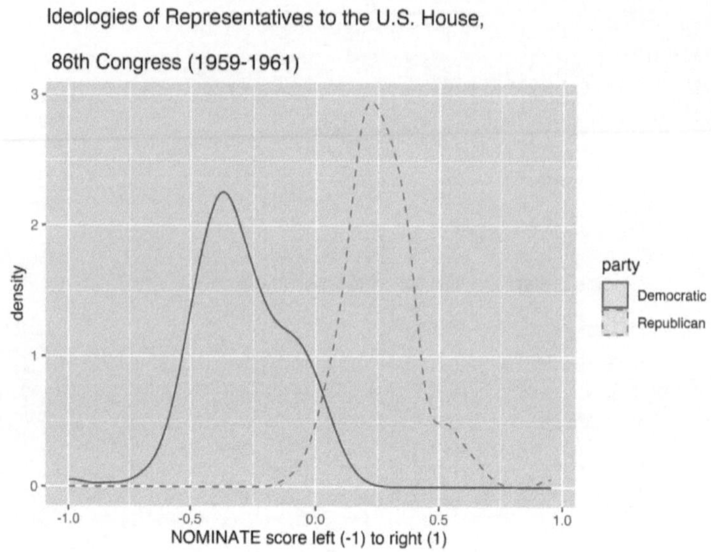

FIGURE 4.6 Density plot of legislator ideology scores in the U.S. House, 86th Congress (1959–1961), displayed by party caucus. Created with **ggplot2**, this plot compares the distribution of ideologies for Democrats (solid line) and Republicans (dashed line), showing the distinct central tendencies and spread of each party's ideological positions.

4.5 Scatterplots

For a visual comparison of two continuous measures, a scatterplot is by far the most common graphic. We will construct a simple scatterplot for logged GDP per capita and fertility rate at the country level. Since we have already observed this figure, we will simply construct the *ggplot2* code for it, equivalent to the `qplot()` code from previous chapters. The geometry for creating a scatterplot is `geom_point()`; it requires two aesthetic mappings onto X and Y.

The data will be from *countries.csv*:

```
countries<-read_csv(file="countries.csv")
```

Two familiar variables are logged GDP per capita by fertility rate.

```
ggplot(data=countries) +
    geom_point(aes(x=log(gdp_pc_pp), y=fert_rate))
```

Try running this code. Notice the log transformation around GDP per capita, `log(gdp_pc_pp)`. One additional modification to the aesthetic would be to color the points by region, by adding `color=region` to the aesthetic of `geom_point()` — observe how the scatterplot changes `geom_point(aes(x=log(gdp_pc_pp), y=fert_rate, color=region))`.

An aesthetic mapping for a third variable on a scatterplot: bubble chart

A bubble chart is a type of scatterplot that displays three — not two — variables. While the X and Y axes encode two different variables, a third one is encoded into the size of the points plotted in the scatterplot (resembling bubbles). The example in Figure 4.7 displays a figure of this type.

```
ggplot(data=countries) +
geom_point(aes(x=log(gdp_pc_pp), y=fert_rate,
      size=total_population))
```

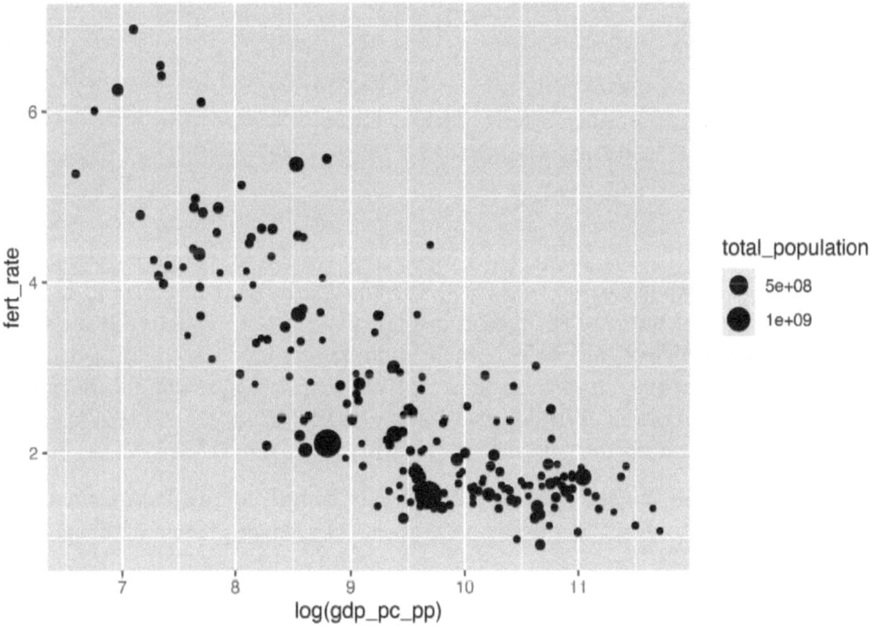

FIGURE 4.7 Bubble chart created with **ggplot2**, displaying the relationship between fertility rate and log GDP per capita. Each point represents a country, with bubble size proportional to total population. This visualization highlights how larger-population countries are distributed across the GDP–fertility relationship.

From the data, the three relevant variables are *total_population*, *fert_rate*, and *gdp_pc_pp*. Adding `size=total_population` to the aesthetic results in variation in the points displayed on the figure.

There are two additional considerations once the points are plotted by values of an additional variable. One is the magnitude of the change from smallest to largest, and the other is the display of the population figures. The magnitude of the points is set by `scale_size()`, with limits to indicate how much larger the biggest point should be relative to the smallest. A difference of 10 to 1 would be set with `scale_size(range=c(1,10))`. In addition, to make the population figures more legible in the graph legend, we could plot population in millions: `size=(total_population/1000000)`.

In addition, the bubbles overlap each other, like ink splatters. The different points in Figure 4.7 are often indistinguishable. So we will make the points a little transparent. Transparency is controlled by what is known as an alpha channel; in R, an alpha setting is controlled by a number, ranging from 0 (transparent) to 1 (opaque). In the figure, to add transparency to the points, we simply add `alpha=` to `geom_point()`. The code below for Figure 4.8 shows the result of setting the magnitude of point size, the scale of population, and alpha transparency.

```
ggplot(data=countries) +
geom_point(aes(x=log(gdp_pc_pp), y=fert_rate,
      size=(total_population/1000000)), alpha=.3) +
  scale_size(range=c(1,10))
```

Once the plots are transparent, the overlapping display of countries is visible. Once we specify the size, `ggplot()` will automatically include a legend in the figure. The default name of the legend in Figure 4.8 is the name of the variable encoded into the size of the circles, which could be modified. It would be helpful to add a title, caption, and more descriptive X and Y axes labels. Because `geom_point()` was modified by `size=`, we can specify what to call this feature of the data with an option for the `labs()` function, `size="Total population"`.

Filling in some additional labels results in the code and Figure 4.9. We move the legend to the bottom of the graphic with `theme(legend.position = "bottom")`:

```
ggplot(data=countries) +
geom_point(aes(x=log(gdp_pc_pp), y=fert_rate,
      size=total_population/1000000), alpha=.3) +
  labs(x="log GDP per capita, 2010 U.S. dollars",
      y="fertility rate (average births per woman)",
      size="Total population in millions") +
  scale_size(range=c(1,10)) +
```

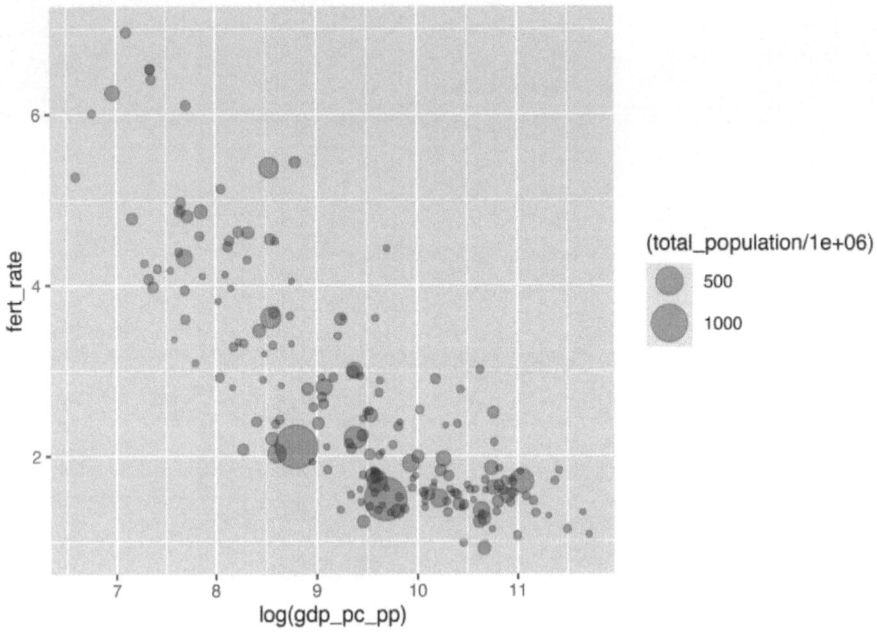

FIGURE 4.8 Bubble chart of fertility rate by log GDP per capita, with bubble size proportional to total population. This chart introduces transparency (alpha shading) in the points to reduce overplotting and improve the visibility of overlapping bubbles, especially for countries with similar GDP and fertility levels.

```
theme(legend.position = "bottom")
```

Next we add labels (country names) to particular points that we wish to highlight. We will use a function `geom_text_repel()` from another package that works within `ggplot()`, **ggrepel** (Slowikowski, 2023). The function will need to be installed if it is not already, `install.packages("ggrepel")`. Then attach it:

```
library(ggrepel)
```

The use of the function here is to have a geometry added to a **ggplot2** graphic, so that names of countries appear in a legible form on the figure, positioned sufficiently close but still legible next to the point each one represents. The function name is `geom_text_repel()`, and it requires the same aesthetics as `geom_point()`, with one additional `label=` argument for the

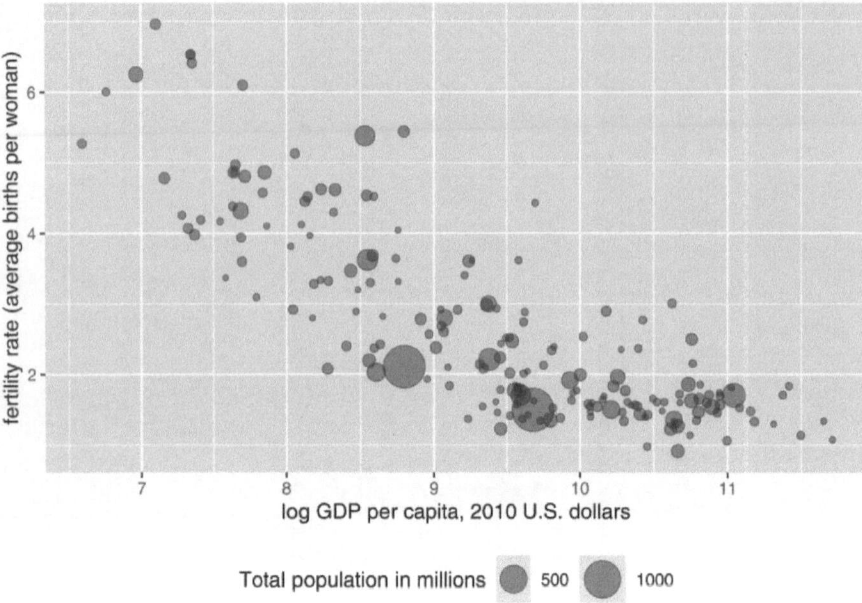

FIGURE 4.9 Bubble chart of fertility rate by log GDP per capita, with bubble size proportional to total population and alpha transparency applied to improve visibility of overlapping points. This version of the chart refines the axis labels and legend text for greater clarity and audience readability.

name of the countries. The geometry is added as a separate layer to the graphic — as if plotting selected text over the points, so we have to re-specify the `aes()` part of the graphic. If we wanted to plot all of the country names, we would simply type `geom_text_repel(aes(x=log(NY.GDP.PCAP.KD), y=SP.DYN.TFRT.IN, label=country))`. The *country* variable is the first one in the countries dataframe. Nonetheless, plotting all the country names would make for an unintelligible plot. So we want to specify a subset of the data.

To subset the part of the data we wish to plot points on, we would use the `subset()` function on the dataframe, to select observations within the dataframe meeting a particular condition. We could specify a subset of fertility rates greater than six. In `geom_text_repel()`, this data subset would look like `data=subset(wbdata1, SP.DYN.TFRT.IN > 6)`.

Putting it all together, along with another alpha parameter to have the country names a little lighter than the points, we have the following function:

```
geom_text_repel(data=subset(wbdata1, SP.DYN.TFRT.IN > 6),
                aes(x=log(NY.GDP.PCAP.KD),
                y=SP.DYN.TFRT.IN, label=country),alpha=.6)
```

Added as a separate line to Figure 4.10:

```
ggplot(data=countries) +
geom_point(aes(x=log(gdp_pc_pp), y=fert_rate,
      size=total_population/1000000), alpha=.3) +
  labs(x="log GDP per capita, 2010 U.S. dollars",
      y="fertility rate (average births per woman)",
      size="Total population in millions") +
  scale_size(range=c(1,10), guide="legend")  +
 geom_text_repel(data=subset(countries, fert_rate > 6),
       aes(x=log(gdp_pc_pp), y=fert_rate,
           label=country), alpha=.6)  +
   scale_size(range=c(1,10)) +
    theme(legend.position = "bottom")
```

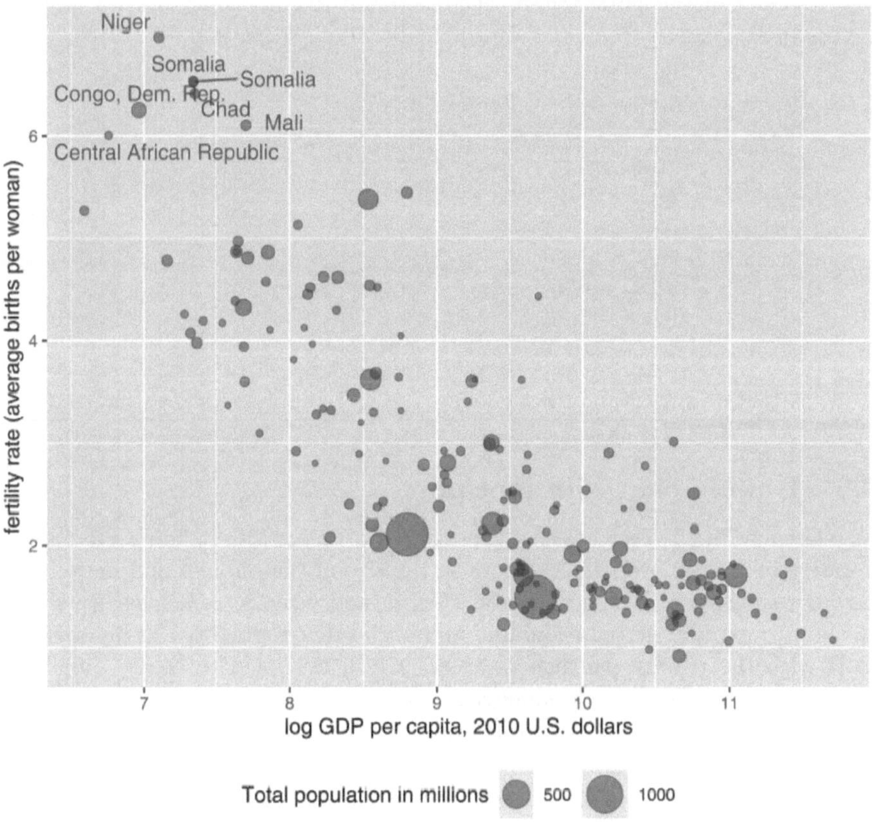

FIGURE 4.10 Bubble chart of fertility rate by log GDP per capita, with bubble size proportional to total population, and country names based on the condition of fertility rates higher than six average births per woman.

Finally, we could add a linear trend line summarizing the relationship between Y and X on the figure. The trend line requires adding a `stat()` layer. In this case, the stat we want to use is `stat_smooth()`, which stands for a smoothed line, either nonlinear or linear. Because it is a separate layer, we need to specify `aes()` for the text layer: `stat_smooth(se=FALSE, method="lm", aes(x=log(gdp_pc_pp), y=fert_rate), color="red")`. There are a couple of other options to add. We do not have a sample, so we do not need to see a measure of sampling uncertainty around the line, thus we set `se=FALSE`. In the function, `method="lm"` where the "lm" stands for linear model, while `color="red"` simply draws the line in red. Leaving `method="lm"` out of the layer will result in a nonlinear best fitting curve to the data. Adding this last line to the data results in Figure 4.11:

```
ggplot(data=countries) +
geom_point(aes(x=log(gdp_pc_pp), y=fert_rate,
     size=total_population/1000000), alpha=.3) +
  labs(x="log GDP per capita, 2010 U.S. dollars",
     y="fertility rate (average births per woman)",
     size="Total population in millions") +
 geom_text_repel(data=subset(countries, fert_rate > 6),
       aes(x=log(gdp_pc_pp), y=fert_rate,
           label=country), alpha=.6)  +
     theme(legend.position = "bottom") +
  stat_smooth(se=FALSE, method="lm",
             aes(x=log(gdp_pc_pp), y=fert_rate), color="red")
```

4.6 Bubble charts in motion

As referenced in Chapter 1, a classic use of bubble plots is in building a dynamic iteration of change over time — both in the size of the bubbles and movement through the X and Y space. These "Gapminder" plots, conceived by Hans Rosling to explain global economic and social development, can be created in R with the **plotly** package (Sievert, 2020), designed to create dynamic graphics such as these. A Gapminder plot is essentially a series of static bubble charts compiled together, one after another, indexed over a series of years. Showing change over time, the plots require typically four or five measured variables from data: X, Y, point size, time period, and perhaps a point color to represent region. The data must consist of a time series cross-sectional dataset, containing the same cross-sectional observations repeated over years.

Because the plots are dynamic, the visualizations must be embedded within an HTML file. The graphics can be rendered in *RStudio* or a web browser,

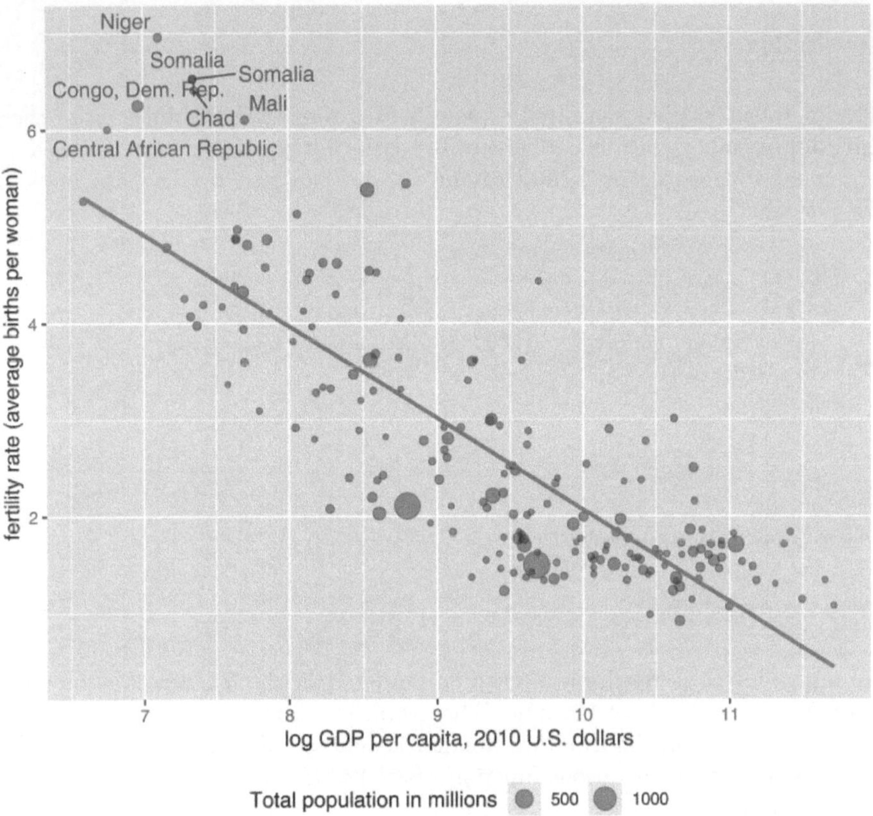

FIGURE 4.11 Bubble chart of fertility rate by log GDP per capita, with bubble size proportional to total population. A trend line summarizes the negative relationship between GDP per capita and fertility rate.

but not exported to a PDF. Below the code and dynamic output consists of a screengrab of the dynamic graphic. Run the code to observe how the dynamic graphic is created.

Putting an existing bubble chart in motion, such as Figure 4.10, begins with saving it as a graphic object, with a few modifications, such as changing the country labels.

The function *ggplotly()* from the **plotly** package will render a static **ggplot2** graphic and add dynamic elements.

First install the *plotly* package if needed (`install.packages("plotly")`), then attach it:

```
library(plotly)
```

The modifications to the bubble plot involve removing the lines from the **ggrepel** package, which is not compatible with **plotly**, adding a `color=region` argument to `geom_point()`, and saving the bubble plot to the data object *dynbubble1*:

```
dynbubble1<- ggplot(data=countries) +
  geom_point(aes(x=log(gdp_pc_pp), y=fert_rate, color=region,
      size=total_population/1000000), alpha=.3) +
  scale_size(range=c(1,10), guide="none")
  labs(x="log GDP per capita, 2010 U.S. dollars",
      y="fertility rate (average births per woman)",
      size="Total population in millions")
```

Place the saved graphic object in the `ggplotly()` function:

```
ggplotly(dynbubble1)
```

Running this code results in a graphic rendered in the Viewer pane on the lower right-hand side of RStudio. The graph is interactive, with the cursor revealing additional information about the points. The interactivity of the graph is changed with various functions. See the Resources section at the end of the chapter for details.

To set the graphic in motion, we first need to read in a dataset compiling the same cross-sectional measurements, year over year. Import the *countries_tsxs.csv* CSV file:

```
countriestsxs<-read_csv(file="countries_tsxs.csv")
```

We add an additional global aesthetic `aes(frame=year, ids=country)` to the `ggplot()` function at the beginning of the graphic, and change the data source to *countriestsxs*. For higher contrast colors, the line `scale_colour_viridis_d()` appears at the bottom of the graphic:

```
graph_dynamic<- ggplot(data=countriestsxs,
                       aes(frame=year, ids=country)) +
  geom_point(aes(x=log(gdp_pc_pp), y=fert_rate, color=region,
      size=total_population/1000000), alpha=.3) +
  scale_size(range=c(1,10), guide="none")
  labs(x="log GDP per capita, 2010 U.S. dollars",
      y="fertility rate (average births per woman)",
```

```
    size="Total population in millions")  +
  scale_colour_viridis_d()
```

```
ggplotly(graph_dynamic)
```

Once *graph_dynamic* is loaded, the result is an interactive bubble plot like Figure 4.12 appearing in the Viewer.

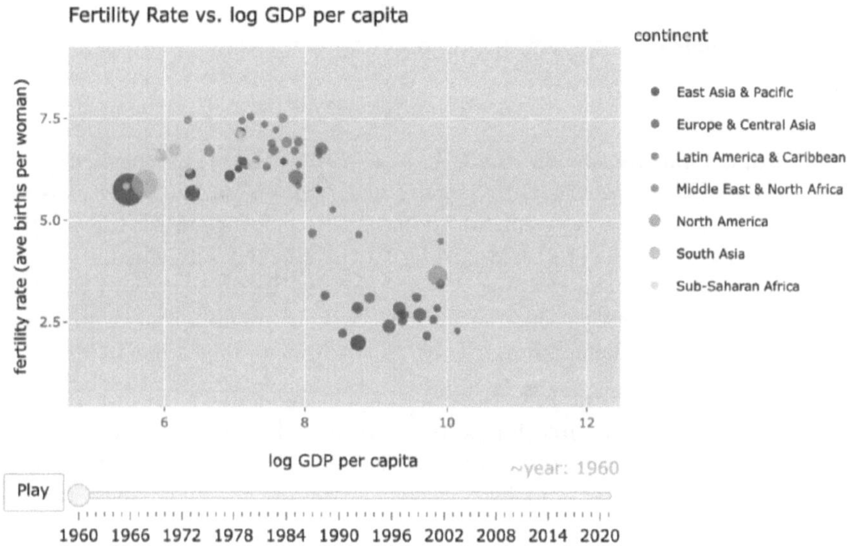

FIGURE 4.12 Interactive, dynamic bubble chart created with the **plotly** package, showing the relationship between fertility rate and log GDP per capita over time (1960–2020). Each point represents a country, with bubble size proportional to population and color indicating region. The animation slider allows exploration of changes in these relationships over time.

In the figure, a slider and "Play" button allows the graphic to become dynamic as the relationships change over time.

Resources

The political scientist Edward Tufte published *The Visual Display of Quantitative Information* (2001), one of the pillars of the subject's study.

See Healy (2019) for an in-depth review of creating data visualizations with **ggplot2**.

In the context of skills in data science, Baumer et al. (2017) consider data visualization from an abstract theory and practice of **ggplot2** functions.

Chang's *R Graphics Cookbook* (2018) illustrates and explains R code for about every data visualization type imaginable, as does the R Graph Gallery https://r-graph-gallery.com/. See a gallery of **plotly** graphics and **R** code at https://plotly.com/r/.

4.7 Exercises

(1) Import *86_house_dw.csv*. Create a histogram of ideology (*nominate_dim1*) in the U.S. House, grouped by party caucus. Use `facet_wrap(~party)` as an additional line of below `geom_histogram()`. Compare the faceted histograms to the faceted box-whisker plot introduced in the chapter. What patterns in ideology and party are visible from the histogram that are not as apparent from the box-whiskers plot?

(2) Import *poll_ft.csv*. Pick three feeling thermometer scores for political figures and three for political groups. For each one, create a visualization of the distribution of feelings in the electorate toward each group. (Create at least one each of a density plot, histogram, and box-whisker plot among the six.) Describe the results. Pick two groups or figures showing evidence of polarization (bimodality) and use `facet_wrap(~)` to compare feelings by individual group characteristics, such as gender or party identification. From the chosen characteristic, which groups are more polarizing and which are less so? Describe the evidence across the visualizations.

(3) Collect some data from the World Bank, using your data collection and organization skills from Chapters 2 and 3. After collecting and organizing the data, present summary statistics and at least one dynamic visualization (or a series of static) visualizations, to show how country level carbon dioxide (CO_2) emissions have changed over time, as economies develop. You will want to search for (CO_2) emissions and GDP measures in the World Bank development indicators database. Explain your choice of measures and organization of the dataset. Present the graphic(s) and/or link to a webpage for the dynamic graphic. In two paragraphs, interpret the summary statistics and graphics, explaining how emissions and economies

have changed together, paying particular attention to any unusual or countries of particular interest to you.

(4) Construct a scatterplot on two continuous variables, grouped by color. If necessary, apply a transformation to linearity on one or both variables. Add a linear trend line fit across all observations. If applicable, comment on the extent to which the line should be fit excluding outliers, within a paragraph explaining what you observe within the scatterplot.

(5) Use the **plotly** package to construct a dynamic graphic displaying how greenhouse gas emissions and GDP, at country level, have covaried over time. Describe the changes in the graphic over time.

(6) Prepare a presentation, explanatory quality visualization to highlight some feature of the data from any of the datasets within this chapter. Explain how the graphic is prepared as a presentation quality visualization rather than an exploratory graphic. To highlight the intended features of the data, if necessary, look through help sources to add additional code to improve the presentation quality of the graphic.

5

Data Wrangling: Cleaning and Transforming Data

Chapter 5 reviews functions for manipulating raw data to prepare it for analysis or visualization.

Learning objectives and chapter resources

By the end of this chapter, you should be able to (1) apply key functions to sort, filter, arrange, and restructure data tables for analysis and presentation; (2) use these techniques to process county-level voting returns from an official data source; (3) create clean data tables for visualization; and (4) document the steps involved in transforming raw data into a usable format. The material in the chapter requires the **dplyr** (Wickham et al., 2023b), **ggplot2** (Wickham et al., 2023a), and **tibble** (Müller and Wickham, 2023) packages, which should already be installed as part of the *tidyverse*. The material uses the following datasets: *house_memb_117.csv* and *2016GEN_MI_CENR_BY_COUNTY.xls* from https://faculty.gvsu.edu/kilburnw/inpolr.html.

5.1 Wrangling and tidying data

Data is rarely analysis-ready or machine readable when obtained from its source. Transforming raw data into a format suitable for analysis — known as "data wrangling" — is a critical step in nearly any research project. And just as important is having code to make the process reproducible, both for transparency and for repeating the steps in the event of an error. This chapter provides tools to take messy or unwieldy data tables and to clean, structure, extract, and prepare data for visualization and further analysis. It is organized into three sections:

TABLE 5.1 Functions from the dplyr package for altering rows and columns of a dataset.

function	Description
filter()	Select rows of a dataframe based on specified conditions.
arrange()	Arrange rows of a dataframe based on specified column(s) in ascending or descending order.
select()	Select and keep specific columns from a dataframe.
rename()	Rename columns in a dataframe.
mutate()	Create new columns or modify existing columns.
group_by()	Group the data by values in one or more columns, enabling operations within each group.
summarise()	Calculate summary statistics (often used with group_by()).

First, in brief we will learn about the use of indices to subset data tables by rows and columns. While these are essential to learn, the **dplyr** package provides a more flexible set of functions for sequencing operations into others such as filtering and arranging rows and columns. Second, we will learn about the core functions of the **dplyr** package, displayed by function name in Table 5.1 followed by a brief explanation. We will learn to use these functions with the `%>%` 'pipe' operator, for combining functions by passing the output of one to another. Third, we will apply these skills to cleaning and preparing for visualization Michigan County level voting records from an official source.

Attach the **tidyverse** to access functions from the packages **dplyr** and **ggplot2**.

```
library(tidyverse)
```

Save the two Chapter datafiles to your working directory; use the Session menus to change it as needed. Use `read_csv()` to read in *house_memb_117.csv* and store it as the data object *house_memb_117*:

```
house_memb_117<-read_csv(file="house_memb_117.csv")
```

The result, *house_memb_117*, is stored as a `tibble`, a particular format for printing a dataframe to the Console. Enter *house_memb_117* to observe it. The file *house_memb_117.csv* is a table of characteristics of representatives to the 117th session of the U.S. House of Representatives, from the canonical source of data on members of Congress and their votes, Voteview.com (Lewis et al., 2023). The file consists of 455 rows and 9 columns. (See the Appendix for details.)

Apart from identifiers, the data includes two measures of legislator ideology, *nominate_dim1* and *nominate_dim2* (Poole, 2005; Poole and Rosenthal, 2006). Developed by the political scientist Keith Poole, 'NOMINATE' (an acronym) scores estimate ideology on the principal line of political conflict in Congress, an economic liberalism-conservatism (*dim1*) and a second (*dim2*) capturing the legislator's position on the remaining bundle of issues within a particular session of Congress. See Everson et al. (2016) for an explanation of the meaning and estimation of the scores. Given the file, imagine that we are interested in extracting observations on legislator ideology by State or party caucus, or perhaps sort the legislators by ideology.

Selecting rows and columns with indices

Indices are a fundamental way to select specific rows and columns in a dataframe. Below, the code selects four columns (*state_abbrev*, *party_code*, *bioname*, and *nominate_dim1*) from *house_memb_117*:

```
house_memb_117[, c(4, 5, 6, 8) ]
```

```
## # A tibble: 455 x 4
##    state_abbrev party_code bioname                nomin~1
##    <chr>        <chr>      <chr>                    <dbl>
##  1 AL           Republican ROGERS, Mike Dennis      0.363
##  2 AL           Democratic SEWELL, Terri           -0.396
##  3 AL           Republican BROOKS, Mo               0.652
##  4 AL           Republican PALMER, Gary James       0.678
##  5 AL           Republican CARL, Jerry L.           0.52
##  6 AL           Republican MOORE, Barry             0.642
##  7 AL           Republican ADERHOLT, Robert         0.386
##  8 AK           Republican YOUNG, Donald Edwin      0.283
##  9 AK           Democratic PELTOLA, Mary Sattl~    -0.156
## 10 AS           Republican RADEWAGEN, Aumua Am~     0.336
## # ... with 445 more rows, and abbreviated variable
## #   name 1: nominate_dim1
```

Indices specify positions within the dataframe using the format `[r, c]`, where `r` represents rows, and `c` represents columns. When left blank, all rows or columns are selected. For example, `house_memb_117[, 6]` selects all rows from the sixth column, *bioname*. Moving the number 6 to the `r` position selects all columns for the sixth observation, `house_memb_117[6, ]`. Multiple columns can be selected with the `c()` function: `house_memb_117[, c(1, 2, 3, 4, 5)]` selects columns 1 through 5, while `house_memb_117[, c(1:5)]` achieves the same result with the sequence 1 through 5 specified by `1:5`.

In the example above, `house_memb_117[, c(4, 5, 6, 8)]` returns all rows but only columns 4, 5, 6, and 8. When executed, the result is displayed as a tibble, with a header `# A tibble: 455 × 4`, showing that the dataset contains 455 rows and 4 columns. The first 10 rows are displayed along with column names, types (`<chr>` for character, `<dbl>` for numeric, `<fct>` for factor), and a comment indicating the remaining rows: `# ... with 445 more rows`.

5.2 Functions for organizing rows and columns

Indices are essential for working with dataframes. The tools from **dplyr** are different: rather than directly identifying rows and columns with indices to specify a position, **dplyr** works with functions. As individual functions, each follows a clear pattern. The functions require first the name of a dataset, in this case, `house_memb_117`, followed by a set of conditions that select rows or columns of the dataset. The result is typically returned as a tibble. Below are examples of key dplyr functions:

Highlight a row of the dataset: `filter()`

The *filter()* function selects rows of a data table based on row values. For example, if we wanted to view ideology scores on the elected representatives from New York:[1]

```
filter(house_memb_117, state_abbrev=="NY")
```

```
## # A tibble: 29 x 4
##    state_abbrev party_code bioname              nomin~1
##    <chr>        <chr>      <chr>                  <dbl>
##  1 NY           Democratic HIGGINS, Brian        -0.348
##  2 NY           Democratic CLARKE, Yvette Diane  -0.61
##  3 NY           Democratic TONKO, Paul           -0.418
##  4 NY           Republican REED, Thomas W. II     0.269
##  5 NY           Democratic MENG, Grace           -0.379
##  6 NY           Democratic JEFFRIES, Hakeem      -0.489
##  7 NY           Democratic MALONEY, Sean Patri~  -0.239
##  8 NY           Republican ZELDIN, Lee M          0.397
##  9 NY           Democratic RICE, Kathleen Maura  -0.28
## 10 NY           Republican STEFANIK, Elise M      0.269
## # ... with 19 more rows, and abbreviated variable name
## #   1: nominate_dim1
```

[1] In this printed text, since all of the columns will not fit on a page, I select a subset of the columns, such as in this case, I entered `filter(house_memb_117[, c(4, 5, 6, 8)], state_abbrev=="NY")`

Notice in the printout, the list of representative names followed by the ideology score, ranging from (-1 for more liberal) to (+1 for more conservative). For example, current House minority leader Representative Hakeem Jeffries is scored a -.489, while Representative Elise Stefanik is scored a .269.

Saving data subsets

The `filter()` function, like other **dplyr** functions, does not save a new dataset. Instead, it only returns the lines of the dataframe that match the conditions, printing the dataframe out as a tibble.

To save a filtered dataset, such as a New York subset of House member data, we use the assignment operator. To create a data table or 'tibble' *NY* and assign it the contents of the filter:

```
NY<-filter(house_memb_117, state_abbrev=="NY")
```

The result is a dataframe consisting of New York legislators. While the actual dataset includes all 9 variables, subscripts could select just the party, name, and ideology score for printing: *party_code*, *bioname*, *nominate_dim1*: `[, c(5,6,8)]`:

```
NY[, c(5,6,8)]
```

```
## # A tibble: 29 x 3
##    party_code bioname                  nominate_dim1
##    <chr>      <chr>                            <dbl>
##  1 Democratic HIGGINS, Brian                  -0.348
##  2 Democratic CLARKE, Yvette Diane            -0.61
##  3 Democratic TONKO, Paul                     -0.418
##  4 Republican REED, Thomas W. II               0.269
##  5 Democratic MENG, Grace                     -0.379
##  6 Democratic JEFFRIES, Hakeem                -0.489
##  7 Democratic MALONEY, Sean Patrick           -0.239
##  8 Republican ZELDIN, Lee M                    0.397
##  9 Democratic RICE, Kathleen Maura            -0.28
## 10 Republican STEFANIK, Elise M                0.269
## # ... with 19 more rows
```

Note that inserting parentheses around the entire expression results in both saving the dataframe to a new object and returning the lines that fit the conditions. So the line `(NY<-filter(house_memb_117, state_abbrev=="NY"))` will both save the contents of `filter()` to a new dataframe, *NY*, and will return the result.

Filtering rows by condition

Filtering and other applicable functions, allow you to use familiar comparison operators. While we used `==` equal to in the prior examples , other operators are less than or equal to `< =` , greater than or equal to `> =` , not equal to `!=`, and of course less than `<` and greater than `>`. Other functions can be combined with `filter()`, such as `is.na()` for identifying observations with missing values; `filter(house_memb_117, is.na(born))` would find any row with a missing birth year.

We can use simple Boolean logic to construct `filter()` functions that satisfy multiple conditions. For example, when conditions in `filter()` are listed sequentially, the result of the filter operation is the dataset when all conditions are evaluated as true. So the two statements below are equivalent:

```
filter(house_memb_117, state_abbrev=="NY" , born > 1969)
filter(house_memb_117, state_abbrev=="NY" & born > 1969)
```

These statements query for representatives who are from NY and born after 1969. The symbol for 'and' in the line above means that only for those lines when the two conditions `state_abbrev=="NY" & born > 1969` are true does the `filter()` function return a line of the dataset. The symbol `|` as an operator for 'or', substituted for &, would return all `NY` representatives or those born after 1969, regardless of State.

Combining the two, we can find the names of representatives from New York and New Jersey that fit the two conditions.

```
filter(house_memb_117, state_abbrev=="NY" |
                       state_abbrev=="NJ", born > 1969)
```

```
## # A tibble: 17 x 9
##    chamber icpsr distri~1 state~2 party~3 bioname  born
##    <chr>   <dbl>    <dbl> <chr>   <chr>   <chr>   <dbl>
##  1 House   21723        5 NJ      Democr~ GOTTHE~  1975
##  2 House   21937        3 NJ      Democr~ KIM, A~  1982
##  3 House   21964       11 NJ      Democr~ SHERRI~  1972
##  4 House   21101       23 NY      Republ~ REED, ~  1971
##  5 House   21342        6 NY      Democr~ MENG, ~  1975
##  6 House   21343        8 NY      Democr~ JEFFRI~  1970
##  7 House   21539        1 NY      Republ~ ZELDIN~  1980
##  8 House   21541       21 NY      Republ~ STEFAN~  1984
##  9 House   21916       19 NY      Democr~ DELGAD~  1977
## 10 House   21949       14 NY      Democr~ OCASIO~  1989
## 11 House   22105       16 NY      Democr~ BOWMAN~  1976
## 12 House   22117        2 NY      Republ~ GARBAR~  1984
```

```
## 13 House   22127        17 NY      Democr~ JONES,~  1987
## 14 House   22133        11 NY      Republ~ MALLIO~  1980
## 15 House   22154        15 NY      Democr~ TORRES~  1988
## 16 House   22169        19 NY      Democr~ RYAN, ~  1982
## 17 House   22170        23 NY      Republ~ SEMPOL~  1983
## # ... with 2 more variables: nominate_dim1 <dbl>,
## #   nominate_dim2 <dbl>, and abbreviated variable
## #   names 1: district_code, 2: state_abbrev,
## #   3: party_code
```

A tempting shortcut might be to think you can enter the function as `state_abbrev=="NY" | "NJ"`, but that would be incorrect. Instead, to avoid re-typing multiple `state_abbrev==` statements, use the "in" operator `%in%`, as in the following structure:

```
filter(house_memb_117, state_abbrev %in% c("NY", "NJ"), born > 1969)
```

Arranging rows

The `arrange()` function is used to arrange rows, or change the order in which rows are displayed, rather than select particular rows. With the *nominate_dim1* scores, the most obvious use would be to find the most liberal or conservative representatives. For example, `arrange(house_memb_117, nominate_dim1)` would display, in ascending order, the rows of the dataset starting with the lowest (or most liberal) value. To reverse the order to descending, use `desc()`.

```
## # A tibble: 455 x 4
##    state_abbrev party_code bioname                nomin~1
##    <chr>        <chr>      <chr>                    <dbl>
##  1 NM           Republican HERRELL, Yvette          0.936
##  2 SC           Republican NORMAN, Ralph            0.848
##  3 AZ           Republican BIGGS, Andrew S.         0.833
##  4 GA           Republican CLYDE, Andrew S.         0.823
##  5 IL           Republican MILLER, Mary E.          0.806
##  6 GA           Republican GREENE, Marjorie Ta~     0.8
##  7 TX           Republican ROY, Charles             0.8
##  8 VA           Republican GOOD, Bob                0.8
##  9 GA           Republican HICE, Jody Brownlow      0.797
## 10 TX           Republican JACKSON, Ronny           0.773
## # ... with 445 more rows, and abbreviated variable
## #   name 1: nominate_dim1
```

```
arrange(house_memb_117, desc(nominate_dim1))
```

The tibble, indexed to `house_memb_117[, c(4, 5, 6, 8)]`, displays the top ten most conservative lawmakers, at the top Rep. Yvette Herrell (R-NM), followed by some familiar names.

Selecting columns

Without indices, the tibble sorted on ideology displayed all columns by default. While indexing with `house_memb_117[, c(5, 6, 8)]` selects columns by their numeric positions, `select()` makes it clearer and more intuitive by referencing the column names directly. For example, the following function selects three columns from the dataset for display:

```
select(house_memb_117, party_code, bioname, nominate_dim1)
```

```
## # A tibble: 455 x 3
##    party_code bioname                          nominate~1
##    <chr>      <chr>                                 <dbl>
##  1 Republican ROGERS, Mike Dennis                   0.363
##  2 Democratic SEWELL, Terri                        -0.396
##  3 Republican BROOKS, Mo                            0.652
##  4 Republican PALMER, Gary James                    0.678
##  5 Republican CARL, Jerry L.                        0.52
##  6 Republican MOORE, Barry                          0.642
##  7 Republican ADERHOLT, Robert                      0.386
##  8 Republican YOUNG, Donald Edwin                   0.283
##  9 Democratic PELTOLA, Mary Sattler                -0.156
## 10 Republican RADEWAGEN, Aumua Amata Coleman        0.336
## # ... with 445 more rows, and abbreviated variable
## #   name 1: nominate_dim1
```

This use of `select()` is equivalent to ‘`house_memb_117[, c(5,6,8)]`. Or to display a sequential order of columns, separate the columns with a colon: `select(house_memb_117, party_code:nominate_dim1)`. Adding the function *everything()* to the end of the function will simply place the remaining variables at the end of the dataframe, `select(house_memb_117, party_code:nominate_dim1, everything())`.

Renaming columns

To rename a column of a dataframe, use the rename function. In the `rename()` function, the new variable name is listed first, followed by the old variable name. So to rename `nominate_dim1` as `left_right_ideology`:

```
rename(house_memb_117, left_right_ideology = nominate_dim1 )
```

```
## # A tibble: 455 x 3
##    party_code bioname                          left_rig~1
##    <chr>      <chr>                                 <dbl>
##  1 Republican ROGERS, Mike Dennis                   0.363
##  2 Democratic SEWELL, Terri                        -0.396
##  3 Republican BROOKS, Mo                            0.652
##  4 Republican PALMER, Gary James                    0.678
##  5 Republican CARL, Jerry L.                        0.52
##  6 Republican MOORE, Barry                          0.642
##  7 Republican ADERHOLT, Robert                      0.386
##  8 Republican YOUNG, Donald Edwin                   0.283
##  9 Democratic PELTOLA, Mary Sattler                -0.156
## 10 Republican RADEWAGEN, Aumua Amata Coleman        0.336
## # ... with 445 more rows, and abbreviated variable
## #   name 1: left_right_ideology
```

As in the other functions, for the result to be saved to a dataset, the assignment operator needed to either overwrite the existing dataset or save a new version of it, as in `house_memb_117<- rename(house_memb_117, left_right_ideology = nominate_dim1 )`. Multiple renames can be combined, such as `rename(house_memb_117, left_right_ideology = nominate_dim1, state=state_abbrev)`, to rename the ideology and state name.

Creating new columns

The *mutate()* function allows you to create new variables as combinations of other variables. For example, an age in years of each representative in the year 2022 could be calculated from *born*. With `mutate()` we name the new variable *age* and the formula for calculating it, which is the value 2022 minus the variable *born*, appended as the last column in the data:

```
mutate(house_memb_117, age = 2022-born)
```

```
## # A tibble: 455 x 5
##    state_abbrev party_code bioname            born   age
##    <chr>        <chr>      <chr>             <dbl> <dbl>
##  1 AL           Republican ROGERS, Mike De~  1958    64
##  2 AL           Democratic SEWELL, Terri     1965    57
##  3 AL           Republican BROOKS, Mo        1954    68
##  4 AL           Republican PALMER, Gary Ja~  1954    68
##  5 AL           Republican CARL, Jerry L.    1958    64
##  6 AL           Republican MOORE, Barry      1966    56
```

```
## 7 AL         Republican ADERHOLT, Robert  1965    57
## 8 AK         Republican YOUNG, Donald E~  1933    89
## 9 AK         Democratic PELTOLA, Mary S~  1973    49
## 10 AS        Republican RADEWAGEN, Aumu~  1947    75
## # ... with 445 more rows
```

There are many useful functions within `mutate()`. One is the ranking function.

We might be interested, for example, in ranking representatives in ideology. The `min_rank()` function will do this. By default, it assigns smallest values the smallest ranks; with the `desc()` function, we can assign smaller ranks to higher values. For example, assigning smallest ranks to more conservative (higher ideology scores):

```
mutate(house_memb_117, conservative_rank =
         min_rank(desc(nominate_dim1)))
```

The result is a table displaying the most conservative legislators, rank-ordered from most conservative starting at rank position 1. Further limiting the display of columns to fit on the printed page results in:

```
## # A tibble: 455 x 4
##    bioname                           born nomin~1 conse~2
##    <chr>                            <dbl>   <dbl>   <int>
##  1 ROGERS, Mike Dennis               1958   0.363     178
##  2 SEWELL, Terri                     1965  -0.396     349
##  3 BROOKS, Mo                        1954   0.652      41
##  4 PALMER, Gary James                1954   0.678      31
##  5 CARL, Jerry L.                    1958   0.52      103
##  6 MOORE, Barry                      1966   0.642      48
##  7 ADERHOLT, Robert                  1965   0.386     170
##  8 YOUNG, Donald Edwin               1933   0.283     207
##  9 PELTOLA, Mary Sattler             1973  -0.156     227
## 10 RADEWAGEN, Aumua Amata Coleman    1947   0.336     188
## # ... with 445 more rows, and abbreviated variable
## #   names 1: nominate_dim1, 2: conservative_rank
```

Without the `desc()` function, `min_rank()` assigns higher ranks to smaller values.

Saving new datasets

Without an assignment operator, the functions such as `mutate()` and `select()` will not alter the dataset saved within the Environment pane. To create a subsetted data table (or to overwrite the existing data table), the assignment operator is necessary. With the assignment operator for the example with *age*, the variable is written to the end of the dataframe. For example the following lines add the variable *age* to the dataframe:

```
house_memb_117<-mutate(house_memb_117, age = 2022-born)
```

To save only a subset of variables – including the newly created *age* variable – we would use the `select()` function to identify specific variables. For example the two lines below create the *age* variable with `mutate()`, appending *age* to *house_memb_117_subset*, which is a subset of the five variables identified in the next line, saved again as *house_memb_117_subset.*

```
house_memb_117_subset<-mutate(house_memb_117, age = 2022-born)
house_memb_117_subset<-select(house_memb_117_subset, state_abbrev,
                              party_code, bioname, born, age)
```

The new subset data table contains only the five variables:

```
house_memb_117_subset
```

```
## # A tibble: 455 x 5
##    state_abbrev party_code bioname             born   age
##    <chr>        <chr>      <chr>              <dbl> <dbl>
##  1 AL           Republican ROGERS, Mike De~   1958    64
##  2 AL           Democratic SEWELL, Terri      1965    57
##  3 AL           Republican BROOKS, Mo         1954    68
##  4 AL           Republican PALMER, Gary Ja~   1954    68
##  5 AL           Republican CARL, Jerry L.     1958    64
##  6 AL           Republican MOORE, Barry       1966    56
##  7 AL           Republican ADERHOLT, Robert   1965    57
##  8 AK           Republican YOUNG, Donald E~   1933    89
##  9 AK           Democratic PELTOLA, Mary S~   1973    49
## 10 AS           Republican RADEWAGEN, Aumu~   1947    75
## # ... with 445 more rows
```

5.3 Combining functions

Often in managing unwieldy datasets, we combine different functions to accomplish a series of steps, such as mutating or creating a new variable, selecting columns, filtering the order of rows, and printing out or saving a subset. Doing so in an efficient way, however, requires us to combine multiple functions in a sequential order. The "pipe operator", the characters `%>%` (Bache and Wickham, 2020), allows multiple functions to be chained together.

Conceptually, the pipe operator translates to "and then". So for example, we could filter a dataset, and then select particular rows, and then arrange the rows. The pipe operator simplifies the process of combining functions in a highly readable way, sequential steps separated by `%>%`. As a core element of the **tidyverse**, the piper operator is available once **dplyr** or the entire **tidyverse** is attached. It is worth noting that a recent update to **R** added a native pipe, `|>`, which works nearly identically, although the code in this text relies on `%>%`. Line by line we will use `%>%` to combine functions to do the following:

(1) Filter on representatives from New York.
(2) Select the columns for *party_code*, *bioname*, and *nominate_dim1*.
(3) Arrange the dataframe rows by conservatism, from most conservative in row 1 to least at the end.

(4) Print out the data table.

To do so, we start with the `house_memb_117` dataframe and then add the functions line by line, adding `%>%` until the last command:

```
house_memb_117 %>%
  filter(state_abbrev=="NY") %>%
  select(party_code, bioname, nominate_dim1) %>%
  arrange(desc(nominate_dim1))
```

```
## # A tibble: 29 x 3
##    party_code bioname               nominate_dim1
##    <chr>      <chr>                         <dbl>
##  1 Republican SEMPOLINSKI, Joseph           0.545
##  2 Republican TENNEY, Claudia               0.453
##  3 Republican ZELDIN, Lee M                 0.397
##  4 Republican JACOBS, Chris                 0.306
##  5 Republican MALLIOTAKIS, Nicole           0.304
##  6 Republican GARBARINO, Andrew R.          0.274
##  7 Republican REED, Thomas W. II            0.269
##  8 Republican STEFANIK, Elise M             0.269
##  9 Republican KATKO, John                   0.185
## 10 Democratic RYAN, Patrick                -0.206
## # ... with 19 more rows
```

Notice that at the end of the tibble, a comment states there are 19 more rows, in addition to the printed 10, for a total of 29 rows and 3 columns. To print the entire tibble (all of the rows) we add `print(n=39)` as an additional row. Below the `print()` function prints the top five:

```
house_memb_117 %>%
  filter(state_abbrev=="NY") %>%
  select(party_code, bioname, nominate_dim1) %>%
  arrange(desc(nominate_dim1)) %>%
  print(n=5)
```

```
## # A tibble: 29 x 3
##   party_code bioname                 nominate_dim1
##   <chr>      <chr>                           <dbl>
## 1 Republican SEMPOLINSKI, Joseph             0.545
## 2 Republican TENNEY, Claudia                 0.453
## 3 Republican ZELDIN, Lee M                   0.397
## 4 Republican JACOBS, Chris                   0.306
## 5 Republican MALLIOTAKIS, Nicole             0.304
## # ... with 24 more rows
```

Notice that in each of the functions, such as `filter()`, it is not necessary to include the name of the dataframe. Through the pipe operator, the dataframe to the left of the first pipe operator `house_memb_117 %>%` is passed on to each of the functions below it. To save the table to a dataframe, for example named *ny_reps*, we would change the first line to `_ny_reps<-house_memb_117 %>%`.

Grouped summary statistics

It is often useful to sort through data by groups within it. The `group_by()` function, sets a grouping structure for the dataframe that is taken into account when functions like are subsequently applied. It is often useful in combination with the `summarize()` function, for calculating summary statistics on groups. For example, we may be interested in average ideological orientation of State delegations, or averages within parties within States. Or perhaps we would like to count the number of representatives from each State, by political party.

We would first group the data with `group_by()` then apply `summarize()`. Because the `summarize()` function reduces the data down to a single summary statistic, it requires both a name of the new summary statistic in the data table, and a function to calculate the summary statistic, such as `mean()`, `sum()`, `range()`, `max()` or `min()`.

For example, in `summarize(mean_ideology = mean(nominate_dim1))`, `mean_ideology` will be the column name in the summary dataframe, and `mean(nominate_dim1)` is the function that calculates the average ideology score.

Let's observe how it works within partisan groups of New York legislators.

```
house_memb_117 %>%
  filter(state_abbrev=="NY") %>%
  group_by(party_code) %>%
  summarize(mean_ideology = mean(nominate_dim1))
```

```
## # A tibble: 2 x 2
##   party_code mean_ideology
##   <chr>              <dbl>
## 1 Democratic        -0.381
## 2 Republican         0.334
```

The `summarize()` function reduces the dataframe to a set of means, ideology by party group. The average Democrat and Republican representative from New York is roughly similar, although the mean Democratic ideology is slightly larger in magnitude than Republican.

The `summarize()` function can take multiple arguments, allowing you to calculate several summary statistics at once. For example, we could calculate both the mean and standard deviation within parties:

```
house_memb_117 %>%
  filter(state_abbrev=="NY") %>%
  group_by(party_code) %>%
  summarize(
    mean_ideology = mean(nominate_dim1),
    sd_ideology = sd(nominate_dim1))
```

```
## # A tibble: 2 x 3
##   party_code mean_ideology sd_ideology
##   <chr>              <dbl>       <dbl>
## 1 Democratic        -0.381       0.123
## 2 Republican         0.334       0.111
```

The `sd_ideology` expression follows after `mean_ideology`. Note that in these examples of calculating summary statistics, there are no missing values in the data. Where there are missing values **R** requires specific instructions on what to do with the missing values, in this case it would be to remove missing values from any calculations. For example, the statement for the standard deviation would need to be `sd_ideology = sd(nominate_dim1, na.rm = TRUE)`.

One additional useful function to add to `summarize()` is `n()`, which does not require any arguments. The function will count up the number of rows defined by the `group_by()` function. For example, to add counts of the `party_code` groups:

```
house_memb_117 %>%
  filter(state_abbrev=="NY") %>%
  group_by(party_code) %>%
  summarize(
    mean_ideology = mean(nominate_dim1),
    sd_ideology = sd(nominate_dim1),
    count_n = n())
```

```
## # A tibble: 2 x 4
##   party_code mean_ideology sd_ideology count_n
##   <chr>              <dbl>       <dbl>   <int>
## 1 Democratic        -0.381       0.123      20
## 2 Republican         0.334       0.111       9
```

Overall, with a few functions, starting with those listed in Table 5.1, datasets can be transformed in nearly limitless ways. The following section demonstrate the use of **dplyr** functions for shaping up raw datasets, wrangling and cleaning data for use in analysis.

5.4 Cleaning and transforming county-level votes

Another important application of data wrangling is in creating tidy datasets from a source, such as a US Secretary of State's election portal, aggregated to a geographic level. In this example, we look at election results from the November 2016 general election, aggregated to the Michigan County level. We will extract County-level presidential votes and organize a tidy dataset. From there, we will construct visualizations of the election results across Michigan counties with a specific data visualization, a "Cleveland dotplot" (Cleveland and McGill, 1984).

Working with county-level election results

Save the *2016GEN_MI_CENR_BY_COUNTY.xls* datafile to your working directory; use the Session menus to change it.[2]

[2]To download the data directly see the Michigan Secretary of State website, https://www.michigan.gov/sos/elections/election-results-and-data. Scrolling through past election results, find the 2016 November general election, which will link to a webpage showing County-level results for all races on the ballot, https://mielections.us/election/results/2016GEN_CENR.html. At the top of this page, click "Data" then the link to "TAB-delimited by County", which will start the process of downloading a file with the election results.

The data is imported with the `read_tsv()` (tab-separated value) function from the **dplyr** package. Place the file in your working directory and run the function. The datafile from the Secretary of State's Office (*2016GEN_MI_CENR_BY_COUNTY.xls*) has the file extension *.xls*, suggesting it is a Microsoft Excel formatted file, but it is actually saved as a tab delimited text file:

```
MI_2016<-read_tsv(file="2016GEN_MI_CENR_BY_COUNTY.xls")
```

When you import the data, you will see a list of columns 'parsed' from the data, starting with the `ElectionDate` and ending with '`Nominated(N)/Elected(E)`. Each row of the dataset corresponds to a single candidate for any elected office appearing on a ballot within each County. So for each County, there are as many rows as there are candidates appearing on the ballot for election to any office, which is why there are 6,082 rows:

```
MI_2016
```

```
## # A tibble: 6,082 x 19
##    ElectionDate Offic~1 Distr~2 Statu~3 Count~4 Count~5
##    <chr>        <chr>   <chr>   <chr>     <dbl> <chr>
##  1 2016-11-08   01      00000   0             1 ALCONA
##  2 2016-11-08   01      00000   0             1 ALCONA
##  3 2016-11-08   01      00000   0             1 ALCONA
##  4 2016-11-08   01      00000   0             1 ALCONA
##  5 2016-11-08   01      00000   0             1 ALCONA
##  6 2016-11-08   01      00000   0             1 ALCONA
##  7 2016-11-08   01      00000   0             1 ALCONA
##  8 2016-11-08   01      00000   0             1 ALCONA
##  9 2016-11-08   01      00000   0             1 ALCONA
## 10 2016-11-08   01      00000   0             1 ALCONA
## # ... with 6,072 more rows, 13 more variables:
## #   OfficeDescription <chr>, PartyOrder <chr>,
## #   PartyName <chr>, PartyDescription <chr>,
## #   CandidateID <dbl>, CandidateLastName <chr>,
## #   CandidateFirstName <chr>,
## #   CandidateMiddleName <chr>,
## #   CandidateFormerName <lgl>, ...
```

We will extract the vote totals within each county for the presidential election, and then organize a Michigan County level presidential elections dataset, to visualize candidate support across counties. So for this purpose the file from the State includes mostly rows of irrelevant election results. So the first step is to delete all rows that do not relate to the presidential election. We use the `filter()` function to select only the results — the rows — for the presidential election. The `OfficeDescription` field in the dataset contains indicators for different elections. Open the data spreadsheet in the Environment pane to review the offices. Or table out the office descriptions with `table(MI_2016$OfficeDescription)`.

The presidential election results are stored as `President of the United States 4 Year Term (1) Position`, in the first set of counties. The other offices follow. To find any set of elections, such as all results for U.S. House elections, we could use a search function to look for all instances of the term `Congress` in the `OfficeDescription` field. For example, the function `str_detect()` will find and return the locations in the dataset of a search string. For example, to find the rows where "Congress" appears, we could enter `str_detect(MI_2016$OfficeDescription, "Congress")`. This function would give us a result, either `TRUE` or `FALSE` for each row. Let's see what that summary would look like:

```
summary(str_detect(MI_2016$OfficeDescription, "Congress"))
```

```
##      Mode   FALSE    TRUE    NA's
## logical    5651     430       1
```

It shows that 430 rows reference a race for Congress. If we wanted to select only these rows, we would use `str_detect()` inside the `filter()` function. These two lines subset the `MI_2016` dataset by filtering the `OfficeDescription` field to select only the rows where the term "Congress" appears: `MI_2016 %>% filter(str_detect(OfficeDescription, "Congress"))`.

Following this example, we filter on the presidential election and save a new version of the data – *MI_prez_2016* with just the variables we needed to analyze vote returns over Counties: `CountyName`, `CandidateLastName`, `CandidateVotes`, `CountyCode`. `CountyCode` is a Federal Information Processing Standard (FIPS) indicator for each County[3]

```
MI_prez_2016<-MI_2016 %>%
  filter(OfficeDescription==
    "President of the United States 4 Year Term (1) Position") %>%
    select(CountyName, CandidateLastName, CandidateVotes, CountyCode)
```

```
MI_prez_2016
```

```
## # A tibble: 1,079 x 4
##    CountyName CandidateLastName CandidateVotes County~1
##    <chr>      <chr>                      <dbl>    <dbl>
```

[3]County-level FIPS codes are a standardized set of numerical codes used to uniquely identify counties within the United States. More information about county-level FIPS codes, including their structure and usage, can be found on the U.S. Census Bureau's website, https://www.census.gov/quickfacts/fact/note/US/fips.

```
## 1 ALCONA      Trump                      4201        1
## 2 ALCONA      Clinton                    1732        1
## 3 ALCONA      Johnson                     164        1
## 4 ALCONA      Castle                       28        1
## 5 ALCONA      Stein                        54        1
## 6 ALCONA      Soltysik                      0        1
## 7 ALCONA      Fox                           0        1
## 8 ALCONA      Hartnell                      0        1
## 9 ALCONA      Hoefling                      0        1
## 10 ALCONA     Kotlikoff                     0        1
## # ... with 1,069 more rows, and abbreviated variable
## #   name 1: CountyCode
```

Reshaping the data, long to wide format

In *MI_prez_2016* there is one column of vote counts (`CandidateVotes`), and for every County a row containing an entry for each presidential candidate listed on the ballot in that County. This way of organizing the data is called 'long' format, because the dataset is longer than it is wide; instead of having a different column (or variable) for each candidate's votes, there is one column to record votes and separate rows for each candidate.

To simplify analysis, we want to restructure the data to 'wide' format. In this format, the dataset will have one row per county (exactly 83 rows) and a separate column for each candidate's vote totals. This wide structure is particularly useful for calculations, such as summing vote totals over all third-party candidates, because it facilitates column based operations.

The function `pivot_wider()` will translate the data from long to wide data format. The required arguments for `pivot_wider()` are (1) the names of each new column in wide format, and (2) the values to be stored in these columns. From long format, the names in `CandidateLastName` will be passed to wide format, while the values in `CandidateVotes` will be stored in the new wide format columns.

```
MI_prez_2016_wide <- MI_prez_2016 %>%
    pivot_wider(names_from = CandidateLastName,
      values_from=CandidateVotes)

MI_prez_2016_wide
```

```
## # A tibble: 83 x 15
##   CountyN~1 Count~2 Trump Clinton Johnson Castle Stein
##   <chr>       <dbl> <dbl>   <dbl>   <dbl>  <dbl> <dbl>
## 1 ALCONA          1  4201    1732     164     28    54
## 2 ALGER           2  2585    1663     177     19    67
```

```
## 3 ALLEGAN          3 34183   18050    2513    289   596
## 4 ALPENA           4  9090    4877     498     53   160
## 5 ANTRIM           5  8469    4448     459     47   146
## 6 ARENAC           6  4950    2384     270     26    60
## 7 BARAGA           7  2158    1156     106     15    49
## 8 BARRY            8 19202    9114    1424    181   346
## 9 BAY              9 28328   21642    2189    208   544
## 10 BENZIE         10  5539    4108     382     36   149
## # ... with 73 more rows, 8 more variables:
## #   Soltysik <dbl>, Fox <dbl>, Hartnell <dbl>,
## #   Hoefling <dbl>, Kotlikoff <dbl>, Maturen <dbl>,
## #   McMullin <dbl>, Moorehead <dbl>, and abbreviated
## #   variable names 1: CountyName, 2: CountyCode
```

With the data now in wide format, we create an 'other' vote count column. This variable will record vote totals for all candidates who are neither Trump nor Clinton. Within `mutate()` , the *other* variable is the sum of each individual candidate column:

```
MI_prez_2016_wide<-MI_prez_2016_wide %>%
  mutate(other=Castle+ Fox + Hartnell  + Hoefling + Kotlikoff + Maturen
        + McMullin +  Moorehead + Soltysik + Stein) %>%
  select(CountyName, CountyCode, Trump, Clinton, other) %>%
  print(n=5)
```

```
## # A tibble: 83 x 5
##   CountyName CountyCode Trump Clinton other
##   <chr>           <dbl> <dbl>   <dbl> <dbl>
## 1 ALCONA              1  4201    1732   101
## 2 ALGER               2  2585    1663    93
## 3 ALLEGAN             3 34183   18050  1040
## 4 ALPENA              4  9090    4877   233
## 5 ANTRIM              5  8469    4448   206
## # ... with 78 more rows
```

Then we `select()` only the necessary columns for recording presidential vote totals across counties, identifiers *CountyName* and *CountyCode*, followed by *Trump*, *Clinton*, and *other*. Given the raw votes, we could create vote percentage totals for each candidate:

```
MI_prez_2016_wide<-MI_prez_2016_wide %>%
 mutate(Trump_percent=Trump/(Trump + Clinton + other)*100) %>%
 mutate(Clinton_percent=Clinton/(Trump + Clinton + other)*100)
```

```
## # A tibble: 83 x 7
```

```
##    County~1 Count~2 Trump Clinton other Trump~3 Clint~4
##    <chr>      <dbl> <dbl>   <dbl> <dbl>   <dbl>   <dbl>
##  1 ALCONA         1  4201    1732   101    69.6    28.7
##  2 ALGER          2  2585    1663    93    59.5    38.3
##  3 ALLEGAN        3 34183   18050  1040    64.2    33.9
##  4 ALPENA         4  9090    4877   233    64.0    34.3
##  5 ANTRIM         5  8469    4448   206    64.5    33.9
##  6 ARENAC         6  4950    2384    92    66.7    32.1
##  7 BARAGA         7  2158    1156    70    63.8    34.2
##  8 BARRY          8 19202    9114   589    66.4    31.5
##  9 BAY            9 28328   21642   818    55.8    42.6
## 10 BENZIE        10  5539    4108   199    56.3    41.7
## # ... with 73 more rows, and abbreviated variable
## #   names 1: CountyName, 2: CountyCode,
## #   3: Trump_percent, 4: Clinton_percent
```

The dataset now consists of an organized, wide table of presidential candidate support, one row for each county, and candidate support across multiple columns.

5.5 Wrangling election data and dotplotting results

The 'Cleveland dotplot', named after its creator William S. Cleveland, is a data visualization form for comparing quantitative values across categories of a qualitative variable. In this case, we are interested in visualizing levels of candidate support across each Michigan county. Similar to a bar graph, where the lengths of bars would be drawn to scale the percentage of the vote won by each candidate across counties, a Cleveland dotplot replaces bars with dots, arranged along a horizontal or vertical axis. The length of a bar is replaced by position of a dot. By eliminating the area of bars on the graph, the dotplot facilitates comparisons across many categories.

For *MI_prez_2016_wide*, visualizing raw vote totals is not feasible in one graphic because of the disparity between the largest counties, such as Detroit's Wayne County, and any of the smaller counties that would appear indistinguishable. What interests us more than raw votes is the percentage of each candidate's support by county. To construct the dotplot, first in `MI_prez_2016_wide` because the County names are in all caps, we will convert each one to "title case" – uppercase first letters and lowercase remaining letters — with the `str_to_title()` function within `mutate()`:

```
MI_prez_2016_wide<- MI_prez_2016_wide %>%
 mutate(CountyName=str_to_title(CountyName))
```

The `str_to_title()` function (from the **stringr** package (Wickham, 2022) in the **tidyverse**) converts each word, for example, "ST.CLAIR" to "St. Clair". Notice the difference:

```
MI_prez_2016_wide %>%
  print(n=3)
```

```
## # A tibble: 83 x 7
##   CountyN~1 Count~2 Trump Clinton other Trump~3 Clint~4
##   <chr>       <dbl> <dbl>   <dbl> <dbl>   <dbl>   <dbl>
## 1 Alcona          1  4201    1732   101    69.6    28.7
## 2 Alger           2  2585    1663    93    59.5    38.3
## 3 Allegan         3 34183   18050  1040    64.2    33.9
## # ... with 80 more rows, and abbreviated variable
## #   names 1: CountyName, 2: CountyCode,
## #   3: Trump_percent, 4: Clinton_percent
```

A key aspect of the Cleveland dotplot is to order the categories (counties) to make the visual comparisons clearer. For example, rather than organize the counties alphabetically, we will arrange counties by *Trump_percent* or *Clinton_percent*.

Reorder function

The dots are constructed in `ggplot()` by `geom_point()`. In the `aes()` argument, we specify the order of the counties within an additional function. The `reorder()` function sorts the names of the counties by Trump's percentage of the vote (`reorder(CountyName, Trump_percent)`) or Clinton (`reorder(CountyName, Clinton_percent)`). This function is placed within the `aes()` function as the name of the County variable. Figure 5.1 displays a Cleveland dotplot with counties sorted by the Trump vote percent.

```
MI_prez_2016_wide %>%
  ggplot() +
  geom_point(aes(reorder(CountyName, Trump_percent), Trump_percent)) +
  labs(title="Trump Percentage Vote by Michigan County, 2016",
       x="County", y="Vote Percentage") +
  coord_flip()
```

Each dot represents one county, plotted along the X-axis based on the vote percentage. The counties are arranged in ascending order of support on the

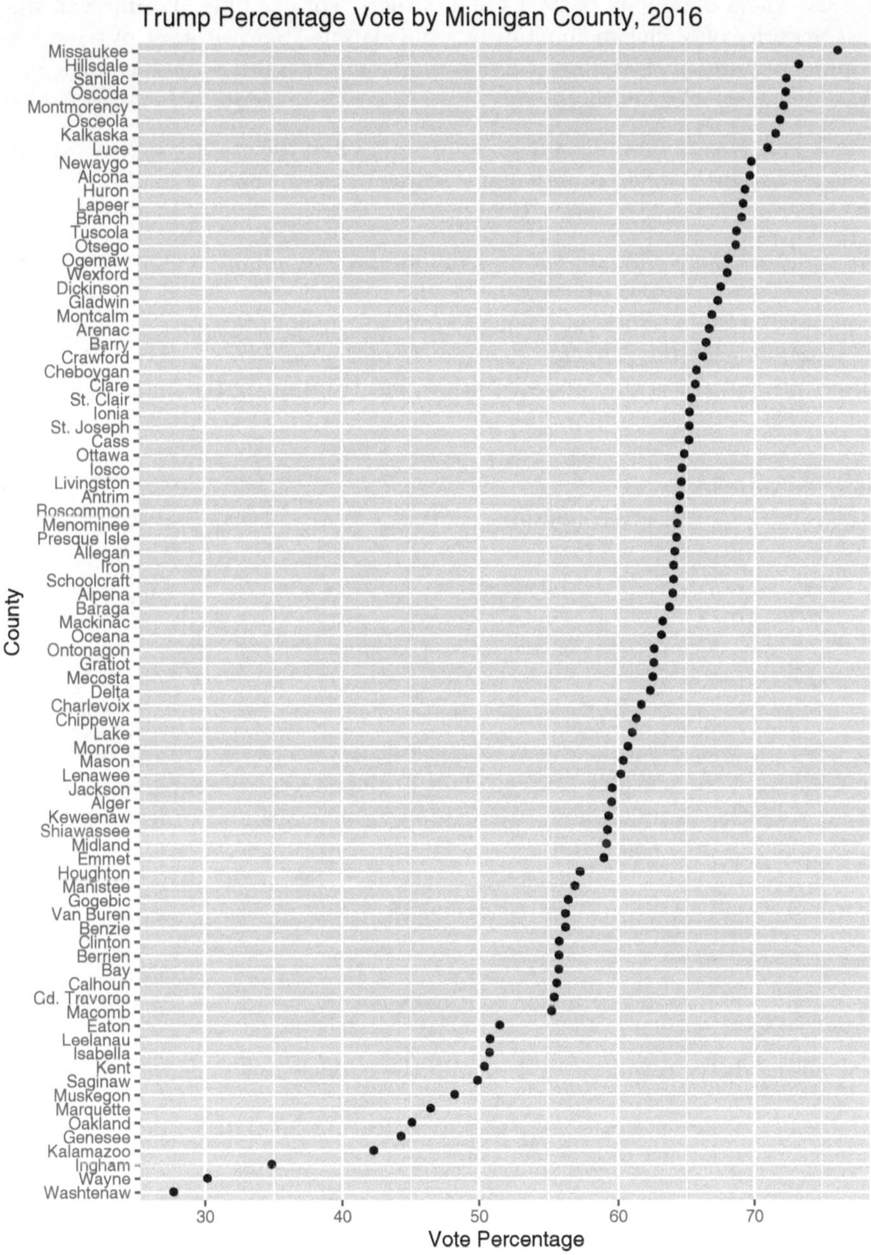

FIGURE 5.1 Cleveland dotplot of Trump's vote percentage by Michigan county in the 2016 presidential election, created with **ggplot2**. County points are ordered by vote percentage, making it easier to compare across counties than in a traditional bar chart.

Y-axis, which highlights patterns and outliers. For example, Trump won the vast majority of Michigan counties by wide margins, but lost most of the more populous counties of Michigan cities by a small margin. Wayne county and Washtenaw are clear outliers.

The X and Y axes are flipped in `coord_flip()` , and the graphic aspect ratio is elongated to make all 83 county names visible. The `reorder()` function changes the order of Michigan counties. Otherwise the syntax is similar to a scatterplot.

We could add two different series, Clinton and Trump, and color each series. Note that *ggplot2* expects data to be in a long format. As a shortcut, or perhaps exploratory graphic, without having to convert the data to long format, it may be easier to plot the percents in two different `geom_point()` series. To save space, this dotplot is not produced below, but try running the lines to observe the result:

```
MI_prez_2016_wide %>%
  ggplot(aes(reorder(CountyName, Trump_percent), Trump_percent)) +
  geom_point(color="red") +
  labs(title="Trump and Clinton Votes by Michigan County, 2016",
       x="County", y="Number of Votes",
          caption="Clinton in blue, Trump in red") +
  coord_flip() +
  geom_point(aes(CountyName, Clinton_percent), color="blue")
```

In Chapter 6 we consider election data in the context of maps, and among other subjects, examine how to create election maps and other aspects of geographically referenced data.

Resources

The official **dplyr** reference guide details the package functions with examples: https://dplyr.tidyverse.org/reference/index.html.

5.6 Exercises

Answer the questions with R code supported by a narrative explanation within an RMarkdown document.

(1) Construct a series of box-whiskers plots for the ideology scores within State delegations to the US House. Pick a subset of five States to compare and wrangle the data, using `filter()` to identify the States of interest. Save the data as a dataframe or pass on the wrangled data to **ggplot2** for the construction of the box-whiskers plots.

(2) Based on the Congressional ideology (Voteview.com) data, write a three-paragraph report on the ideology of the Congressional delegation from Florida. Be sure to include a discussion of how the delegation relates to each Chamber's overall median and within party caucus median ideology score. Be sure to include appropriate summary statistics with your answer.

(3) Construct a data visualization consisting of two histograms to compare the ideological orientations of representatives to the U.S. House by party caucus, in one of the following states: Texas, California, or New York. Wrangle the data to create a subset for the state, and for the histograms use color and transparency to highlight differences between the two caucuses.

(4) In Michigan, which US House legislators had the most liberal and most conservative ideology scores? Create two data tables to display district number, name, party affiliation, and ideology (left - right) score.

(5) Pick a US State, identify the Congressional delegation from it. Create a Cleveland dotplot to visualize the most conservative to most liberal legislators, from a given State, by their ideology score.

6

Maps and Spatial Data

Chapter 6 reviews the use of geographic packages for drawing and analyzing data embedded in maps.

Learning objectives and chapter resources

By the end of this chapter, you should be able to (1) explain key characteristics of geospatial data for creating political maps; (2) identify and collect geospatial datafiles relevant to the study of politics; (3) apply tools in visualizing and analyzing geospatial data, analyzing the extent of gerrymandering in U.S. legislative districts. The material in the chapter requires the following packages to be installed: **leaflet** (Cheng et al., 2023), **leaflet.extras** (Gatscha et al., 2024), **sf** (Pebesma, 2018; Pebesma and Bivand, 2023), **ggspatial** (Dunnington, 2023), **cartogram** (Jeworutzki, 2023), and **tidycensus** (Walker and Herman, 2024). For data preparation and labeling maps: **dplyr** (Wickham et al., 2023b) and **ggplot2** (Wickham et al., 2023a) from the **tidyverse** (Wickham, 2023b), which should already be installed. The material uses the following datasets *michigan12_16.csv*, *Michigan_State_Senate_Districts_2011_plan.geojson*, *ZAF_ADM4.geojson*, *Crime_Incidents_in_2024.geojson*, and *Counties___v17a.shp* from https://faculty.gvsu.edu/kilburnw/inpolr.html.

6.1 Maps as data

Maps and the visualization of data associated with political boundary lines are of inherent interest to the study of politics. The chapter begins with a brief overview of key terminology in the analysis of data with geographic characteristics. From there we apply these principles in the creation of maps visualizing political phenomena, such as election results, leading to the study of 'gerrymandering', the practice of drawing legislative districts to benefit one party over another.

The use of geospatial data requires some unusual assumptions about how the data was collected and what visually it represents. For example, these assumptions involve underlying models of how latitude and longitude coordinates are determined, and how a particular geographic area of the three-dimensional earth gets represented on a two-dimensional image. The discussion below is necessarily brief, providing an overview of essential ideas for identifying, cleaning, and combining geographic data. For a more in-depth discussion and application to more advanced methods of analysis, see Lovelace, Nowosad, and Muenchow (2019) and Walker (2023).

The term "geospatial" or "spatial data" refers to a particular type of data, information to identify the geographic location and characteristics of natural and man-made features around the world, from rivers and mountain ranges to political boundaries and election results. Geospatial information is organized into two broad types, vector and raster.

Vector data is the type more common in political science research; it is used to represent particular geometries such as points, lines, and polygons that in turn represent political phenomena of interest. For example, points could represent the location of people or cities, lines for roads, and polygons for the shape of political entities such as legislative districts. Raster data is relatively uncommon in the study of politics. A "raster" refers to a grid or matrix of cells (imagine screen pixels), each containing a value that represents a specific attribute, such as elevation or temperature. To make sense of the data, each of these geometric entities usually has "attribute data" attached, such as the name of the city, road, or political boundary. Attribute data are characteristics of the geographies that are not part of defining the geography itself, such as the number of registered Republican or Democratic voters in a legislative district.

Key terms in geospatial data

The terms and underlying assumptions about a "datum", "ellipse", and "projection" require brief explanation. A "datum" is a reference model of the earth's shape, used as the basis for the coordinate system of latitude and longitude. The datum is based on a particular mathematical model of the "ellipse", the three-dimensional representation of the earth's shape upon which the coordinates are developed. From a starting point in this model, the northerly or southerly location (latitude) and easterly or westerly location (longitude) coordinates are developed.

In working with geospatial data, you will often see reference to "North American Datum of 1983 (NAD83)" or more likely the "World Geodetic System of 1984 (WGS84)". The WGS 84 determines the reference coordinate system used by Global Positioning System (GPS) tools. Most geospatial data available for download on the web will use this datum. Ideally, data collected from a different datum intended to be combined should be placed on the same datum, but this step is really only necessary for a high degree of precision or map resolution.

The cardinal directions North, South, East, and West correspond to particular coordinate designations. For latitude, coordinates in the Northern hemisphere coordinates are recorded with an N and Southern coordinates with an S; for longitude, Eastern hemisphere with an E and Western hemisphere with a W. In place of cardinal directions, positive coordinate numbers are used for N and negative coordinate numbers for S — similarly, an E with a positive coordinate number and W with a negative coordinate number.

Map projection

A projection refers to a particular way of taking geographic features on the curved three-dimensional surface of Earth and 'projecting' or translating it onto a two-dimensional flat image. Like taking a curved piece of paper with an image of North America and flattening it, parts of the paper will have to fold over onto itself, distorting the original curved image. Different projections will distort different aspects of the map (like area, shape, or distance) to a greater or lesser extent. For an analysis involving measurements like distance or geographic area, or when working with large areas, it is usually necessary to consider projection. Without a proper projection, measurements can be inaccurate without accounting for the curvature of the Earth.

To illustrate differences between projections, Figure 6.1 presents two different projections of the continental USA. The first is a Mercator projection. The Mercator projection, in the upper panel of Figure 6.1, is a projection prioritizing navigation (think driving directions), but notice the distorted straight line across the USA-Canada border and the distorted shapes of States near it. Below it is the "Albers Equal-Area Conic Projection", which more closely preserves area and shape of the States at the expense of navigation. While less useful for navigating, the map projection features States with a more familiar shape and accurate area.

Geospatial datafile formats

In searching for your own spatial data — vector data for displaying points, lines, or polygons — there are two commonly found formats.

(1) A "Shapefile" refers to a set of files in a format originally developed by ESRI, a geographic information systems company. The Shapefile itself consists of not just one *.shp* ('shape') file, but also additional helper files including *.shx*, *.dbf* (for storing attribute data), and *.prj*. When working with Shapefiles, these three ancillary files are necessary and should be downloaded with the *.shp* file, all of which are usually made available in a *.zip* file archive. Shapefiles are

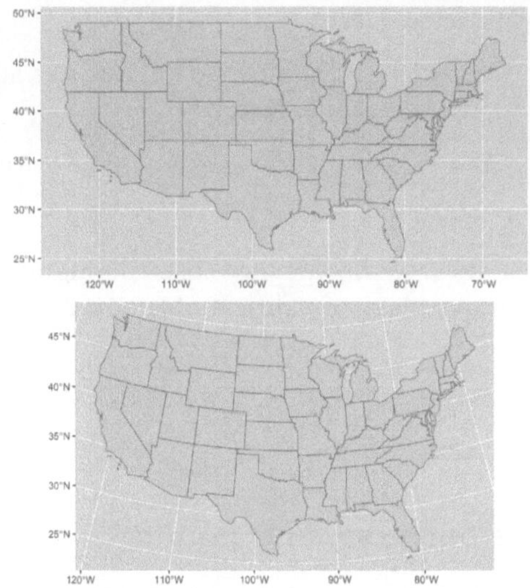

FIGURE 6.1 Comparison of two map projections for the continental United States. The upper panel uses the Mercator projection, while the lower panel uses the Albers Equal-Area Conic projection. These maps illustrate how projection choice affects the visual representation of geography.

relatively easy to find for the creation of vector maps, such as legislative boundary lines and voting precincts.

(2) A "GeoJSON" file refers to the organization of the spatial data into a JSON file, the same as reviewed in Chapter 3. A single *.geojson* file may contain both geographic and attribute data sometimes referred to as "properties" of the geographic datafile. Because it is a single self-contained file and relatively smaller in file size, data analysts often prefer to work with GeoJSON over the "Shapefile" format.

Key steps and the big picture of maps and spatial data analysis

Constructing maps to analyze data tied to geographic features typically involves four basic steps.

(1) Locating and downloading relevant geospatial datafiles, such as a *.geojson* file outlining legislative districts. If needed, make changes to the map projection; if combining multiple map files together as map layers (discussed further below), verify that each layer references the same datum and projection.

(2) Locating or organizing any additional attribute data, characteristics of the geographic features (such as votes cast within a State) that

are the focus of the analysis.

(3) Merging the geospatial and attribute data, associating the attribute data with relevant geographic features.

(4) Transforming and visualizing the attribute data within the geography.

In the sections that follow, we first review constructing maps, starting with the creation of interactive maps of political phenomena, then we will create static image maps.

6.2 Visualizing polygons and points

In working through Chapter 6, note that mapping packages are large in file size and time-consuming to download. Drawing maps requires much more computing power, along with much more patience until the maps show up in the Plots or Viewer pane of RStudio. So to make these analyses more feasible, the maps and geospatial data in the chapter usually focus on just one particular U.S. State, Michigan, or small areas with relatively few geographic features.

The functionality of interactive maps, such as maps that allow immediate panning or zooming on different areas or altering the data displayed on a map, is created through the **leaflet** (Cheng et al., 2023) package and accessed through the "Viewer" pane (lower right of RStudio). The map interactivity is limited to this Viewer pane or, to view interactive maps outside of RStudio, the maps can be exported to an HTML file and hosted on a web server. The interactive maps cannot be embedded within a Microsoft Word document or PDF. So there is a tradeoff: while interactive maps allow for exploration of geographic features, this interactivity is limited to RStudio or webpages.

A basic interactive map using **leaflet** starts with the `leaflet()` function, which will require a geographic location to display in the map. One way to set the location of the map is by simply specifying latitude and longitude coordinates. The `setView()` function asks for both latitude (`lat=`) and longitude (`lng=`), and a level of focus or 'zoom', `zoom =`. A simple interactive map can be drawn with these two lines, followed by a third function, `addTiles()` which will draw the 'tiles' or image squares of the map.

If necessary, install **leaflet** (`install.packages("leaflet")`). To draw a map centered around Cape Town, South Africa, longitude is set to 18.6 degrees and latitude to -34 degrees:

```
library(leaflet)

leaflet() %>%
  setView(lng = 18.6, lat = -34, zoom = 10) %>%
  addTiles()
```

The map will appear in the Viewer pane on the lower right-hand side of RStudio as in Figure 6.2, based on map data from OpenStreetMaps[1]. Using the mouse to pan the map in different directions, and the plus and minus signs to zoom in and out, the map can focus on any arbitrary area. The limiting scope of an interactive map of this kind is the attribute data provided by a researcher. As an additional layer to the map, a researcher can add any additional geospatial and attribute data of political interest.

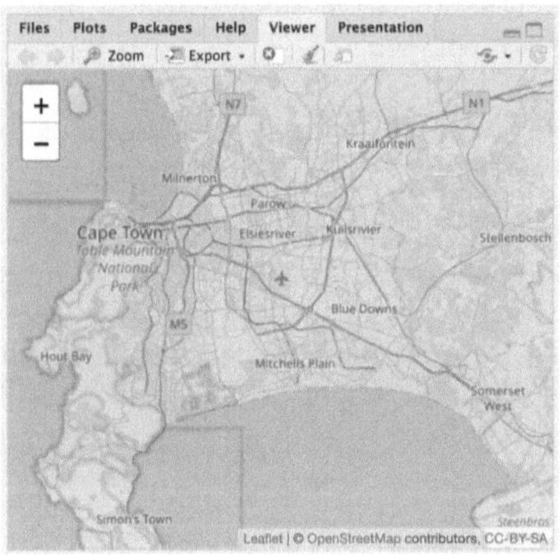

FIGURE 6.2 Interactive **leaflet** map of Cape Town, South Africa, with OpenStreetMap tiles as the base layer, as displayed in the Viewer pane.

Visualizing political boundaries

For example, some political boundaries are viewable through the interactive map, but perhaps over Cape Town a researcher plans to visualize other more detailed political boundaries and characteristics of each. The GeoQuery app at AidData.org[2] makes available a wide range of geospatial data for state administrative units; one such set of units is the lowest level of administrative units for South Africa, recorded in the .geojson file *ZAF_ADM4.geojson* (Goodman et al., 2019).

[1]See https://www.openstreetmap.org
[2]See https://www.aiddata.org/.

From the **sf** package, the `st_read()` function will interpret the geospatial data, and with the assignment operator `<-` we create a map object `sa_admin_boundaries`. Soon after importing the data, we will clean it with the **dplyr** package.

```
library(tidyverse)
library(sf)
```

```
sa_admin_boundaries<-st_read("ZAF_ADM4.geojson")
```

```
## Reading layer `ZAF_ADM4' from data source
##   `/Users/kilburnw/Documents/ZAF_ADM4.geojson'
##   using driver `GeoJSON'
## Simple feature collection with 4392 features and 8 fields
## Geometry type: MULTIPOLYGON
## Dimension:     XY
## Bounding box:  xmin: 16.45 ymin: -34.83 xmax: 32.94 ymax: -22.13
## Geodetic CRS:  WGS 84
```

After reading in the *.geojson* file, the function reports on the contents: (1) a map or 'Simple feature collection' of 4392 features, the administrative units throughout South Africa; (2) eight fields, or eight columns for each of the row administrative units within the file, plus an `id` or identification column; (3) the geometry type of multiple shapes or polygons, described in a coordinate system within a 'bounding box' or map with those four corners.

The goal in this example is to map out each of these administrative units on the interactive map of Cape Town. The map object `sa_admin_boundaries` contains the administrative units for all of South Africa. After entering `glimpse(sa_admin_boundaries)` you will see the column `shapeName` names the polygon based on location in a particular village or city.

To identify the rows corresponding to Cape Town, we can `filter()` on the rows in which the words "Cape Town" appear, overwriting the map data `sa_admin_boundaries` with only those rows. In Chapter 5 we used the `filter()` command to test whether a particular variable contained an exact character string or numeric value, but in this case *shapeName* contains both the name of the location and the number of the administrative unit, so `filter(shapeName=="Cape Town")` would not work. Instead, we want to find any rows containing at least the string "Cape Town". The `str_detect()` function will find rows containing a string (returning a TRUE if the string is present in a row), and within `filter()` those rows will be selected for inclusion in the *sa_admin_boundaries* dataframe. Thus, we filter on rows where `str_detect(shapeName, "City of Cape Town")` are `TRUE`:

```
sa_admin_boundaries<- sa_admin_boundaries %>%
  filter(str_detect(shapeName, "City of Cape Town"))
```

Because we have imported the map boundary file with `st_read()`, **leaflet** will automatically read the *geometry* column and locate the map boundaries, centering the boundaries in the map Viewer. The boundaries are polygons, so we add the boundary map with the function `addPolygons()`:

```
leaflet() %>%
 addPolygons(data=sa_admin_boundaries) %>%
 addTiles()
```

The administrative ward polygons should appear as in Figure 6.3. At the default scale, the boundary lines obscure the map features; zoom in to see how the lines divide up the city.

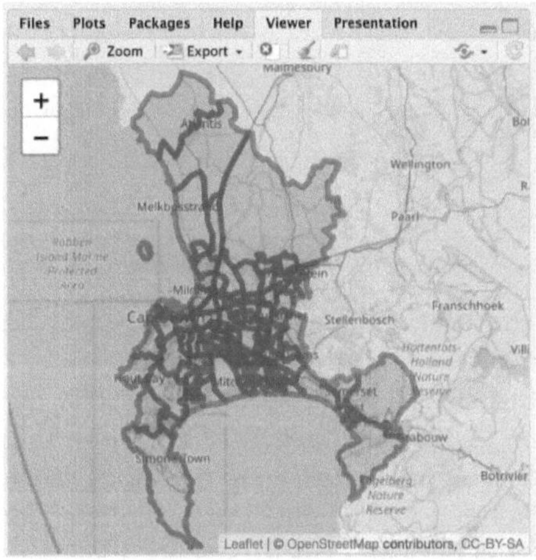

FIGURE 6.3 Interactive **leaflet** map of Cape Town, South Africa, displaying polygon boundaries of administrative units, with OpenStreetMap tiles as the base layer.

Earlier we briefly reviewed the concept of map projection, transforming spatial data from the three-dimensional Earth to a two-dimensional map. Although we are combining two different sources of geospatial data in Figure 6.3, when adding map layers to `leaflet()` it is usually not necessary to alter a map projection. With the additional layer located in latitude and longitude on the common WGS84 datum, `leaflet()` will line up the map layers correctly. Leaflet, like other 'map tile' apps such as Google Maps, uses a specific projection known as "Pseudo-Mercator" or "EPSG 3857". For a relatively small area such as wards in Cape Town, distortions introduced by the Mercator projection are generally minimal and often unnoticeable. Nonetheless, for calculations involving area and distance where precision matters, the projection should be reconsidered for maximal accuracy.

By default, `leaflet()` draws the polygon shapes with a slight hue to show the extent of the polygon coverage over the map. Thus beyond the borders of each administrative unit, the polygons do not contain any additional data. We use variation in the shading of the borders to display additional data. For example, AidData.org provides a measure of fine particulate air pollution averaged to each polygon.

Visualizing boundary line attributes

The air pollution data measures PM2.5 concentration, measured in the year 2011, averaged to each administrative unit (Brauer et al., 2016). (The 2.5 in PM 2.5 refers to the size of the pollutant, in micrometers, such as from wildfire smoke, power plants or factory emissions.) When selected for downloading from AidData, the air pollution data is stored as a data table CSV, separated from the geospatial data. As characteristics of the polygons, the air pollution data can be treated as 'attribute data', and needs to be merged with the administrative unit boundaries. The first step is to import it and subset it to Cape Town.

```
sa_air_pollution<-read_csv(file="air_pollution_data.csv")
```

The air pollution dataset, stored within RStudio as the dataframe *sa_air_pollution* consists of eight columns. Review the contents with `glimpse(sa_air_pollution)`. The air pollution level is recorded in column *ambient_air_pollution_2013_fus_calibrated.2011.mean*. Like the data for administrative units, the column *shapeName* describes the geographic location, with each row corresponding to the different boundaries for an administrative ward. We will subset the data to Cape Town, as done previously:

```
sa_air_pollution<- sa_air_pollution %>%
  filter(str_detect(shapeName, "City of Cape Town"))
```

To visualize the air pollution data, we need to merge it with the polygon data. The column *ShapeName* matches the same *ShapeName* column in the map data, so we will use that one for merging and displaying the differences in air pollution levels. All rows of the datafiles match with only one row of the other. While there are a few different merges that can be used, we will use `inner_join()` to merge the air pollution data onto the *sa_admin_boundaries* dataset:

```
sa_admin_boundaries <- sa_admin_boundaries %>%
  inner_join(sa_air_pollution, by="shapeName")
```

Run `glimpse(sa_admin_boundaries)` to see how the dataset has changed. The air pollution data is appended onto the administrative boundaries. Duplicate column names are marked by the suffix *.x* (from *sa_admin_boundaries*) and *.y* (from *sa_air_pollution*). Notice the lengthy name for the measure of air pollution. We will change it to something simpler, such as *air_pollution_level*:

```
sa_admin_boundaries<-sa_admin_boundaries %>%
 rename(air_pollution_level =
          ambient_air_pollution_2013_fus_calibrated.2011.mean)
```

To plot the variation in polygon color to display levels of air pollution, we need to first specify a color palette to display the variation in color. The function for doing so is `colorNumeric()`.

```
pal <- colorNumeric(palette="Blues",
        domain = sa_admin_boundaries$air_pollution_level)
```

Then we plot the data as follows, after filtering out missing values, storing the complete data in *filtered_boundaries*:

```
filtered_boundaries <- sa_admin_boundaries %>%
  filter(!is.na(air_pollution_level))

leaflet(filtered_boundaries) %>%
 addPolygons(fillColor=~pal(air_pollution_level), fillOpacity = .8,
             weight = 0.5, ) %>%
 addTiles() %>%
 addLegend(pal = pal, values = ~air_pollution_level,
          title = "2.5 air pollution levels", position = "bottomright")
```

The leaflet map will appear similar to Figure 6.4. Within `addPolygons()` the argument `fillColor()` links to the color palette for the levels of *air_pollution_level*. The boundary lines are further reduced to half their original size with `weight=.5`.

Visualizing points and point densities

Rather than embedded within boundary lines, data can be represented by points and visualized on a map. Each point must be measured in latitude and longitude. For a small number of points, the coordinates can be organized

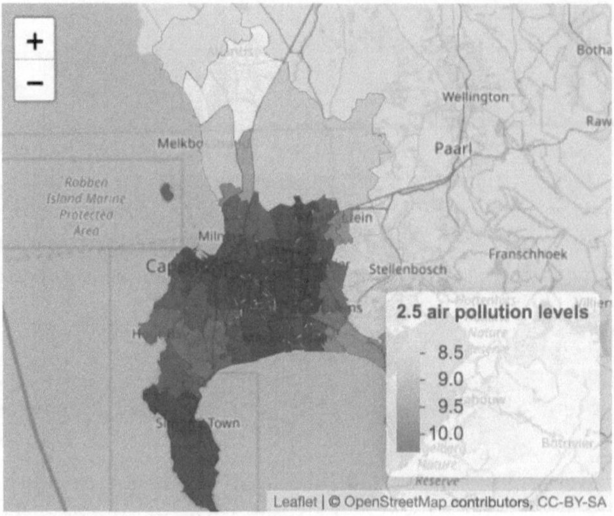

FIGURE 6.4 Interactive **leaflet** map of Cape Town, South Africa, displaying polygon boundaries of administrative units, shaded by fine particulate air pollution levels (PM2.5), with OpenStreetMap tiles as the base layer.

into a small dataset. (The process of translating street address locations into coordinates — geocoding — requires additional packages or other resources, described in the Resources section.)

In identifying point coordinates, you may return more precision than is needed. For example, Google Maps identified the location of the USA Consulate in Cape Town in decimal degree form at "-34.0768110805937,18.4296571685817", which is a location with far more precision than necessary. As a rough rule of thumb for accuracy or precision, at one decimal place a location is accurate to within 10 kilometers, at two decimal places 1 kilometer, at three decimal places 100 meters, and at four decimal places 10 meters. So within 100 meters, we could say the USA Consulate is located 34.077 degrees south of the equator (-34.077) and 18.430 degrees east of the Prime Meridian (18.430). Thus for most purposes two to three decimal places is sufficient.

Given the coordinates, just one descriptive field for each point may be sufficient to plot the points on a map. The `tribble()` function from the **tibble** (Müller and Wickham, 2023) package facilitates row-wise data entry. As a simple example, we could plot the location of three state consulates for the USA, China, and Russia. Latitude and longitude should be organized into two different columns, and we will need at least a third column to record the name of each one. In the `tribble()` function, the data elements are organized row-wise with the first row reserved for variable names preceded by a tilde ~ symbol. In this example, we record the latitude in a variable *lat*, longitude in *long*, and the name of the consulate in a variable *name*. The simplest storage type decision is to record character strings in quotes, while the coordinates will automatically be recorded in numeric form.

The `tribble()` contents of this function are assigned to a small dataset, *three_consulates*.

```
three_consulates<-tribble(
  ~lat, ~long, ~name,
  -33.979,18.445,"Chinese Consulate",
  -33.920,18.424, "Russian Consulate",
  -34.077,18.430, "American Consulate")
```

In entering the data with `tribble()`, each row (until the last one) ends with a comma, and the closed parenthesis completes the function. As a dataset it appears as a 3 by 3 table:

```
three_consulates
```

```
## # A tibble: 3 x 3
##     lat  long name
##   <dbl> <dbl> <chr>
## 1 -34.0  18.4 Chinese Consulate
## 2 -33.9  18.4 Russian Consulate
## 3 -34.1  18.4 American Consulate
```

For legibility, the resulting **tibble** prints only the first decimal of coordinates, but the full coordinates are saved within the dataset. The `three_consulates` layer can be added to any map as an additional layer. Figure 6.5 displays the consulates over the base map; the function `addCircleMarkers()` identifies the *three_consulates* dataframe coordinates and a label. The points could be added to the map of administrative unit polygons as an additional layer after `addPolygons()`.

```
leaflet( ) %>%
  addTiles() %>%
  addCircleMarkers(lat=three_consulates$lat, lng=three_consulates$long,
     label = three_consulates$name)
```

Visualizing point densities

Data represented by points could be visualized discretely, as individual points, or if points are sufficiently numerous, as densities or 'heatmaps' that show trends over geographic areas. For example, data.gov hosts crime reports from

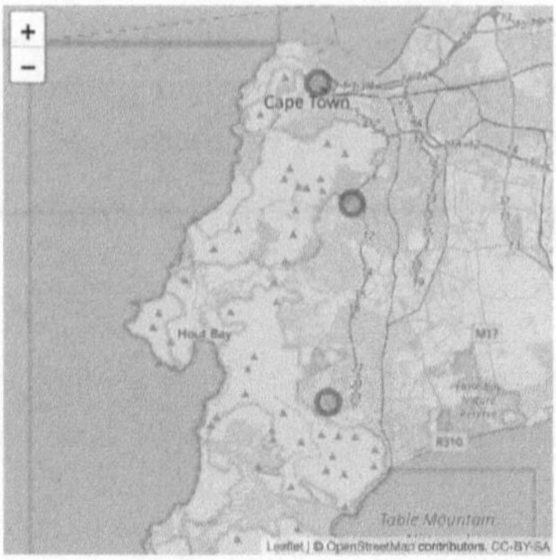

FIGURE 6.5 Interactive **leaflet** map of Cape Town, South Africa, with OpenStreetMap tiles as the base layer, displaying three point markers based on latitude and longitude coordinates.

the Washington, DC, Metropolitan Police (Open Data DC, 2024). The *.geojson* file *Crime_Incidents_in_2024.geojson* contains crimes reported through the fall of 2024, each one geocoded to a particular location.

```
dc_crime<-st_read("Crime_Incidents_in_2024.geojson")
```

The variable *OFFENSE* records types of crimes. See the codebook for details. One level of *OFFENSE* is *HOMICIDE*. Subset the data to homicides:

```
homicide_data <- dc_crime %>%
  filter(OFFENSE == "HOMICIDE")
```

Coordinates are recorded in variables *LONGITUDE* and *LATITUDE*. Each geocoded homicide can be plotted individually through `addCircleMarkers()`, or the density of each event through `addHeatmap()` from the **leaflet.extras** package (Gatscha et al., 2024). To plot the points, we first set the boundaries of the **leaflet** map, centered around downtown DC, then `addCircleMarkers()`:

```
leaflet() %>%
  setView(lng = -77.0369, lat = 38.9072, zoom = 12) %>%
  addTiles() %>%
addCircleMarkers(data = homicide_data, lng = ~LONGITUDE,
    lat = ~LATITUDE, color = "red", radius = 2)
```

The upper panel in Figure 6.6 displays the point map, each circle corresponding to a single homicide. The function argument `radius=2` sets the point radius to two pixels. The points appear to cluster around the areas south of Anacostia and Columbia Heights. A heatmap helps to communicate the trends in the clusters. After installing the **leaflet.extras** package if necessary (`install.packages("leaflet.extras")`), we can plot the points through `addHeatmap()`:

```
library(leaflet.extras)

leaflet() %>%
  setView(lng = -77.0369, lat = 38.9072, zoom = 12) %>%
  addTiles() %>%
  addHeatmap(data = homicide_data, lng = ~LONGITUDE, lat = ~LATITUDE,
     blur = 20, max = 0.05, radius = 15)
```

The heatmap in the lower panel of Figure 6.6 focuses on clusters rather than individual events, and it does reveal a focal point south of Anacostia. Within `addheatmap()`, the three last arguments — `blur=`, `max=`, and `radius=` — affect how the map represents density patterns. The `radius` sets the size of the area around each point (measured in pixels) that contributes to the heatmap; larger values result in smoother, more generalized density patterns, while smaller values keep the density tightly localized to each point's immediate vicinity. The `blur` option controls how sharply the density transitions from the center of a point to the edges of its radius. The `max` option sets the scaling for the density intensity, defining how bright or "hot" the most dense areas will appear. Try adjusting the arguments to observe how each affects the map.

These maps in **leaflet** are meant to be interactive, rather than printed through an RMarkdown file as an image to a Word or PDF document. Creating these images requires the creation of maps as plotted graphics.

Creating static polygon and point maps

Static maps based on shapes or points imported from a ShapeFile or .geojson file require the **sf** package, but the inclusion of a base layer necessitates **ggspatial**, along with functions from the **tidyverse** for data wrangling and visualization, and optionally labelling with **ggrepel**.

```
library(ggspatial)
```

The same .geojson file of administrative boundaries, read into R with `st_read()` and saved as *sa_admin_boundaries* can be plotted in a static **ggplot2** graphic. The geometry in the `ggplot()` graphic, `geom_sf()`, plots any "simple features" map object (such as a ShapeFile or geoJSON file) imported into R with `st_read()`. The function will automatically find the *geometry* column (in `data=sa_admin_boundaries`) for plotting. Note that drawing static maps via `ggplot()` takes a little more computing time than the interactive maps; so

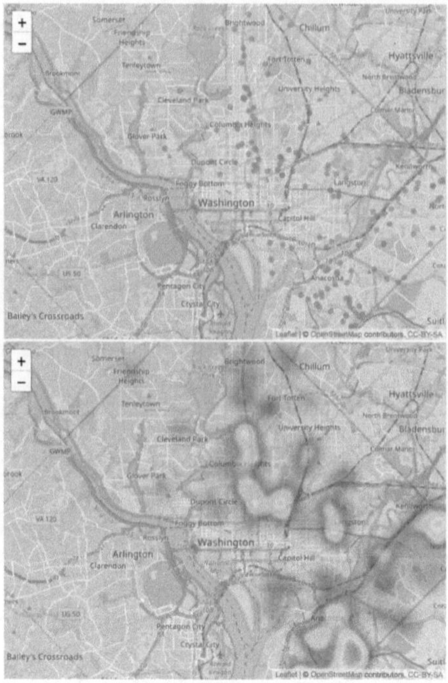

FIGURE 6.6 Interactive **leaflet** map of Washington, DC, displaying homicides in 2024, with OpenStreetMap as the base layer. The upper panel shows individual homicide locations as points, while the lower panel presents a heatmap of homicide concentrations.

after you enter the commands below, you may have to wait a minute or so to see the next prompt `>`, and then perhaps 1-2 minutes for the actual map to appear in the "Plots" pane.

The simplest plot of the administrative boundaries would consist of two lines `ggplot() + geom_sf(data=sa_admin_boundaries)`. For completeness, however, static maps of this kind usually include an orientation symbol and a scale legend. Among other functions, the **ggspatial** package provides these features. The two components are added with the functions `annotation_north_arrow()` and `annotation_scale()`. The function includes a `style= north_arrow_fancy_orienteering` for a particular orientation style, placing it at the bottom right of the graph (`location= "br"`). The intended width of the legend scale on the plot area is controlled with `width_hint=` as a proportion of 1.

To plot both the boundary lines for wards and the air pollution level, we specify a fill layer within the aesthetic `aes()` function. The `geom_sf()` function still requires that the dataset `data=` is explicitly identified. Figure 6.7 shows the administrative wards filled by average level of air pollution.

```
ggplot() +
  geom_sf(aes(fill=air_pollution_level), data=sa_admin_boundaries) +
  theme(legend.position = "bottom") +
  annotation_north_arrow(location = "br",
        style = north_arrow_fancy_orienteering) +
  annotation_scale(location = "br", width_hint = 0.4) +
  scale_x_continuous(breaks = seq(18, 19, by = .5))
```

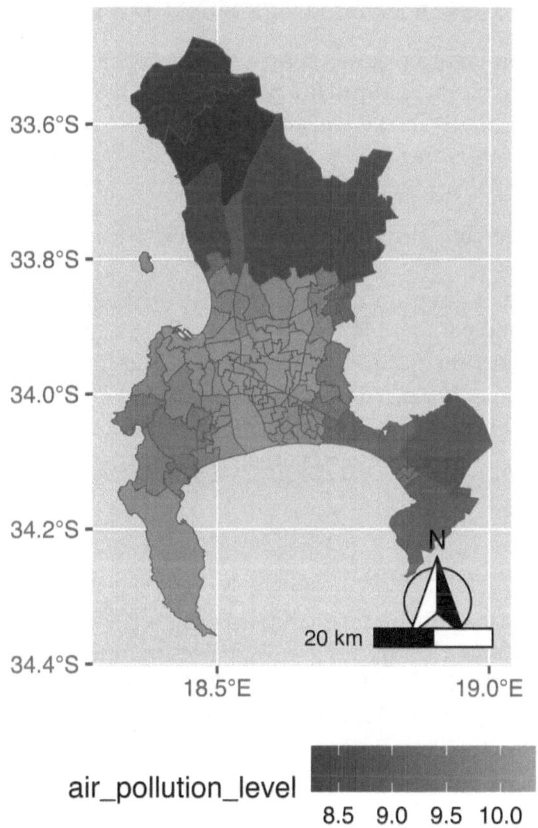

FIGURE 6.7 Heat map of Cape Town, South Africa, showing average PM2.5 air pollution concentration by administrative unit in 2011, visualized with **ggplot2**. Shading represents pollution levels, with darker areas indicating higher concentrations.

Perhaps the most common use of maps is in organizing election results by county.

6.3 Mapping elections in red, blue, and purple

The creation of 'choroplethic' heatmaps for elections — such as the red-blue USA election maps in every presidential election — relies upon the same principles of map-making we have seen in drawing a map of administrative wards in Cape Town. For example, in elections, the geographic unit (such as a State, County, or precinct) is shaded either by a categorical variable (win/loss for a candidate) or continuous variable (a candidate's margin of victory).

To illustrate how to create election maps, we approach it first with the same `geom_sf()` function in the **ggspatial** package. Second we will consider the limitations of these maps in the construction of "cartogram"-style maps with the help of the **cartogram** package (Jeworutzki, 2023). Finally we will review how interactive maps can be created to investigate gerrymandering.

The basic steps to create these maps are as follows:

(1) Decide upon a geography for the map – Counties? States? precincts?
(2) Find a boundary line file that outlines the geography.
(3) Find the election results – aggregated to the intended geography; prepare a data table.
(4) Merge the election results with the map file, on the key field that links the map geographies together.
(5) Display the resulting map + election results.

A map of Michigan counties

Our goal is to map the outcome and margins of victory for presidential candidates Trump and Clinton in the 2016 election, at the Michigan County level. The first step is to identify a suitable boundary line file. Many units of government have geographic data portals with open access geospatial data. A simple internet search will find these portals. Within each one, boundary lines files can easily be downloaded for analysis.

The Michigan County boundary line file is *Counties__v17a.shp*. From a working directory, use the **sf** package to read in file, with `st_read()`:

```
michigan<-st_read("Counties__v17a.shp")
```

```
## Reading layer `Counties__v17a' from data source
##   `/Users/kilburnw/Documents/Counties__v17a.shp'
##   using driver `ESRI Shapefile'
## Simple feature collection with 83 features and 15 fields
## Geometry type: MULTIPOLYGON
```

```
## Dimension:     XY
## Bounding box:  xmin: 161300 ymin: 128100 xmax: 792200 ymax: 859200
## Projected CRS: NAD83 / Michigan Oblique Mercator
```

The geospatial dataset consists of 83 rows (one for each Michigan county) and 15 variables describing each row. Like the geoJSON file for Cape Town, the geometry is *MULTIPOLYGON* referencing the shapes for each Michigan county. Enter `glimpse(michigan)` to observe different features of the dataset, in particular *NAME* and *LABEL* that describe the counties, the first row of which are for *Alcona* and *Alcona County*. In addition to name, the counties are identified by numeric FIPS codes, from States down to the smallest Census Bureau tracts. Both columns contain the same information, the only difference being the leading zeros. (States are identified by a two-digit code that precedes the FIPS code shown here; for example, Michigan's State level FIPS code is 26, thus the FIPS code identifying Alcona County as a county in Michigan would be *26001*).

Before moving on to the next step, we will plot the `michigan` simple feature dataset:

```
ggplot() +
  geom_sf(data=michigan)
```

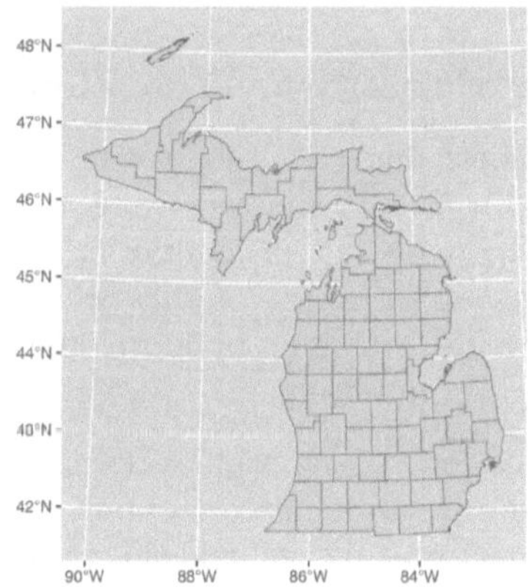

FIGURE 6.8 Map of Michigan county boundaries created with **ggplot2** to plot spatial data.

As expected, Figure 6.8 shows the outlines of the Michigan counties from the simple features data *michigan*. The next step is to link election results to the counties, by merging an election dataframe to the *michigan* counties map data. The election data, *michigan12_16.csv*, contains County-level presidential vote counts for 2012 and 2016. It is organized with Counties on rows and votes, along with various other county characteristics on the columns.

Using the `read_csv()` we will read in the CSV file and store it as `MI1216`.

```
MI1216<-read_csv(file="michigan12_16.csv")
```

Like the *michigan* map data, the *MI1216* dataset contains 83 rows, one for each Michigan county. There are two variables that record the name of each county, as well as a FIPS code column. To merge the two datasets, one method would be to use a `join`, such as`inner_join()`. The two matching key variables from each dataset are *FIPSNUM* in *michigan* and *fips* in *MI1216*. `michigan<-michigan %>% inner_join(MI1216, by=c("FIPSNUM", "fips")`

Because the order of the rows in each dataset matches each other, an alternative to a join is simply to bind the columns from one dataset to the other, with the `cbind()` function. Since the datasets mirror each other in row structure — Michigan counties in rows in the same alphabetical order — we can simply merge the *MI1216* data onto it adding all the columns from *MI1216* to the side of *michigan*. This is easy to do with the function `cbind()`.

```
michigan<-cbind(michigan, MI1216)
```

The dataset *michigan* now contains the geographic data as well as the attribute elections data. To verify, try `glimpse(michigan)`.

A choroplethic map of county-level election results

First we will plot a choropleth map of Michigan with some 2016 election results. We will plot the map with `ggplot()` and code counties the traditional red-blue colors. Then in a second map, we will fill counties by a color gradient including purple to identify the 'swing' counties. So now we can fill in the map with colors from the 2016 election, assigning blue to counties where Clinton won, and red to counties where Trump won.

We need to specify for the map which counties each candidate won. One way to do this would be to specify an indicator column that takes the values 1 if Clinton won and 0 if Trump won. We will add this measure to the dataset with the `ifelse()` function. The function will evaluate a condition for each row of the dataset, whether `clinton_vote` is greater than `trump_vote`: `clinton_vote > trump_vote`. Then counties where the condition is true will receive a score of 1 and 0 if false.

```
michigan <- michigan %>%
    mutate(indicator=ifelse(clinton_vote > trump_vote, 1, 0))
```

We could leave this indicator as is, with 0 and 1 for Trump and Clinton — but we could also associate candidate names with these numbers. Doing so requires us to convert the numeric scores into another variable type, a `factor` variable, which combines a numeric score with an associated verbal label, such as "Trump" and "Clinton" in this case. To convert it, we use `mutate()` with an additional function, `as_factor()` that will convert the numeric score into a factor measure. All we have to do is specify the verbal labels for each numeric code, after converting it to a factor with `as_factor()`:

```
michigan<- michigan %>%
   mutate(indicator = as_factor(indicator))
michigan<- michigan %>%
   mutate(indicator = fct_recode(indicator,
   "Trump" = "0", "Clinton" = "1"))
```

To create an election map that simply displays a red-colored county for Trump and blue for Clinton, we need to add a fill color. That fill will be determined by the `indicator variable`, and the color we set manually with `scale_fill_manual()`. To add the fill, we specify it as `geom_sf(aes(fill=indicator))`. Since the indicator variable has only two levels, "Trump" and "Clinton", we will specify two colors in the order red and blue, and then add a name for the legend: `scale_fill_manual(values = c("red", "blue"), name= "Winning candidate")`.

A simple red-blue election map could be visualized with the following lines of code:

```
ggplot(data=michigan) +
  geom_sf(aes(fill=indicator)) +
  labs(title="Michigan County level vote",
       subtitle="2016 presidential election") +
  scale_fill_manual(values = c("red", "blue"),
                    name= "popular vote winner")
```

The map for this code would include X and Y axes scale labels with the latitude and longitude coordinates. To remove the coordinates, add `theme_void()`, and then reposition the legend to the bottom of the map, `theme(legend.position = "bottom")`, so that there is less white space above and below it. Figure 6.9 illustrates this adjusted map of the 2016 County level results. Note that in the grayscale, text printed version of the map, the "blue" Counties appear darker than red.

```
ggplot(data=michigan) +
  geom_sf(aes(fill=indicator)) +
  labs(title="Michigan Counties, 2016 presidential election") +
  scale_fill_manual(values = c("red", "blue"),
```

```
                    name= "popular vote winner")+
  theme_void() +
  theme(legend.position = "bottom")
```

FIGURE 6.9 Traditional red-blue election map of Michigan counties visualized with **ggplot2**. On the grayscale printed version of the map, blue counties appear darker than red.

Blending counties: red, blue, and purple

Of course, a common criticism of maps such as these is that the map obscures the votes received by the loser in each county. After all, Trump's margin of victory was less than 10,000 votes, so we could create a map displaying the shift between red and blue — to purple — among the counties that were relatively close. The key to understanding how this map would be constructed is to imagine that within each state we are plotting the size of the margin of victory won by either Trump or Clinton along a color spectrum ranging from red — through purple — to blue.

Instead of specifying red and blue colors with `scale_fill_manual()`, we will use a function that creates a gradient of colors between the two. This function is `scale_fill_gradient2()`, where the `2` stands for a divergent contrast between the two colors. By default, the function uses the color white as the midpoint, so instead we will specify purple as the midpoint in the transition.

The next decision is whether to plot Trump's percentage minus Clinton's, or vice versa. That choice affects which color is at the minimum of the gradient and which is specified at the maximum. If it is Trump's percent of the vote minus Clinton's, then the minimum, or low would be blue and the high red: larger values of this measure would correspond to a more red state, while lower would be blue, and then purple is the midpoint: `scale_fill_gradient2(low="red", mid="purple", high="blue")`.

We create the margin of victory measure:

```
michigan <- michigan %>%
  mutate(margin=trump_percent - clinton_percent)
```

And then create the map displayed in Figure 6.10:

```
ggplot(data=michigan) +
  geom_sf(aes(fill=margin)) +
  labs(title="Michigan Counties, 2016 presidential election",
       subtitle="percentage margin of victory by Trump (red)
                 or Clinton (blue)") +
  scale_fill_gradient2(low="blue", mid="purple", high="red") +
  theme_void() +
  theme(legend.position = "bottom")
```

Notice in the legend that ggplot automatically chooses the breakpoints, the range of percentage scores associated with a particular color. In this case, it chose four bins, or ranges, in increments of 25 percentage points. We can specify our own preferred breakpoints; perhaps in units of ten, `breaks=c(-30, -20, -10, 0, 10, 20, 30)`. This argument would be added to `scale_fill_gradient2()` after `high="red"`. Another aesthetic change would be to move the legend to the side, under the upper peninsula, `theme(legend.position = "left")`. Other maps show these with a neutral color such as white as the midpoint, so that the map emphasizes red or blue shades. Figure 6.11 displays such a map:

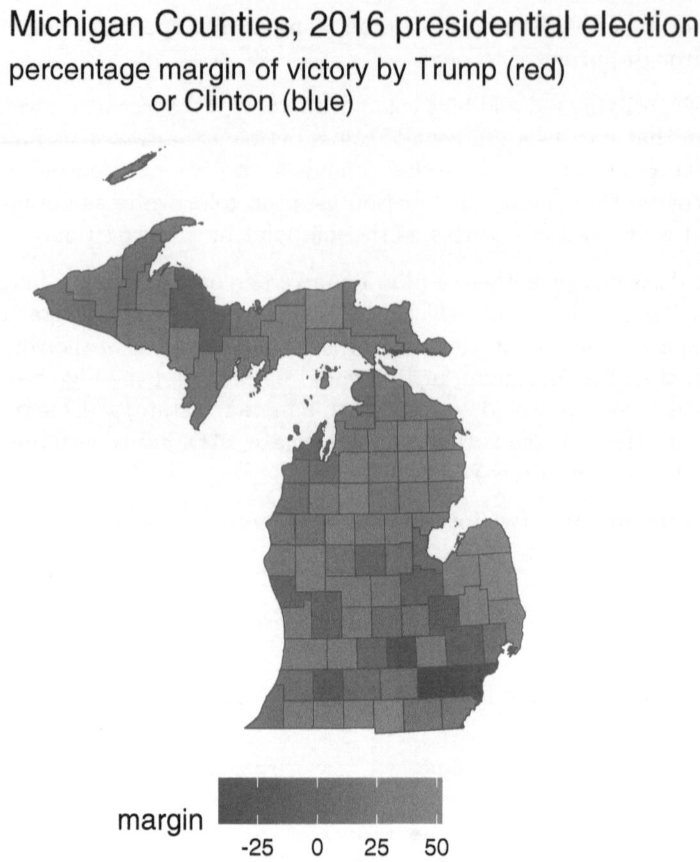

FIGURE 6.10 County level election map displaying a percentage margin of victory from red through purple to blue, visualized with **ggplot2**. On a graycale printed page, these color transitions are not visible.

```
ggplot(data=michigan) +
  geom_sf(aes(fill=margin)) +
  labs(title="Michigan Counties, 2016 presidential election",
      subtitle="percentage margin of victory by
      Trump (red) to Clinton (blue)") +
  scale_fill_gradient2(low="blue", mid="white", high="red",
                       breaks=c(-30, -20, -10, 0, 10, 20, 30)) +
  theme_void() +
  theme(legend.position = "left")
```

For an introduction to other mapping packages, see Machlis (2019).

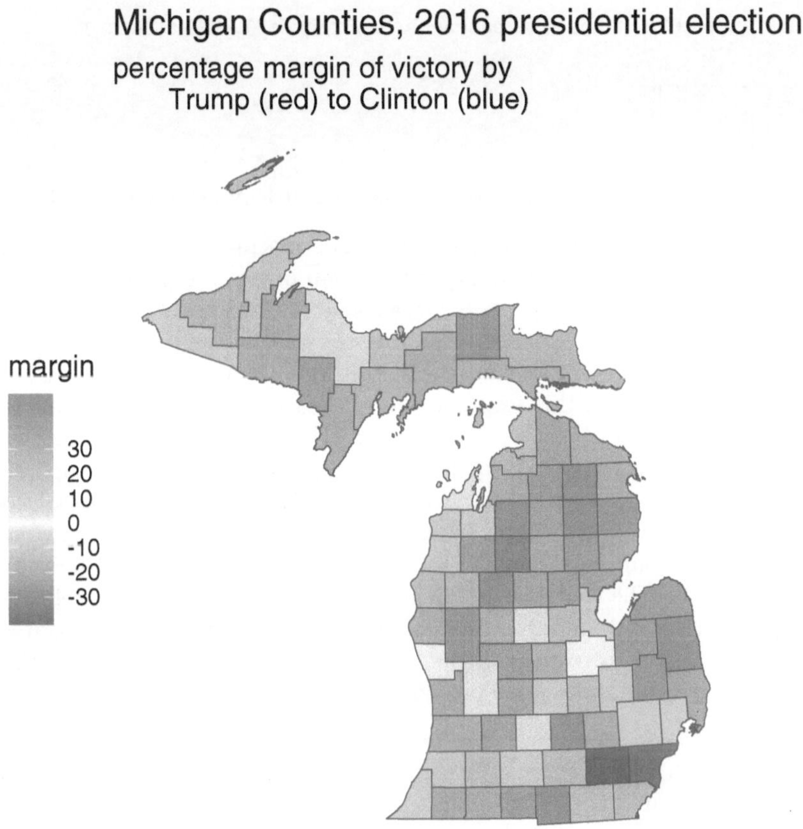

FIGURE 6.11 County level election map displaying a percentage margin of victory from red through white to blue, in increments of ten, visualized with **ggplot2**. On a graycale printed page, these color transitions are not visible.

6.4 Cartogram corrections to election maps

A cartogram is a type of map in which the area of a geographic unit as it appears on the map is intentionally distorted by another feature of the geographic units, such as population. In the case of a Michigan County election map, this transformation ensures that counties with relatively lower populations (or votes cast) appear proportionally smaller. By scaling geographic areas in this way, cartograms prevent the common misconception that larger land areas correspond to larger contributions to the total vote, a bias inherent in traditional maps. In **R**, the **cartogram** package (Jeworutzki, 2023) will transform existing election maps into the cartogram style visualization. We will transform the Michigan County red-blue election map. Install the package **cartogram** if you have not already `install.packages("cartogram")`, and attach it.

```
library(cartogram)
```

Cartogram of the Michigan county presidential vote in 2016

The starting point for the cartogram is an existing map constructed via the **sf** package. In **cartogram**, starting with a map object the main function for creating the cartogram map is `cartogram_cont()`. The necessary arguments are the map object, the variable for scaling the geographic boundaries, and a maximum number of iterations for the algorithm to run in order to create the new boundary lines. In this example the cartogram uses the same `michigan` simple features data object used to construct the election heat maps with **ggplot2()**. The cartogram will be used to draw County shapes in proportion to their total vote — how much each County contributed to the total vote statewide. So the County shapes for lower population counties will be drawn smaller, higher counties larger, and the color of each County will be red or blue accordingly. The choropleth map shows that upper Michigan is relatively red, along with the thumb, but most of these counties are relatively sparsely populated. So the cartogram algorithm redraws the County lines, making these counties smaller and more populous counties larger.

One potential obstacle in drawing the cartogram is that the algorithm that scales boundary lines requires an acceptable map projection as the starting point for finding the geographical 'center' of each geographic unit. The `st_transform()` function for altering the projection may be necessary if the **cartogram** function returns an error indicating a different projection must be used.[3]

To begin we specify the projection:

```
michigan<-st_transform(michigan, CRS=epsg:3857)
```

Through `st_transform()`, we specified the CRS as `epsg:3857` and assigned it to the input map *michigan*. Next we will draw a cartogram map, with the function `cartogram_cont()`. We specify the output (`mi_cart`), the input map (`michigan`), and the measure with which to draw the cartogram (`total_votes`). The cartogram algorithm works iteratively, trying to find an arrangement of the area marked by county lines that satisfies the condition of drawing the county lines relative to `total_votes`, counties with more votes to contribute to

[3]As a last resort, if no other map projection works, try `epsg:4326`, which is a pseudo-projection, an "equirectangular" projection (see https://en.wikipedia.org/wiki/Equirectangular_projection) or `epsg:3857`the usual projection for transportation maps.

the total are drawn relatively larger, and smaller vote counties are drawn with a smaller geographic area. It finds one solution, then tries to improve upon it. Setting `itermax=` determines how many of these iterations to go through; four is fine. Changes beyond that tend to be subtle.

```
mi_cart<-cartogram_cont(michigan, "total_votes", itermax=4)
```

The cartogram algorithm returns back a dataset like `michigan` in every respect except the county line coordinates, which are now scaled according to the `cartogram_cont()` algorithm. The new cartogram map is stored as *mi_cart*. We will redraw the election map with `ggplot()` and `geom_sf()`, essentially repeating the same `ggplot()` syntax from Figure 6.9 applied to *mi_cart* as the input data rather than *michigan*. (To draw a cartogram style map of Figures 6.10 or 6.11, we would rely on the syntax for each of those maps.)

Figure 6.12 displays the cartogram representation of Figure 6.9, Michigan County level votes in the 2016 presidential election. Compare the Cartogram to the prior election map to observe the differences. TK In the cartogram, the upper and interior solidly 'red' parts of the State are drawn much smaller, while the more populous counties surrounding and containing Detroit on the southeast side of the State are drawn proportionally larger. Trump won the State's popular vote in part by running more competitively in these counties, which appear purple in the color gradient map; perhaps a cartogram map of margin of victory would explain the election result better than the red-blue version.

```
ggplot(data=mi_cart) +
  geom_sf(aes(fill=indicator)) +
  labs(title="Michigan Counties, 2016 presidential election") +
  scale_fill_manual(values = c("red", "blue"),
                    name= "popular vote winner")+
  theme_void() +
  theme(legend.position = "bottom")
```

6.5 Measuring gerrymandering in legislative districts

The term 'gerrymandering' refers to the practice of drawing odd or unusually shaped legislative districts, typically for the purpose of partisan advantage. The study of gerrymandering has a lengthy history in political science; for an overview see McGhee (2020). Two common ways of assessing gerrymandering is through the analysis of a legislative district shape, the degree of compactness,

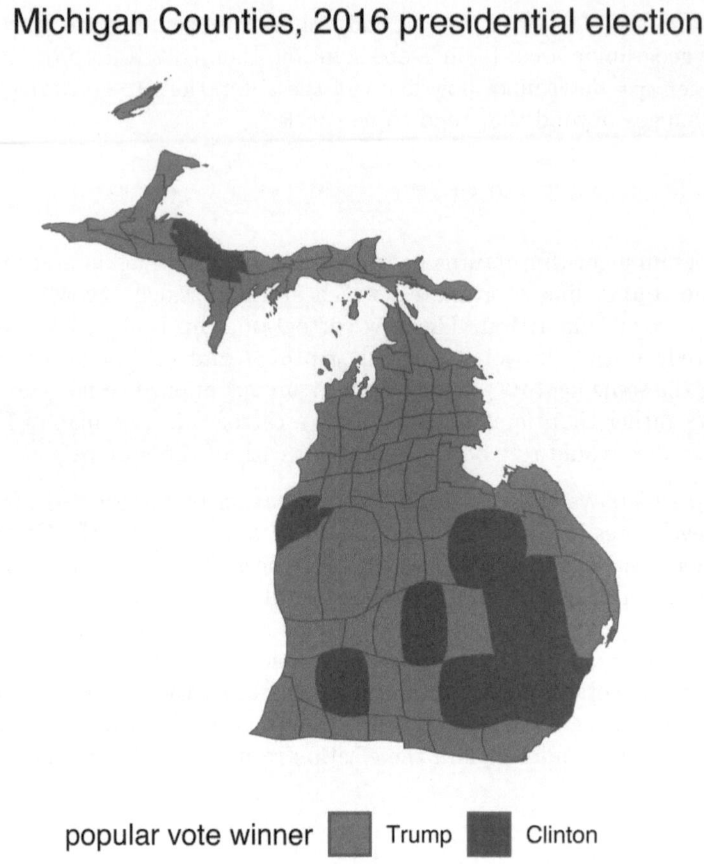

FIGURE 6.12 Cartogram representation of the 2016 presidential vote in Michigan Counties, visualized with the **cartogram** package.

and the skew or lopsidedness of the vote. These gerrymandering metrics can be assessed with some extensions to the spatial operations we have already learned and some skills in data wrangling. This section assumes the **sf** and **leaflet** packages are already installed, and that the **tidyverse** packages are previously attached.

```
library(sf)
library(leaflet)
```

In this example, we will assess legislative district shape, compactness, for the 2011 redistricting plan for the Michigan State Senate. Like any other **.geojson** file, we read the file with `st_read()` from the **sf** package.

```
mi_senate_2011<-
  st_read("Michigan_State_Senate_Districts_2011_plan.geojson")
```

In addition to identifying information about the legislator within each district, the **mi_senate_2011** file contains border lines of districts in the *geometry* column. Like previous map files reviewed in this chapter, **leaflet** will automatically read the *geometry* column and locate the map boundaries, centering the boundaries in the map Viewer. Because the boundaries are polygons, we add the boundary map with the function `addPolygons()`:

```
leaflet() %>%
 addPolygons(data=mi_senate_2011) %>%
 addTiles()
```

The map will appear as in Figure 6.13, within the Viewer pane, displaying the outline of the districts.

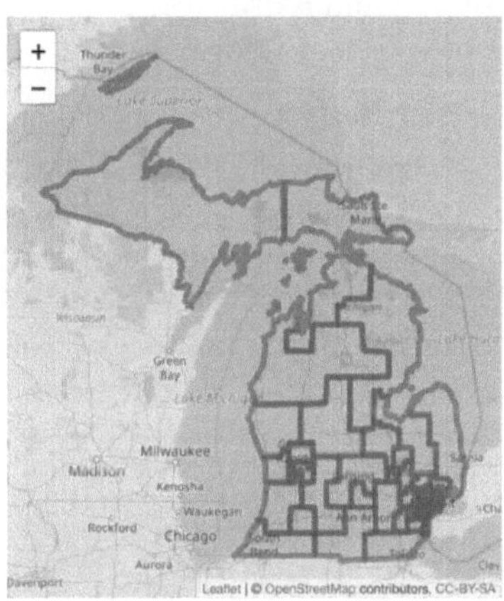

FIGURE 6.13 Interactive **leaflet** map of the 2011 Michigan Senate districts, with OpenStreetMap tiles as the base layer.

To further explore the map, features from the map data can be added as a popup label, by adding the `popup=` argument to `addPolygons()`. For example, to display the legislator name contained within the dataset *LEGISLATOR* column, the function would appear as `addPolygons(data=mi_senate_2011, popup=~LEGISLATOR)`. There are multiple columns in the dataset that describe the districts: "PARTY" and "NAME". The `paste()` function combines the two fields; the map display supports HTML tags for formatting the popup menu items. We could create a new column in the *mi_senate_2011* dataset:

```
mi_senate_2011<-mi_senate_2011 %>%
    mutate(popup_content = paste("<strong>Name:</strong> ", NAME,
            "<br><strong>Legislator:</strong> ", LEGISLATOR,
            "<br><strong>Party:</strong> ", PARTY))
```

The tags enclose the text. For example, `<strong>` and `</strong>` enclose `Name` and present the text in boldface. The tag `<br>` places a line break before the text, such as before "Legislator". Then we can plot the map with more descriptive information contained in the popup menu:

```
leaflet() %>%
 addPolygons(data=mi_senate_2011, popup=~popup_content) %>%
 addTiles()
```

The map of the districts will be redrawn, and clicking on any one of the districts will reveal the popup containing information about the three legislative districts. Try creating the *popup_content* and redraw the map with the `popup=` argument and observe the behavior of the map information.

We will calculate some gerrymandering metrics based on assessing compactness. The idea of compactness measurement is that a district shape is compared against a hypothetical compact shape, such as a circle, square or polygon. One of the most common compactness measures is the "Convex Hull", a comparison of the shape of the district to a polygon that encloses the district. To illustrate, we start with the 29th district, which encompasses the city of Grand Rapids and areas east and southeast.

To ease visualization of the compactness measure, we will filter on this district from the data and save it as a single district datafile, *district_29*:

```
district_29 <- mi_senate_2011 %>%
  filter(NAME==29)
```

Try visualizing the district with `leaflet() %>% addPolygons(data=district_29, popup=~popup_content) %>% addTiles()`.

Convex Hull score

From the **sf** package we use the function `st_convex_hull()` to create the convex hull around the district:

```
district_29<-district_29 %>%
  mutate(convex_hull = st_convex_hull(geometry))
```

The function searches inside *district_29* for a column labelled *geometry*, drawing a minimally encompassing polygon around the district. The convex hull appears around the district as the dashed line in Figure 6.14. Two layers of `addPolygons()` draw the district and the convex hull.

```
leaflet() %>%
  addPolygons(data=district_29, popup=~popup_content) %>%
  addPolygons(data = district_29$convex_hull, color = "red", weight = 2,
              dashArray = "5, 5",
              fillOpacity = 0.3, fillColor = "red") %>%
  addTiles()
```

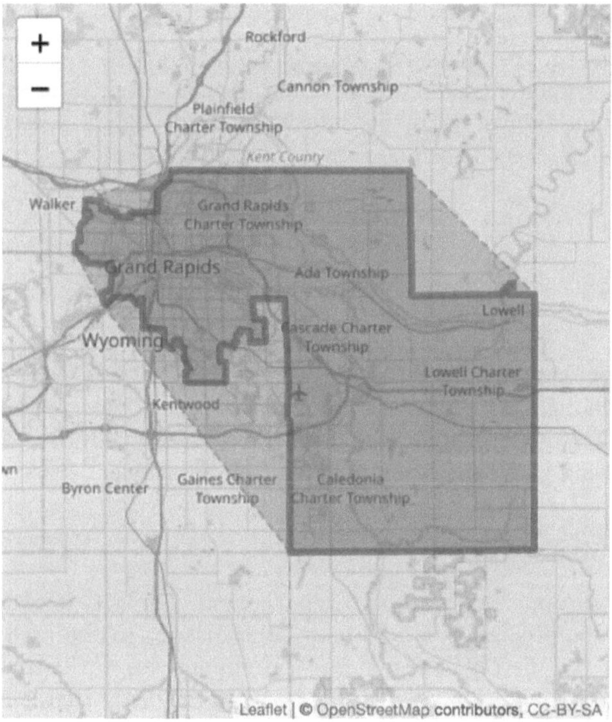

FIGURE 6.14 Interactive **leaflet** map of Michigan's 29th Senate district (2011 plan), with OpenStreetMap tiles as the base layer. The district boundary is outlined with a convex hull surrounding it.

The convex hull metric is simply the ratio of the district area to the area of the hypothetical convex hull enclosing it. The closer the ratio is to 1, the more compact the district. The function `st_area()` will calculate the area of both the district and the convex hull.

```
boundary_area <- st_area(district_29)
convex_hull_area <- st_area(district_29$convex_hull)
```

The `st_area()` function automatically searches for a "geometry"column. For calculating *convex_hull_area*, since the *convex_hull* column is labelled *convex_hull* it is specified as part of the *district_29* dataset. The default scale from the `st_area()` function is square meters; the stored units and area in `boundary_area` are a little over 600 million squared meters.

```
boundary_area
```

```
## 631850157 [m^2]
```

Converting the squared meters to squared kilometers (dividing by one million: 1000 x 1000 for squared meters) results in an estimate from `st_area()` of 631.85 squared kilometers. Notice in the data file *district_29* the recorded area in the column *SQKM* is 632.5. The convex hull measure is

```
boundary_area / convex_hull_area
```

```
## 0.7894 [1]
```

The result, .789, would be interpreted as the legislative district area being 79 percent of the convex hull. Based on this measure, the district is relatively compact, although any district's compactness must be interpreted in the context of the redistricting plan and other districts. Studies of the compactness of a redistricting plan take account of multiple measures.

Polsby-Popper score

In the Polsby-Popper measure the hypothetical compact reference shape is a cirle. The score compares the area of the district to the area of a circle (surrounding the district) defined by a circumference equal to the district perimeter (Polsby and Popper, 1991). The more the district resembles a circle, the more compact and the closer the score will be to 1. A district shape that is less circular, irregular and elongated would have a lower Polsby-Popper score.

The formula comparing the area of the district to the area of the circle is A_d/A_c. To avoid actually calculating the area of such a circle, the formula is expressed in values measured from the legislative district as substitutes for the circle. We will compare the result from this formula with an estimate of the circle area from `st_area()`.

Polsby-Popper compactness $=\frac{4\pi A_d}{P_d^2}$

where A_d is the area of the district, and P_d is the *perimeter* of the district. In the formula, the ratio of $\frac{A_d}{P_d^2}$ is multiplied by 4 times pi, π.

Taking a closer look at the formula, recall that the perimeter of a circle is $(2\pi r)$. In the Polsby-Popper metric, we know the perimeter of the circle (equal to the perimeter of the district), but we do not know the radius of this circle. We need to find the area of the circle with the knowledge of what the perimeter should be. The radius r can be expressed in terms of the perimeter P, $r = \frac{P}{2\pi}$. The area of a circle is πr^2. We can substitute this perimeter value into the formula for the radius, resulting in area$= \pi(\frac{P}{2\pi})^2$. Applying the exponent 2 reduces down to the area of the circle $= A = \frac{P^2}{4\pi}$. The compactness measure is the area of the district divided by the area of the circle with the same circumference as the district perimeter, or $\frac{A_d}{\frac{P^2}{4\pi}}$. The 4π moves up to the numerator, resulting in the Polsby-Popper formula.

To estimate the Polsby-Popper score, we rely on `st_area()` to find the area of the district. Next we find the perimeter by tracing the length around the boundary of the district, with `st_boundary()` nested within `st_length()`.

```
district_29 <- district_29 %>%
    mutate(boundary_area = st_area(district_29)) %>%
    mutate(boundary_perimeter = st_length(st_boundary(district_29)))
```

We previously calculated *boundary_area* with the `st_area()` function. The perimeter is stored as column *boundary_perimeter* in the *district_29* dataset. To visualize the circle defined by the district perimeter compared to the district, we draw an enclosing circle around the district — the circle with a circumference equal to the perimeter of the district, centering the circle over the district. This 'centering' is accomplished by locating the geographic centroid (or center point) of the district and drawing a circle with an appropriate radius from that centroid, for the circumference equal to the district perimeter. For the visualization, we calculate the radius based on the formula for the circumference: $C = 2\pi r$.

```
district_29 <- district_29 %>%
  mutate(radius = boundary_perimeter / (2 * pi))
```

Next to find the centroid of the district, we use the function `st_centroid()`:

```
district_29<-district_29 %>%
  mutate(centroid = st_centroid(district_29))
```

To see where the centroid is located, in the lines of `leaflet()` code, try inserting the line `addMarkers(data=district_29centroid) %>%` in front of the last line addTiles(), in the previous set of commands to visualize the district in the Viewer pane. Now that the radius and centroid are calculated, the circle can be drawn around the district:

```
district_29<-district_29 %>%
  mutate(circle = st_buffer(centroid, dist = radius))
```

The circle centered over the district appears as in Figure 6.15:

```
leaflet() %>%
 addPolygons(data=district_29, popup=~popup_content) %>%
 addPolygons(data=district_29$circle, color = "red", weight = 2,
             dashArray = "5, 5",
             fillOpacity = 0.3, fillColor = "red") %>%
 addTiles()
```

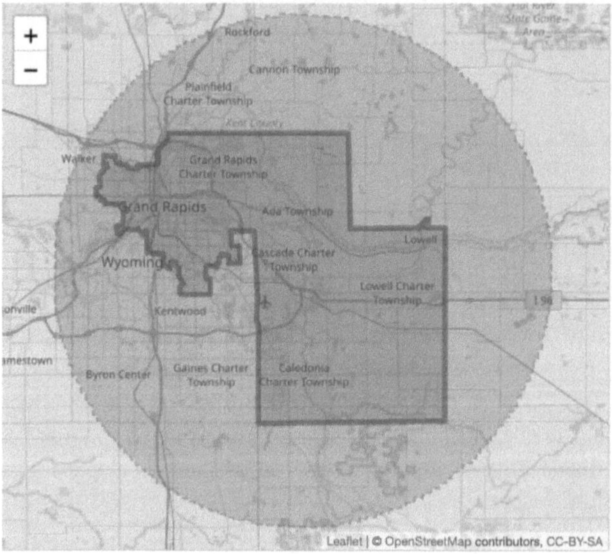

FIGURE 6.15 Interactive **leaflet** map of Michigan's 29th Senate district (2011 plan), with OpenStreetMap tiles as the base layer. The district boundary is outlined with the Polsby-Popper circle surrounding it.

The circle does not appear perfectly smooth around the perimeter — it has jagged lines resulting from the 'buffer' algorithm. We will compare the buffer calculation to the perimeter-based calculation. In **R** the name pi is reserved for π:

```
district_29<-district_29 %>%
  mutate(polsby = (4 * pi * boundary_area) / (boundary_perimeter^2))
```

```
district_29$polsby
```

```
## 0.3102 [1]
```

The result, 0.3102, means the area of the district is 31 percent of the area of the circle. Next we approximate this calculation with the buffer area of the circle. We already have the area of the district. The denominator of the measure is the area of the circle calculated from `st_area()`:

```
district_29$boundary_area/st_area(district_29$circle)
```

```
## 0.3067 [1]
```

The approximation from the buffer circle is .3067, fairly close to the calculation based on the formula.

Compactness calculations across a redistricting plan

Researchers assess compactness across each district within a redistricting plan. Given a map of districts, we can do so by grouping the calculations by district:

```
mi_senate_2011 <- mi_senate_2011 %>%
  group_by(NAME) %>%
  mutate(convex_hull = st_convex_hull(geometry)) %>%
  mutate(convex_hull_area =st_area(convex_hull)) %>%
  mutate(boundary_area = st_area(geometry)) %>%
  mutate(hull_score = boundary_area / convex_hull_area)
```

The `mutate()` function creates the new variables and `mi_senate_2011` now contains a set of the scores, `hull_score`, for each district. We can list the scores below:

```
mi_senate_2011 %>%
 select(NAME, LEGISLATOR, PARTY, hull_score)
```

```
## Simple feature collection with 38 features and 4 fields
## Geometry type: MULTIPOLYGON
## Dimension:     XY
## Bounding box:  xmin: -90.42 ymin: 41.7 xmax: -82.41 ymax: 48.26
## Geodetic CRS:  WGS 84
```

```
## # A tibble: 38 x 5
## # Groups:   NAME [38]
##    NAME  LEGISLATOR             PARTY hull_score
##    <chr> <chr>                  <chr>        [1]
##  1 27    "Jim Ananich"          D          0.861
##  2 23    "Curtis Hertel, Jr." D          0.866
##  3 20    "Sean McCann"          D          0.998
##  4 29    "Winnie Brinks"        D          0.789
##  5 28    "Mark Huizenga "       R          0.745
##  6 08    "Doug Wozniak "        R          0.580
##  7 10    "Michael MacDonald " R          0.764
##  8 09    "Paul Wojno"           D          0.846
##  9 13    "Mallory McMorrow"     D          0.795
## 10 11    "Jeremy Moss"          D          0.785
## # ... with 28 more rows, and 1 more variable:
## #   geometry <MULTIPOLYGON [°]>
```

We can store these columns into a separate dataset, *hull_scores_dataset_mi_senate_2011.*

```
hull_scores_dataset_mi_senate_2011<- as_tibble(mi_senate_2011) %>%
   select(NAME, LEGISLATOR, PARTY, hull_score)
```

Observe the first ten scores:

```
hull_scores_dataset_mi_senate_2011 %>%
print(n=10)
```

```
## # A tibble: 38 x 4
##    NAME  LEGISLATOR             PARTY hull_score
##    <chr> <chr>                  <chr>        [1]
##  1 27    "Jim Ananich"          D          0.861
##  2 23    "Curtis Hertel, Jr." D          0.866
##  3 20    "Sean McCann"          D          0.998
##  4 29    "Winnie Brinks"        D          0.789
##  5 28    "Mark Huizenga "       R          0.745
##  6 08    "Doug Wozniak "        R          0.580
##  7 10    "Michael MacDonald " R          0.764
##  8 09    "Paul Wojno"           D          0.846
##  9 13    "Mallory McMorrow"     D          0.795
## 10 11    "Jeremy Moss"          D          0.785
## # ... with 28 more rows
```

For example, calculate summary statistics by party:

```
hull_scores_dataset_mi_senate_2011 %>%
group_by(PARTY) %>%
summarise(mean=mean(hull_score))
```

```
## # A tibble: 2 x 2
##   PARTY  mean
##   <chr>   [1]
## 1 D     0.731
## 2 R     0.742
```

On average the districts are similar in convex hull compactness. In spread, the standard deviations are similar:

```
hull_scores_dataset_mi_senate_2011 %>%
group_by(PARTY) %>%
summarise(stand_dev=sd(hull_score))
```

```
## # A tibble: 2 x 2
##   PARTY stand_dev
##   <chr>     <dbl>
## 1 D         0.166
## 2 R         0.130
```

6.6 Accessing US Census data

Analyzing gerrymandering often requires assessing the demographic characteristics of electoral districts. The U.S. Census Bureau provides two primary types of data: (1) demographic characteristics of geographic areas, and (2) geographic boundary files (lines, points, and shapes) that define these areas. Several R packages facilitate access to Census data; here, I illustrate the use of the tidycensus package (Walker and Herman, 2024; Walker, 2023).

While the Census offers a wide range of data products, this example focuses on downloading 2022 American Community Survey (ACS) 5-year estimates at the state level[4]. These datasets can be readily merged with boundary files for legislative district maps, as described below.

Data are downloaded directly from the Census Bureau, which requires registration to access their data servers. The first step is to register for a free access code, specifically an API key: https://api.census.gov/data/key_signup.html.

[4]See https://www.census.gov/data/developers/data-sets/acs-5year.2022.html#list-tab-1806015614

After registering and receiving the key, install and attach the **tidycensus** package. (This section of the text also makes use of **tidyverse** functions).

```
library(tidycensus)
```

The API key is stored in your **R** environment with the `census_api_key()` function. You must reload the environment with `readRenviron("~/.Renviron"))` for the key to be available in your **R** session.

```
# not a working API key below
census_api_key("5c9a42d90596a472d3", install=TRUE)
readRenviron("~/.Renviron")
```

Perhaps the most difficult part of working with Census Bureau data is identifying specific measures within any one data product. The sheer volume of variables count into the tens of thousands, a scale much larger than global development indicators from World Bank data. The **tidycensus** package provides a function `load_variables()` to download available measures from a specified product, in this case the 5-year estimates from the most recent ACS survey:

```
variable_names<-load_variables("acs5", year="2022")
```

```
variable_names
```

```
## # A tibble: 28,152 x 4
##    name        label                         concept geogr~1
##    <chr>       <chr>                         <chr>   <chr>
##  1 B01001A_001 Estimate!!Total:              Sex by~ tract
##  2 B01001A_002 Estimate!!Total:!!Male:       Sex by~ tract
##  3 B01001A_003 Estimate!!Total:!!Male:~ Sex by~ tract
##  4 B01001A_004 Estimate!!Total:!!Male:~ Sex by~ tract
##  5 B01001A_005 Estimate!!Total:!!Male:~ Sex by~ tract
##  6 B01001A_006 Estimate!!Total:!!Male:~ Sex by~ tract
##  7 B01001A_007 Estimate!!Total:!!Male:~ Sex by~ tract
##  8 B01001A_008 Estimate!!Total:!!Male:~ Sex by~ tract
##  9 B01001A_009 Estimate!!Total:!!Male:~ Sex by~ tract
## 10 B01001A_010 Estimate!!Total:!!Male:~ Sex by~ tract
## # ... with 28,142 more rows, and abbreviated variable
## #   name 1: geography
```

Without a `year=` argument, the function downloads data from the most recent year available, which as of this writing is 2022. The datatable is saved as data frame *variable_names*, which consists of four columns: (1) name — alphanumeric identifier for each measure; (2) label — a specific explanation of the measure; (3) concept — a broader description of the measure; and (4) geography — the smallest geographic unit for which the measure is available, such as "tract".

A quick way to identify variables is displaying the data table with `View(variable_names)`. The search field within `View()` enables a general search across all spreadsheet. In the upper left-hand corner of the `View()` spreadsheet, the filter button applies a case-insensitive search to any of the chosen fields.

With `filter()` from the **tidyverserse** a smaller table of variables can be saved as a data table, based on a more precise search in specific fields of the data table. For example, to look for population figures, the search can focus on the *concept* column, saving these search results to another dataframe *pop_search*:

```
pop_search<-variable_names %>%
  filter(str_detect(concept, pattern = "Total Population"))
```

In the second line above, the `str_detect()` function from the **stringr** package in the **tidyverse** searches in the column *concept* for any matching patterns (case sensitive) to "Total Population". It identifies 70 matches, stored in *pop_search*, which can be quickly scanned in `View(pop_search)`. Figure 6.16 displays the first few rows of the variable table. Notice that the first row entry under the *concept* column is the entry "Total Population" without any further elaboration, while the rows below it reference population in "... Occupied Housing Units", meaning these rows reference population counts within the specified housing units. The *geography* column shows these counts are available at the block level and higher. The entry in the first row under *name*, *B01003_001* lists the variable name that will be used to collect data on total population within US States.

Filter

name	label	concept	geography
B01003_001	Estimate!!Total	Total Population	block group
B25008A_001	Estimate!!Total:	Total Population in Occupied Housing Units by Tenur...	block group
B25008A_002	Estimate!!Total:!!Owner occupied	Total Population in Occupied Housing Units by Tenur...	block group
B25008A_003	Estimate!!Total:!!Renter occupied	Total Population in Occupied Housing Units by Tenur...	block group

FIGURE 6.16 Excerpt of US Census Bureau variables in data table format, accessed through the spreadsheet viewer.

The function `get_acs()` will retrieve the ACS data, given a specified geographic level and the name of the variable, such as for US States: `get_acs(geography = "state", variables ="B01003_001")`. The Census Bureau organizes data on many geographic levels; for a complete list of geographies, see Walker (2023).

```
get_acs(geography = "state", variables ="B01003_001")
```

```
## Getting data from the 2018-2022 5-year ACS

## # A tibble: 52 x 5
##    GEOID NAME                 variable   estimate   moe
##    <chr> <chr>                <chr>         <dbl> <dbl>
##  1 01    Alabama              B01003_001  5028092    NA
##  2 02    Alaska               B01003_001   734821    NA
##  3 04    Arizona              B01003_001  7172282    NA
##  4 05    Arkansas             B01003_001  3018669    NA
##  5 06    California           B01003_001 39356104    NA
##  6 08    Colorado             B01003_001  5770790    NA
##  7 09    Connecticut          B01003_001  3611317    NA
##  8 10    Delaware             B01003_001   993635    NA
##  9 11    District of Columbia B01003_001   670587    NA
## 10 12    Florida              B01003_001 21634529    NA
## # ... with 42 more rows
```

The 52 observations include all states, Washington, DC and Puerto Rico. Notice the data table appears in long form; additional indicators would be appended to the end. A *variable* column records the name of each variable, while *estimate* records the indicator value. The column *moe* records the margin of error for the sample estimate.

An option for `get_acs()`, `output="wide"` will shift the data table to wide format. Additional variables are added as separate column entries. And mnemonic names can be added to the table in place of the default names. Collecting the names in `c()` modifies the `variables=` argument, such as `variables = c(total_pop="B01003_001", black_pop="B01001B_001")`, which would select and relabel the population total and population selecting solely 'Black or African American' as race.

```
get_acs(geography = "state",
        variables = c(total_pop="B01003_001",
                      black_pop="B01001B_001"), output="wide")
```

```
## Getting data from the 2018-2022 5-year ACS

## # A tibble: 52 x 6
##    GEOID NAME           total~1 total~2 black~3 black~4
##    <chr> <chr>            <dbl>   <dbl>   <dbl>   <dbl>
##  1 01    Alabama         5.03e6      NA 1326341    3804
##  2 02    Alaska          7.35e5      NA   23395     906
##  3 04    Arizona         7.17e6      NA  327077    4147
##  4 05    Arkansas        3.02e6      NA  456693    2289
```

```
## 5 06    California      3.94e7        NA 2202587     9304
## 6 08    Colorado        5.77e6        NA  233712     2900
## 7 09    Connecticut     3.61e6        NA  385407     3664
## 8 10    Delaware        9.94e5        NA  218266     2327
## 9 11    District of C~  6.71e5        NA  297101     1999
## 10 12   Florida         2.16e7        NA 3355708    10029
## # ... with 42 more rows, and abbreviated variable
## #   names 1: total_popE, 2: total_popM, 3: black_popE,
## #   4: black_popM
```

In wide format, the Census variables span across columns, with the selected column names. The function pulls not only the statistic estimates, but also the estimated margins of error for each, noted with E for estimate and M for margin of error appended to the end of each column. The estimates can be selected (along with the geographic identifiers) and saved in a new data table:

```
state_census<-get_acs(geography = "state",
        variables = c(total_pop="B01003_001", black_pop="B01001B_001"),
                        output="wide") %>%
select(GEOID, NAME, total_popE, black_popE)
```

```
## Getting data from the 2018-2022 5-year ACS
```

The table *state_census* records these observations:

```
state_census
```

```
## # A tibble: 52 x 4
##     GEOID NAME                  total_popE black_popE
##     <chr> <chr>                      <dbl>      <dbl>
##  1 01     Alabama                  5028092    1326341
##  2 02     Alaska                    734821      23395
##  3 04     Arizona                  7172282     327077
##  4 05     Arkansas                 3018669     456693
##  5 06     California              39356104    2202587
##  6 08     Colorado                 5770790     233712
##  7 09     Connecticut              3611317     385407
##  8 10     Delaware                  993635     218266
##  9 11     District of Columbia      670587     297101
## 10 12     Florida                 21634529    3355708
## # ... with 42 more rows
```

The US Census Bureau also maintains its own set of geographic Shapefiles[5],

[5]See https://www.census.gov/geographies/mapping-files/time-series/geo/tiger-line-file.html

which can be merged with census data downloaded from **tidycensus** package or, in some cases, downloaded directly through **tidycensus** by adding the argument `geometry=TRUE`. For example, `get_acs(geography = "state", variables = c(total_pop="B01003_001", black_pop="B01001B_001"), output="wide", geometry=TRUE)` would download State (including Washington, DC and Puerto Rico) line coordinates for visualization and further manipulation like other map objects reviewed in this chapter. The complementary **tigris** package (Walker and Rudis, 2024) automates the importation of the Census geographic datafiles; alternatively, the files can be downloaded directly from the Census Bureau website and merged with the data tables on the *GEOID* column.

Resources

Many units of government maintain Geographic Information Systems (GIS) portals for distributing geospatial data; for example, Michigan's site is https://gis-michigan.opendata.arcgis.com/.

The boundary lines for South Africa from this chapter are from GeoQuery at AidData https://www.aiddata.org/geoquery, which hosts many other geospatial datasets.

GeoQuery includes boundary line files from https://www.geoboundaries.org (Runfola et al., 2020), another comprehensive source.

Other comprehensive repositories include the University of California at Berkeley GeoData library, https://geodata.lib.berkeley.edu/. TIGER/Line Shapefiles from the US Census Bureau are available from https://www.census.gov/geographies/mapping-files/time-series/geo/tiger-line-file.html

For Geocoding addresses in the USA, there is the Texas A&M Geoservices https://geoservices.tamu.edu/Services/Geocode/Interactive/.

There are many additional packages for visualizing and manipulating geospatial data, following from the methods introduced within this chapter. For further reading see Lovelace, Nowosad, and Muenchow (2019) (online at https://r.geocompx.org/) and Walker (2023) (online at https://walker-data.com/census-r/index.html).

Additional election and spatial datafiles are available from the Harvard Election Data Archive (https://projects.iq.harvard.edu/eda) and MIT Election Data + Science Lab (https://electionlab.mit.edu/).

6.7 Exercises

For the following exercises, prepare a .Rmd file with the code and responses to the questions below, knitted into an .html file for interactive maps, or PDF or Word document for static maps, as required by the question.

(1) Create a red-blue election map from the 2011 Michigan State Senate districts, either highlighting the winning State Senate candidate in red or blue colors, or display the margin of the vote percent from red through blue. To merge the vote data onto the State Senate legislative district maps, first change the district name in the Senate maps for 2011 from character to numeric `mi_senate_2011 <- mi_senate_2011 %>% mutate(NAME = as.numeric(NAME))`. Then merge the senate wide-formatted data table *MI_state_senate_2014_wide* onto the map data *mi_senate_2011* with `left join()`, merging `by=c("NAME"="DistrictNumber"))`. Create an indicator to display either the winning candidate in red or blue to display with *scale_fill_manual*, or a measure of the vote margin to display with *scale_fill_gradient2*.

(2) Using the polygon data for administrative wards in South Africa, *ZAF_ADM4.geojson* and corresponding air pollution data, subset to a city. With an interactive **leaflet** map, display the variation in air pollution across the wards. Prepare the map as a presentation graphic with clearly labeled parts; in a paragraph, explain what the variation shows and any interesting features you see.

(3) Repeat the steps for the map of South Africa, but apply the steps for a static map created with **ggplot2**. Be sure to include both a base layer and the polygons displaying variation in air pollution.

(4) Identify a US State redistricting plan in a Shapefile (*.shp* file) or GEOjson file; visualize the districts and calculate a compactness measure for each district. Summarize the variation in compactness measures, given a categorical characteristic of the districts such as party affiliation of the election winner.

(5) Pick a US State; download and clean the US County-level presidential election votes by County. Create a County-level map of the presidential election results. Contrast the map with a Cartogram constructed on the same geography. Discuss the Cartogram alongside the prior map, explaining the different features of the Cartogram compared to the traditional election map.

7

Scaling and Clustering for Pattern Detection

Chapter 7 reviews the use of distance metrics for summarizing and exploring relationships between observations.

Learning objectives and chapter resources

By the end of this chapter you should be able to: (1) describe conceptually the use of multi-dimensional scaling (MDS) as a tool for analyzing data; (2) prepare a data table and construct a matrix of Euclidean or Manhattan distance metrics; (3) apply functions for estimating and evaluating nonmetric MDS solutions; (4) interpret a measure of model goodness of fit; (5) contrast the use of distance metrics in MDS with hierarchical clustering; and (6) interpret a dendrogram. The material in the chapter requires the following packages: **smacof** (Mair et al., 2022), which will need to be installed, as well as **ggrepel** (Slowikowski, 2023), and **tidyverse** (Wickham, 2023b). The material uses the following datasets *poll_ft.csv*, *co_na_h_votes_117_wide.csv*, *co_117_euc_distances.csv*, and *s_memb_votes_117_wide.csv* from https://faculty.gvsu.edu/kilburnw/inpolr.html.

Spatial politics and distance metrics

We frequently conceptualize politics in spatial terms. The ubiquity of variations on the phrase 'liberal to conservative spectrum' captures this intuition. Spatial representations of politics are a kind of political science 'brand', encompassing how the discipline explains lines of political conflict through one- or two-dimensional figures (Brady, 2011). An analytic method for taking data and generating insight into spatial politics through visualization is multidimensional scaling (MDS) — a technique closely tied to spatial theories of politics and one that political scientists have actively developed and refined (Armstrong et al., 2020; Jacoby and Armstrong, 2014; Brady et al., 2011).

In Chapter 7, we review the use of MDS, along with a related method, cluster analysis. There are many varieties of these methods; our discussion is limited to an overview of nonmetric MDS (Rabinowitz, 1975) and 'hierarchical' clustering (Bartholomew et al., 2008). While related, each provides distinctly different insight into the same data. We will review the conceptual foundations of MDS and cluster analysis with the creation of 'distance metrics', the primary data input for these procedures, followed by three examples. First, we will apply MDS to the spatial configuration of the American public's feelings toward political candidates and figures. Second, we will apply MDS to explore the spatial relationships between legislators in the U.S. Congress as revealed through roll call voting. Third, we will contrast the insights from MDS with a cluster analysis of the same data sources.

7.1 Introduction to multidimensional scaling

Multidimensional scaling procedures are a graphical, data visualization focused method of analyzing data. Often the purpose is exploratory, to observe patterns of similarity between subjects of political interest, such as the voting behavior of legislators or an electorates' feelings toward political figures. Just about any dataset composed of measures of a common set of observations could be analyzed through MDS, such as country-level measures of political and economic development. Whatever the input source data, the analytical output leads to a dimensional representation of the similarity among the observations.

An example of such a figure – a typical product of an MDS analysis – appears in Figure 7.1. The figure displays in two dimensions the patterns of feeling thermometer scores toward political figures and groups in the 2004 ANES survey; it is a typical 'output' of an MDS analysis, a two-dimensional scatterplot, given the input of individual-level thermometer scores. The figure shows the location of the subjects, where subjects grouped closer together elicited more similar feelings from the electorate. Based on input matrix summarizing the patterns of dissimilarity between the feeling thermometer scores across the figures and groups, the MDS algorithm constructed a set of two-dimensional coordinates for each one, so that the coordinates mirror the dissimilarities as much as possible.

The MDS input data is a square matrix of dissimilarities, measures that quantify the degree of difference between objects of political interest, considered in pairs. A matrix of dissimilarity data from feeling thermometers could be measures that summarize how differently the public feels toward political figures. There are two basic varieties of MDS procedures: metric and nonmetric. A metric MDS procedure makes interval-level assumptions about the input dissimilarities, finding output distances that reflect the numeric

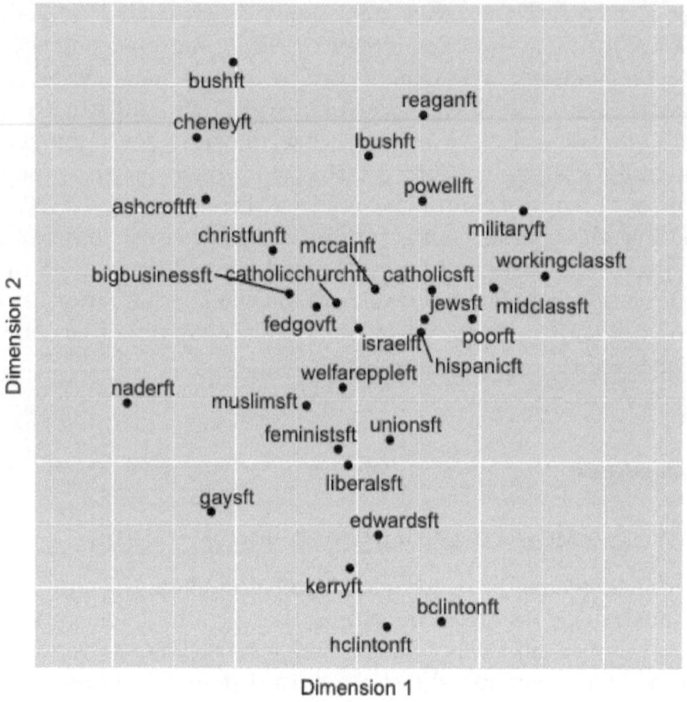

FIGURE 7.1 Nonmetric multidimensional scaling (MDS) plot of feeling thermometer ratings from the 2004 ANES survey, reduced to two dimensions. The plot arranges political figures and social groups based on similarity in respondents' ratings: groups with more similar ratings appear closer together. The familiar partisan and ideological fault lines of American politics emerge, with political figures clustering along ideological directions away from the center, where social groups tend to cluster.

score dissimilarities in the input data. A nonmetric MDS procedure makes only ordinal-level assumptions about the input dissimilarities. In effect the input dissimilarity scores between objects of interest are rank ordered, from least to most similar, and the algorithm finds an optimal translation of the rank-ordered dissimilarities into distances. Below I conceptually review the steps in a nonmetric MDS. A concise and technical introduction to metric and nonmetric is Jacoby (1991) or Jacoby and Ciuk (2018). Nonetheless, both methods start with the construction of dissimilarity matrices. Two commonly used metrics for summarizing dissimilarities are Euclidean and Manhattan distances.

7.2 Euclidean and Manhattan distances

Perhaps the most common dissimilarity measure used in MDS is the Euclidean distance metric. The distance measures the straight line path between two points in a multidimensional space. Figure 7.2 illustrates a Euclidean distance between two points (or objects) in a two-dimensional space, X and Y. The Euclidean distance is based on the familiar Pythagorean theorem, $a^2 + b^2 = c^2$, describing the length of the hypotenuse of a right triangle on these two dimensions. Figure 7.2 displays the Euclidean, straight line distance between the two points, calculated from the squared sum of the distances between each point on the X and Y dimensions. Working through the arithmetic in **R** illustrates how it works.

```
sqrt((6-1)^2 + (5-2)^2)
```

```
## [1] 5.831
```

The distance can be generalized to any number of dimensions and points. If we added a third point and a third dimension, identified p , q, and r:

```
p=c(6,5,7)
q=c(1,2, 10)
r=c(2,5, 12)
```

We can combine rows of the points together (`rbind()`) into a matrix *matx* and apply the distance function `dist()` to calculate the Euclidean distance:

```
mat<-rbind(p,q,r)

mat
```

```
##   [,1] [,2] [,3]
## p    6    5    7
## q    1    2   10
## r    2    5   12
```

```
dist(mat)
```

```
##       p     q
## q 6.557
## r 6.403 3.742
```

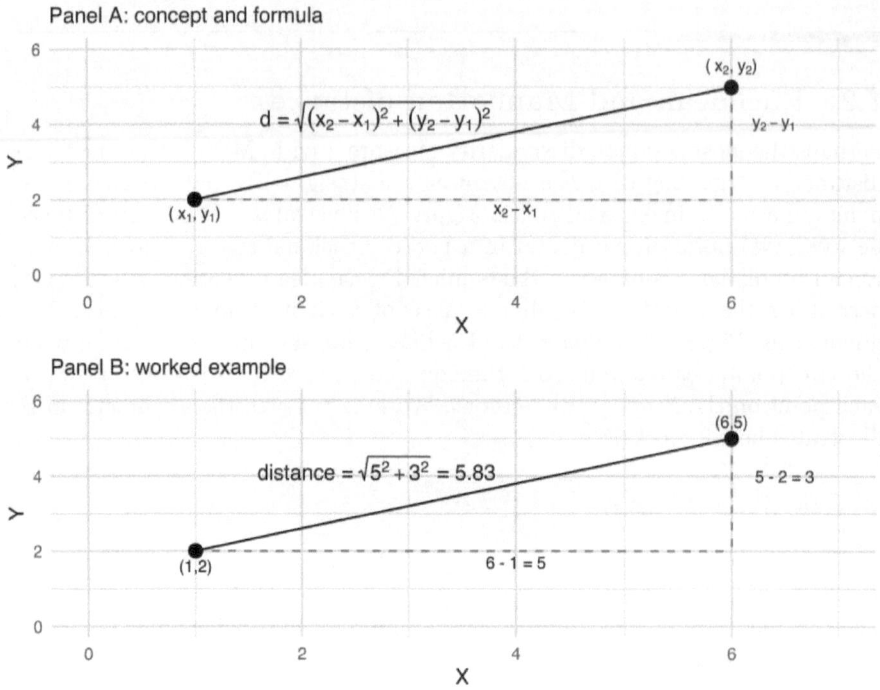

FIGURE 7.2 Euclidean distance formula and calculation of the distance between two points on two dimensions, X and Y.

The result is a Euclidean distance matrix.

Another way of calculating the distance is via a Manhattan or 'city block' distance. Figure 7.3 displays how the Manhattan distance is calculated like walking city blocks, over and up or down. The Manhattan distance is the sum of the absolute difference between the dimensions. For the distance between *p* and *q* on the two dimensions, the difference calculated in **R** is 8: `abs(6-1) + abs(5-2)`. Adding a `distance="manhattan"` argument to `dist()` creates the distances for *matx*:

```
dist(mat, method="manhattan")
```

```
##    p  q
## q 11
## r  9  6
```

Instead of abstract points, in an MDS analysis, distance matrices are created with measurements on political objects of interest — for example, for feeling thermometers, the dimensions (like X and Y) are each a survey respondent's rating of a candidate, or for legislators, the dimensions are the roll call votes.

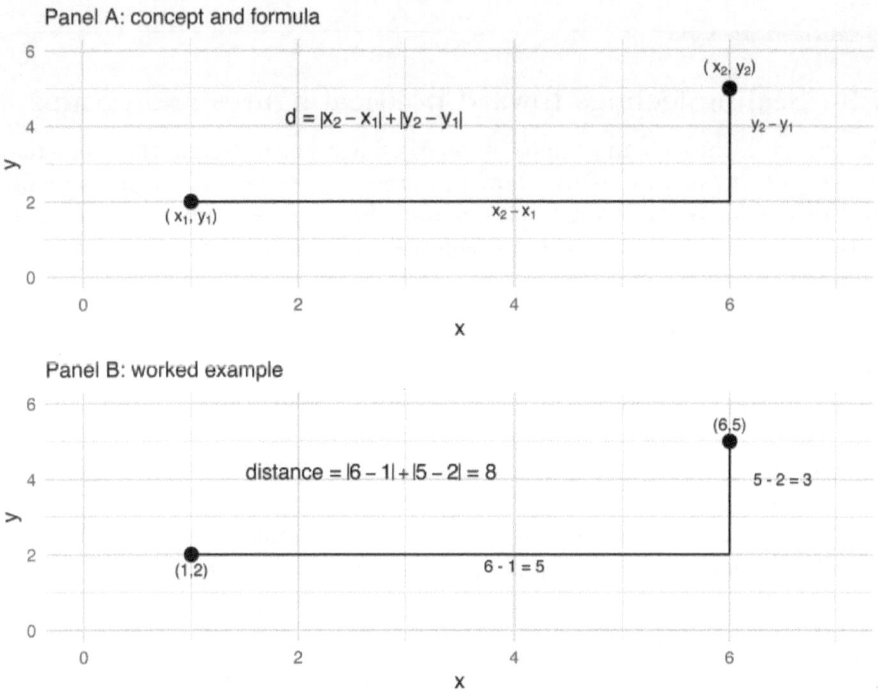

FIGURE 7.3 Manhattan 'city block' distance formula and calculation of the distance between two points on two dimensions, X and Y.

Give a matrix of dissimilarity, the MDS algorithm translates the dissimiliarities into distances.

In a nonmetric MDS, the algorithm accounts for the ordinal arrangement of the dissimilarities starting with "target distances" (or "disparities"), the distances that the algorithm tries to reproduce in the low-dimensional space. The target distances are a new distance matrix, one that preserves the rank order of pairs of objects in their similarity, but not necessarily the exact numerical value of the original matrix of distances. The goal is to find a one- or two-dimensional configuration of the points (as the output visualization) where the actual distances between points closely approximate these target distances. The algorithm focuses on preserving the rank order of the dissimilarities, ensuring that objects perceived as more similar remain closer together in the plot. The fit of the model is evaluated using the function applied by the algorithm to find the configuration, referred to as the "Stress function", which measures the degree of mismatch between the actual distances and the disparities, with lower stress indicating a better fit. The fit statistic is discussed further below. For now, we will work through an example in **R**.

7.3 Scaling feelings toward political figures and groups

We will work through an example of an MDS analysis of feeling thermometer scores *poll_ft.csv*, to illustrate how the procedure works. To begin, we will load two packages, **tidyverse** and **ggrepel** (for labeling graphics), then read in the dataset and store it as object *polfigures*. We use the `read_csv()` function from the **tidyverse** package:

```
library(tidyverse)
library(ggrepel)
```

```
polfigures<-read_csv(file="poll_ft.csv")
```

Once it is stored as *polfigures*, we will remove the columns that do not contain feeling thermometer scores, to simplify the process of applying a distance metric to the data.

```
polfigures<- polfigures %>%
  select(-gender, -campint, -fair, -prezapp, -libcon, -party, -partyid,
         -race, -conservatives, -abor, -educ)
```

```
polfigures
```

```
## # A tibble: 1,212 x 30
##    bushft kerryft naderft cheneyft edwardsft lbushft
##     <dbl>   <dbl>   <dbl>    <dbl>     <dbl>   <dbl>
##  1     70      85       0       50        50      NA
##  2     40      85      NA       40        70      60
##  3    100      50      50      100        50     100
##  4     50      60      60       40        70      60
##  5    100       0      50       85         0     100
##  6     60      30     100       15        30      50
##  7     85      15      40       70        50      70
##  8     50      70      85       60       100      70
##  9     30      70      15       15        60      85
## 10    100       0       0      100         0     100
## # ... with 1,202 more rows, and 24 more variables:
## #   hclintonft <dbl>, bclintonft <dbl>,
## #   powellft <dbl>, ashcroftft <dbl>, mccainft <dbl>,
## #   reaganft <dbl>, hispanicft <dbl>,
```

```
## #   christfunft <dbl>, catholicsft <dbl>,
## #   feministsft <dbl>, fedgovft <dbl>, jewsft <dbl>,
## #   liberalsft <dbl>, midclassft <dbl>, ...
```

Now the dataset consists solely of 30 feeling thermometer scores, queried from the 1,212 respondents to the 2004 ANES survey, a representative cross-section of the U.S. citizenry, 18 years of age or older by the November 2004 election day. (See the Appendix for data details.) The scores for each political figure or social group is rated on the feeling thermometer scale from 0 to 100, the figure of group identified by name, such as *reaganft* for Ronald Reagan. Missing values, *NA*, correspond to "don't know" type responses or any other reason in which a rating was not recorded. Observing the first five rows and columns:

```
polfigures[1:5, 1:5]
```

```
## # A tibble: 5 x 5
##    bushft kerryft naderft cheneyft edwardsft
##     <dbl>   <dbl>   <dbl>    <dbl>     <dbl>
## 1      70      85       0       50        50
## 2      40      85      NA       40        70
## 3     100      50      50      100        50
## 4      50      60      60       40        70
## 5     100       0      50       85         0
```

The `dist()` function to create the summary distance measures between the candidates requires the input matrix to be transposed from its current form. The candidates are currently on the columns, and survey respondent raters are on the rows. The function requires that the 'stimuli', the objects to be assessed for differences, appear on the rows. A transpose function `t()` will switch rows and columns:

```
t(polfigures[1:4, 1:4])
```

```
##          [,1] [,2] [,3] [,4]
## bushft     70   40  100   50
## kerryft    85   85   50   60
## naderft     0   NA   50   60
## cheneyft   50   40  100   40
```

The transpose `t()` function is necessary to transpose the input dataset, moving column (candidates) to rows and each survey respondent to the columns.

Notice the *NA* value *naderft* from the survey respondent in the second column. The `dist()` function handles missing values by simply dropping any pairwise comparisons that involve *NA* values. So for example, across the first four survey respondents or 'dimensions' of comparison, the Euclidean distance between *naderft* and *kerryft* is $\sqrt{(85-0)^2 + (50-50)^2 + (60-60)^2}$, dropping from the calculation the undefined difference between $(85-NA)^2$. From among these four survey respondents, the distance between *naderft* and *kerryft* would be based on comparisons between three respondents, while comparisons between *kerryft* and *bushft* for example would be based on all four. In addition to dropping any pairwise comparisons involving an *NA* value, the `dist()` function also applies a scaling adjustment to any distance calculations in which the number of pairwise comparisons used are less than the total possible number of comparisons.

To observe how the scaling adjustment works, we will apply the `dist()` function to this 4×4 subset of the data.

```
dist(t(polfigures[1:4, 1:4]))
```

```
##          bushft kerryft naderft
## kerryft   69.64
## naderft  100.00   98.15
## cheneyft  22.36   78.42   84.85
```

Notice in the matrix of distances, the Euclidean distance between *kerryft* and *naderft* is 98.14955. This distance differs from the result of 85 from `sqrt((85-0)^2 + (50-50)^2 + (60-60)^2)`. The difference is accounted for by the adjustment factor, $\sqrt{\frac{n}{k}}$, where n is the total possible number of comparisons, the total number of 'dimensions' or respondent ratings if there were no *NA* values. Then k is the number of non-*NA* comparisons. The weight is multiplied by the Euclidean distance; for the distance between *kerryft* and *naderft* for these four survey respondents, the scaled distance would be $\sqrt{\frac{4}{3}} \times 85$, or

```
sqrt((4/3))*85
```

```
## [1] 98.15
```

The result matches the distance in the matrix of six entries on the four columns from the `dist()` function. With the full matrix of scores, we apply the transpose function inside the `dist()` function to create the input distance matrix. The result is stored as *dissimat*:

```
dissimat<-dist(t(polfigures))
```

The result is a 'distance object', *dissimat*, a 30×30 matrix containing (by default) Euclidean distances between each feeling thermometer subject. While the default is Euclidean distance, it can be changed with the `method=` argument, such as `dist(t(polfigures), method='manhattam')`. Researchers usually refer to these matrices as 'dissimilarity matrices' or 'dissimilarity measures' to avoid confusion with the output of an MDS algorithm, a visualization translating these dissimilarities into distances in the visualization.

To view a snippet of the dissimilarity matrix, we first convert it to a matrix and then view the first few rows and columns:

```
dissimat_matrix<-as.matrix(dissimat)

dissimat_matrix[1:5, 1:5]
```

```
##            bushft kerryft naderft cheneyft edwardsft
## bushft        0.0  1878.6    1429    782.9    1790.1
## kerryft    1878.6     0.0    1220   1678.8     651.4
## naderft    1429.5  1220.4       0   1251.3    1208.0
## cheneyft    782.9  1678.8    1251      0.0    1642.7
## edwardsft 1790.1   651.4    1208   1642.7       0.0
```

The subscripts `[1:5, 1:5]` refer to row and column `[r, c]` selections, selecting the first through fifth row. The matrix is symmetric and the main diagonal entries displaying the distance from one political figure on a column to the same political figure on the row is, of course, 0. Across the first row, the distance between *bushft* and *kerryft* is 1878.63. The distance between *cheneyft* and *bushft* is 782.869. Again, these are the scaled distances between pairs of feeling thermometers, calculated across all available pairs of non- *NA* feeling thermometer scores.

The object `dissimat` is the input into the MDS algorithm that translates these dissimilarities into distances. Using the **smacof** package (Mair et al., 2022), the `mds()` function will find a set of output coordinates for the input dissimilarity matrix, for a given number of dimensions. The function requires that we specify the input matrix (in this example, `dissimat`), the number of dimensions (`ndim=`) and for a nonmetric MDS, `type="ordinal"`. Install the package if necessary, `install.packages("smacof")`.

```
library(smacof)
```

The results will be stored as object *nmds_results*:

```
nmds_results<-mds(dissimat, ndim=2,  type="ordinal")
```

Typing the name of the object *nmds_results* reveals details about how the algorithm, in particular a measure of 'badness of fit', which we will review further below. The main result of the `mds()` function is the set of dimensional positions of the feeling thermometers, which is stored as the configuration of points, *nmds_results$conf*. Typing this object out, or viewing the top six rows with `head()`:

```
head(nmds_results$conf)
```

```
##                D1       D2
## bushft     1.0878 -0.51103
## kerryft   -0.9208 -0.02963
## naderft   -0.2673 -0.94219
## cheneyft   0.7874 -0.65896
## edwardsft -0.7895  0.08626
## lbushft    0.7167  0.04366
```

The output displays the positions of the thermometer objects, political figures and groups, on the two dimensions, labeled *D1* and *D2*. Visualizing these configurations is the end goal of the MDS, so the next step is to save the configurations to a dataframe and visualize it:

```
nmds1<-as.data.frame(nmds_results$conf)
```

As a dataframe, it is organized into rows and columns:

```
head(nmds1)
```

```
##                D1       D2
## bushft     1.0878 -0.51103
## kerryft   -0.9208 -0.02963
## naderft   -0.2673 -0.94219
## cheneyft   0.7874 -0.65896
## edwardsft -0.7895  0.08626
## lbushft    0.7167  0.04366
```

Note the dataframe contains row names of the thermometers rather than a column within the dataframe; these are row names, stored as `row.names()` within it. The results plotted as points with the label row names in `ggplot()`. Figure 7.1 displays the MDS algorithm's solution to the problem of finding an arrangement of the feeling thermometers in a two-dimensional X - Y plot, so that the relative distances between each of the points preserves as much as possible the rank order of the disparities calculated from the matrix of Euclidean distances. The `ggplot()` code to generate the figure appears below. The graphic suppresses the numeric placeholders for the scaling solution, as the coordinates of the points do not have inherent meaning on their own.

```
ggplot(nmds1) +
  geom_point(aes(x=D2, y=D1)) +
  geom_text_repel(aes(x=D2, y=D1, label=row.names(nmds1))) +
  scale_y_continuous(limits = c(-1.2, 1.2))+
```

```
scale_x_continuous(limits=c(-1.2, 1.2))+
 labs(x = "Dimension 1", y = "Dimension 2") +
  theme(axis.text.x = element_blank(),
        axis.ticks.x = element_blank() )+
  theme(axis.text.y = element_blank(),
        axis.ticks.y = element_blank() )
```

The figure visualizes the extent to which figures and groups tend to elicit similarly warm or cool feelings; figures such as these date back to the first thermometer items included in the ANES; for example, see Rusk (1972). In interpreting MDS solutions, the key is the relative spacing of the points, the direction or clustering of points — and not to focus exclusively on the movement along the X and Y axes defined in the graphic. Overall, the arrangement in Figure 7.1 reflects a familiar clustering of groups and figures associated with the 'left', 'right' and the Democratic and Republican parties, respectively.

As a map of the electorate's feelings toward figures and groups, clearly there are clusters of Democratic and Republican figures anchoring the upper and lower portions of the graph, perhaps reflecting a partisan ideological dimension of comparison. There are clusters within these regions of the graph, figures and groups associated together, such as the upper-left quadrant clustering Bush, Cheney, Ashcroft, and "Christian Fundamentalists". The social groups tend to cluster in the middle of the graph, perhaps reflecting a lack of polarized feeling toward each. Note that because the interpretation is about the relative position of the points toward each other, the points can be rotated geometrically around the graph if it helps interpretation; scaling texts such as Armstrong and Bakker (2020) and Jacoby (Jacoby, 1991) explain how.

Assessing MDS model fit

Before presenting the analysis of roll call votes, there are two additional concerns that must accompany any MDS analysis: how 'good' of a fit is the dimensional arrangement of observations? And how many dimensions should be visualized? The "Stress" measure of model fit, and an additional figure – a "Shepard diagram" — help to answer these questions.

The Stress statistic, often referred to as "Stress type 1" to differentiate it from other measures is a standard for assessing how well a visual configuration of the observations reflects the dissimilarity input matrix (Bartholomew et al., 2008). While not technically an average, it can be interpreted as the average deviation between the disparities and the fitted distances in the figure. Typing the name of the MDS object, *nmds_results*, reveals the Stress measure and the basic configuration of the model.

```
nmds_results
```

```
##
## Call:
## mds(delta = dissimat, ndim = 2, type = "ordinal")
##
## Model: Symmetric SMACOF
## Number of objects: 30
## Stress-1 value: 0.092
## Number of iterations: 36
```

The Stress measures how well the distances in the solution match the dissimilarities from the input data. It ranges from 0 to 1, where lower values indicate better fit, thus it is a 'badness of fit' measure. If the arrangement of the fitted distances between the thermometers in the one- or two-dimensional space perfectly reflects the rank=ordered distances of the thermometers in the input matrix, then the Stress value would be 0, a perfect fit. From there, some guidelines are .05 is a "good" fit, while .2 or higher is a "poor" fit (Bartholomew et al., 2008). There is a tradeoff between better fit through higher dimensionality versus interpretability. In the example of the feeling thermometers, the Stress value shows a good fit, which could be improved with an additional dimension at the expense of parsimony. Researchers typically plot configurations of one to two dimensions (Jacoby and Ciuk, 2018).

Shepard diagrams

A Shepard diagram is a diagnostic scatterplot for assessing the fit of an MDS (Shepard, 1962). The plot displays information about the fit between the observed dissimilarities or disparities (the input data) and the fitted distances (the output figure). For each pair of points, on the X-axis the figure plots the original dissimilarities or disparities, the 'target distances' that represent the ordinal rank order of similarity between the pairs. The Y-axis plots the fitted distances. If the model fits perfectly — fitted distances match dissimilarities exactly – the points would line up on a straight 45-degree diagonal line from the origin.

The `plot()` function extracts the information from *nmds_results* , specifying `plot.type="Shepard"`.

```
plot(nmds_results , plot.type="Shepard")
```

In Figure 7.4, the diagram displays in gray points measurements on pairs of points, the input dissimilarities on X and the fitted distances in the output on Y. As would be expected from a Stress statistic of .092, the fit is not perfect, but the model fits somewhat well, with points clustered along the 45-degree

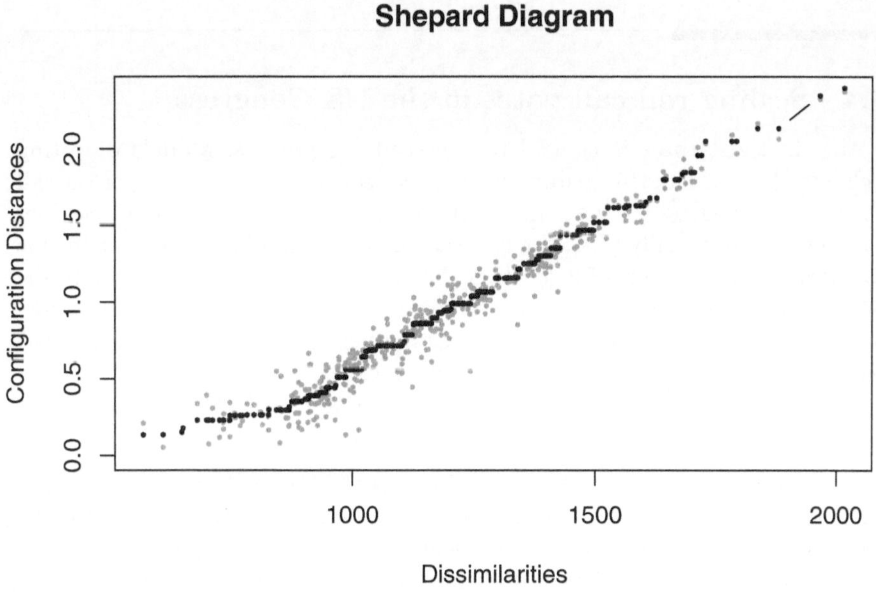

FIGURE 7.4 Shepard diagram for the nonmetric multidimensional scaling (MDS) solution of feeling thermometer ratings from the 2004 ANES, in two dimensions. The plot compares original dissimilarities to the distances in the MDS configuration.

line, although clearly for lower dissimilarities the fit is worse. The spread of points at the lower left side of the graph corresponds to the difficulty in scaling distances over two dimensions for the points with low dissimilarity. With less spread, as dissimilarity increases, fit improves. This contrast makes sense, given that for social groups scaled relatively close together, the algorithm was less successful in finding an optimal arrangement of the points, compared to the more polarizing cluster of political figures toward the edges of the output visualization.

The black points on the Shepard diagram represent the transformed dissimilarities, or disparities, which serve as the target distances for the MDS algorithm. These points reflect the idealized monotonic fit, preserving the rank order of the original dissimilarities. Notice that, from left to right, the black points either increase or remain constant, but never decrease, ensuring the monotonic relationship. Similar to the imaginary 45-degree line in metric MDS, the grey points, which represent the actual distances in the low-dimensional solution, should lie as close as possible to the black points for a good fit. The components of a Shepard diagram can be recovered from the MDS solution object, *mdsgroup_results* contains additional information about the MDS solution, accessible as "variables" within the object `names(mdsgroup_results)`, with `$` and the specific item. See the documentation for the **smacof** package.

7.4 Scaling roll call votes in the US Congress

Voting in legislatures is central to the study of politics. Models of voting behavior have a lengthy history in political science (Poole, 2005). Roll call votes in US Congress are votes in which each member's position ('yea', 'nay', or 'present') is individually recorded, for everything from motions and amendments to votes on final passage of legislation. The recorded votes are entered into the official Congressional Record. Interest groups base their ratings of legislators on these votes. The most prominent estimates of legislator ideology, 'DW-NOMINATE' scores, are based on roll call votes (Poole, 2005). We will analyze votes from an MDS perspective, to visualize patterns of voting – differences between party caucuses and individual members.

Roll call votes are from the 117th US Congress (2021 to 2023). We first work through the process of tidying the data to prepare a matrix of votes for the `mds()` algorithm. While useful for practicing data skills, this step can be skipped, and the matrix of votes can be downloaded directly from the textbook's website. Or, to practice data tidying and MDS analysis skills, download the full datafiles from Voteview.com (Lewis et al., 2023). For each Congress, Voteview organizes the data into three files: one for the voting records, a second for the members, and a third for descriptions of the roll call votes. We will first work with the votes and members' data, then the roll calls to find specific type of legislation. The two datasets used to create the matrix of votes are *HS117_votes.csv* and *HS117_members.csv*, both of which include the House and Senate.

We will first work with Euclidean distances, and to keep the data manageable at first limit it to the Colorado delegation, then examples with the full House and Senate, illustrating properties of the Manhattan distance metric. Either start with the Voteview datafiles, or skip below to the matrix `mds()` input files beginning with *co_117_euc_distances.csv.*

Read in the votes data:

```
HS117_votes<-read_csv(file="HS117_votes.csv")
```

Familiarize yourself with the contents of it. We subset it to four columns:

```
HS117_votes <- HS117_votes %>%
  select(chamber, rollnumber, icpsr, cast_code)
```

Next we recode the numeric values for the 'yeas', 'nays', present, and absent votes. The 'nays' are scored at 0, 'yeas' at 1, and to make it easier to recall, 5 for present and 10 for absent.

```
HS117_votes <- HS117_votes %>%
  mutate(cast_code = case_when(
    cast_code==6 ~ 0,
    cast_code==1 ~ 1,
    cast_code==7 ~ 5,
    cast_code==9 ~ 10))
```

Try `table(HS117_votes$cast_code)` to observe the values for votes cast. Next we read in the data on members:

```
HS117_members<-read_csv(file="HS117_members.csv")
```

The data table contains demographic information about each member, as well as an ideology score (Poole and Rosenthal, 2006). The key field in the members' data is a unique identifier for each member, *icpsr*. The member data needs to be merged with the votes. We want, for each row of votes in *HS117_votes*, to have the member information appended as additional columns matched to *icpsr*. We use a `right_join()` function to merge the votes and members, saving the new merged dataset as *hs_memb_votes_117*.

```
hs_memb_votes_117 <- HS117_members %>%
  select(icpsr, bioname, state_abbrev, party_code) %>%
  right_join(HS117_votes, by="icpsr") %>%
  select(chamber, bioname, party_code, state_abbrev,
         cast_code, rollnumber, icpsr)
```

In the language of mutating joins in the *tidyverse* (Wickham and Grolemund, 2016), *HS117_members* is the X, and *HS117_votes* is the Y. So we would like X matched to Y. There are multiple matches from member information to their row-wise record of votes, identified by *icpsr*. So a warning message will appear letting us know that these multiple matches did in fact occur.

The Voteview data includes mock votes from public positions taken by presidents. Inspect the dataset to observe the presidents in the first few rows. We will remove the presidents:

```
hs_memb_votes_117 <- hs_memb_votes_117 %>%
  filter(state_abbrev!="USA")
```

Then we create a new dataset for House votes: work with votes from the House, extracting the *House* chamber votes to save it as *h_memb_votes_117*.

```
h_memb_votes_117 <- hs_memb_votes_117 %>%
  filter(chamber=="House") %>%
  select(-chamber)
```

In the House, there are 996 possible votes they could have taken. (Enter `table(h_memb_votes_117$bioname)` to observe the count of Representatives and votes. We want to limit the data to just those representatives serving a full term. So we filter the data to all representatives with exactly 996 rows:

```
h_memb_votes_117 <- h_memb_votes_117 %>%
  group_by(bioname) %>%
  filter(n() == 996) %>%
  ungroup()
```

To limit the data and make it more manageable for learning MDS, we subset it to Colorado:

```
co_h_votes_117<- h_memb_votes_117 %>%
  filter(state_abbrev=="CO")
```

Enter *co_h_votes_117* at the prompt to familiarize yourself with it. For labeling the output we need to create a simple label with name, party, and State, which we will use with the entire House and Senate as well.
We use the `separate()` function on the *bioname* column, specifying the `sep=", "` as the place to separate the name. Notice the white space after the comma, which prevents a leading white space appended to the first name; `remove=FALSE`.

```
co_h_votes_117<- co_h_votes_117 %>%
  separate(bioname, c("last_name", "first"), sep=", " )
```

Now the *bioname* column is separated into parts:

```
co_h_votes_117
```

```
## # A tibble: 6,972 x 7
##    last_n~1 first party~2 state~3 cast_~4 rolln~5 icpsr
##    <chr>    <chr>   <dbl> <chr>     <dbl>   <dbl> <dbl>
##  1 LAMBORN  Doug      200 CO            0       1 20704
##  2 LAMBORN  Doug      200 CO            1       2 20704
##  3 LAMBORN  Doug      200 CO            0       3 20704
##  4 LAMBORN  Doug      200 CO            0       4 20704
##  5 LAMBORN  Doug      200 CO            0       5 20704
```

```
## 6 LAMBORN Doug 200 CO 1 6 20704
## 7 LAMBORN Doug 200 CO 0 7 20704
## 8 LAMBORN Doug 200 CO 1 8 20704
## 9 LAMBORN Doug 200 CO 1 9 20704
## 10 LAMBORN Doug 200 CO 1 10 20704
## # ... with 6,962 more rows, and abbreviated variable
## #   names 1: last_name, 2: party_code,
## #   3: state_abbrev, 4: cast_code, 5: rollnumber
```

Next we will change the *party_code* to a "D" or "R" and combine the *last_name*, *state_abbrev*, and *party_code* together, and then drop the remaining *first* and *icpsr* columns:

```
co_h_votes_117<- co_h_votes_117 %>%
  mutate(party_code = case_when(
    party_code == 100 ~ "D",
    party_code == 200  ~ "R")) %>%
  unite(name, c("last_name", "state_abbrev", "party_code")) %>%
  select(-first, -icpsr)
```

At this point we have to decide how to handle the votes cast that are not 'yeas' and 'nays', the problem being that there is no order or at least ordinal scale that links these votes with a vote of 'present' or an absence from voting. –A vote of 'present' or an absence is not meaningfully higher or lower than the 'yeas' and 'nays'.

```
co_h_votes_117  %>%
  count(cast_code) %>%
  mutate(prop. = n / sum(n))
```

```
## # A tibble: 4 x 3
##   cast_code     n    prop.
##       <dbl> <int>    <dbl>
## 1         0  2284 0.328
## 2         1  4571 0.656
## 3         5     1 0.000143
## 4        10   116 0.0166
```

There is only one 'present' vote, and 116 absences, about 1.6 percent of the total. We could count up presences and absences by Representative:

```
co_h_votes_117 %>%
  group_by(name) %>%
  summarize(summary_absences = sum(cast_code == 10, na.rm = TRUE)) %>%
  arrange(desc(summary_absences))
```

```
## # A tibble: 7 x 2
##   name              summary_absences
##   <chr>                        <int>
## 1 BUCK_CO_R                       55
## 2 BOEBERT_CO_R                    24
## 3 LAMBORN_CO_R                    20
## 4 CROW_CO_D                        6
## 5 DeGETTE_CO_D                     5
## 6 PERLMUTTER_CO_D                  5
## 7 NEGUSE_CO_D                      1
```

Or change the *cast_code* to *5* to observe Boebert cast the lone 'present'.

We will calculate the distance metrics between the Representatives while converting the 'present' (*5*) and absent (*10*) votes to *NA* values:

```
co_na_h_votes_117 <- co_h_votes_117 %>%
  mutate(cast_code = case_when(
    cast_code %in% c(5, 10) ~ NA,
    TRUE ~ cast_code))
```

Inspect the dataset to observe the *NA* values. Next we pivot the data, converting it from long format to wide format with `pivot_wider()` from the **tidyverse**.

```
co_na_h_votes_117_wide<-co_na_h_votes_117 %>%
  pivot_wider(values_from = cast_code, names_from = rollnumber)
```

Before continuing, take time to observe the structure of the data `co_na_h_votes_117_wide` in wide format, and think ahead to how the distance metric will handle missing values *NA*. For example, enter `co_na_h_votes_117_wide %>% select(1:11)` and observe that Buck has a missing *NA* value in roll call vote 10. In calculating the distances between the Representatives, all comparisons with Buck will ignore roll call vote 10. For all other distances between Representatives, across the 10 votes, each will be included in the distance calculation. Across all votes, if two Representatives have any *NA* values in the votes being compared, only that specific distance calculation involving an *NA* value will be excluded, with all others included. With the wide dataset, to prepare it for the distance calculations, next we move the *name* variable to the defined names of the rows (in place of numbered row names), which will be necessary for identifying the representatives in the measure of dissimilarity and then the output visualizations.

```
co_na_h_votes_117_wide <- co_na_h_votes_117_wide %>%
  column_to_rownames(var="name")
```

With the Representatives on row names and votes organized across columns, we can calculate and compare two different distance metrics, Euclidean and Manhattan:

```
co_117_euc_distances<-dist(co_na_h_votes_117_wide, method="euclidean")
```

The Euclidean distances are the following:

```
co_117_euc_distances
```

```
##                  LAMBORN_CO_R PERLMUTTER_CO_D BUCK_CO_R
## PERLMUTTER_CO_D        24.746
## BUCK_CO_R              12.960          26.880
## CROW_CO_D              24.551           3.335    26.755
## NEGUSE_CO_D            24.778           3.885    26.882
## BOEBERT_CO_R           14.645          27.845    11.121
## DeGETTE_CO_D           24.850           3.893    26.880
##                  CROW_CO_D NEGUSE_CO_D BOEBERT_CO_R
## PERLMUTTER_CO_D
## BUCK_CO_R
## CROW_CO_D
## NEGUSE_CO_D          3.755
## BOEBERT_CO_R        27.785      27.880
## DeGETTE_CO_D         4.266       2.837       27.831
```

The matrix hints at the furthest ‘right’ Representative being Boebert, who is the most distant from Colorado Democrats. The Democrats have relatively close distances. Compare this matrix with the Manhattan distances:

```
co_117_man_distances<-dist(co_na_h_votes_117_wide, method="manhattan")
```

```
co_117_man_distances
```

```
##                  LAMBORN_CO_R PERLMUTTER_CO_D BUCK_CO_R
## PERLMUTTER_CO_D       612.371
## BUCK_CO_R             167.974         722.526
## CROW_CO_D             602.734          11.123   715.842
## NEGUSE_CO_D           613.945          15.091   722.630
## BOEBERT_CO_R          214.475         775.354   123.686
## DeGETTE_CO_D          617.499          15.152   722.526
```

```
##                 CROW_CO_D NEGUSE_CO_D BOEBERT_CO_R
## PERLMUTTER_CO_D
## BUCK_CO_R
## CROW_CO_D
## NEGUSE_CO_D        14.099
## BOEBERT_CO_R      772.029     777.291
## DeGETTE_CO_D       18.201       8.048      774.552
```

Because the data includes *NA* values, the Manhattan distances are not whole numbers, distances with the scaling factor applied to each calculation. If calculated without the scaling factor, a Manhattan distance has an interesting interpretation. Since the Manhattan distance sums the absolute difference between each roll call vote scored 0 'nay' and 1 'yea', the unadjusted Manhattan distance counts the number of votes where two legislators differ. If two legislators cast a different 'yea' or 'nay' on 300 votes, their Manhattan distance would be 300.

We can compare the scaled up Manhattan distances with the distances calculated on the votes not containing any missing values, by selecting out each of those columns:

```
co_117_man_comp_distances <- co_na_h_votes_117_wide %>%
  select_if(~ all(!is.na(.))) %>%
  dist(method="manhattan")
```

The line `select_if()` function is applied to a "formula" `~ all(!is.na(.))`, selecting any column that does not contain a *NA*. The `.` dot refers to the current column `select_if()` is working through to find *NA* values — and selecting all that do not `is.na(.)` contain any *NA* values.

```
co_117_man_comp_distances
```

```
##                 LAMBORN_CO_R PERLMUTTER_CO_D BUCK_CO_R
## PERLMUTTER_CO_D          554
## BUCK_CO_R                146             644
## CROW_CO_D                545               9       639
## NEGUSE_CO_D              556              12       644
## BOEBERT_CO_R             188             694       110
## DeGETTE_CO_D             558              12       644
##                 CROW_CO_D NEGUSE_CO_D BOEBERT_CO_R
## PERLMUTTER_CO_D
## BUCK_CO_R
## CROW_CO_D
## NEGUSE_CO_D            13
## BOEBERT_CO_R          693         696
```

```
## DeGETTE_CO_D            17           8            694
```

The distances, in whole numbers, measure the number of votes differing between each legislator. We will compare a one-dimensional representation of the voting similarity between the Euclidean and the unscaled up whole number Manhattan distances. First the Euclidean distances, storing the results in *co_h_117_euc_mds*:

```
co_h_117_euc_mds<-mds(co_117_euc_distances, ndim=1, type="ordinal")
```

Stress shows an excellent one-dimensional fit of .0027. (Type `co_h_117_euc_mds$stress`). We store the coordinates *co_h_117_euc_mds$conf* as a dataframe, apply and clean up the labels for each representative, then graph the result:

```
co_mds_df_euc <- as.data.frame(co_h_117_euc_mds$conf) %>%
  rownames_to_column(var = "name") %>%
  mutate(name = str_remove_all(name, "_CO"))
```

In the second line, the row names are moved to a column; in the third, the *name* column is edited with `str_remove_all()` to find and delete each of the State identifiers "`_CO`".

Then *co_mds_df_euc* is fed into `ggplot()`. We use **ggrepel** for the labels, like in the prior graphic. One change in the `ggplot()` is that in place of a variable for a second dimension, we specify `y=0`, since we want to visualize the points on a single dimension. The result appears in Figure 7.5, which show that the Democrats from Colorado voted more cohesively than the Republicans. Within the delegation, DeGette and Boebert had the most dissimilar voting records, and interpreting the distances in ideological terms both were the most liberal and conservative, respectively.

```
ggplot(co_mds_df_euc) +
  geom_point(aes(x = D1, y = 0)) +
  geom_text_repel(aes(x = D1, y = 0, label = name), max.overlaps = 7) +
  labs(x = "Euclidean distance-based ordering of votes", y = " ") +
    theme(axis.text.y = element_blank(),
          axis.ticks.y = element_blank())
```

We can compare the dimensional solution and visualization from the Euclidean distances with the whole number Manhattan distances that count up the number of differences in votes ('yeas' versus 'nays') between them. We would repeat the process used to find and graph the Euclidean MDS solution. While this code is not evaluated, in this example the Manhattan distances would provide a practically identical one-dimensional visualization.

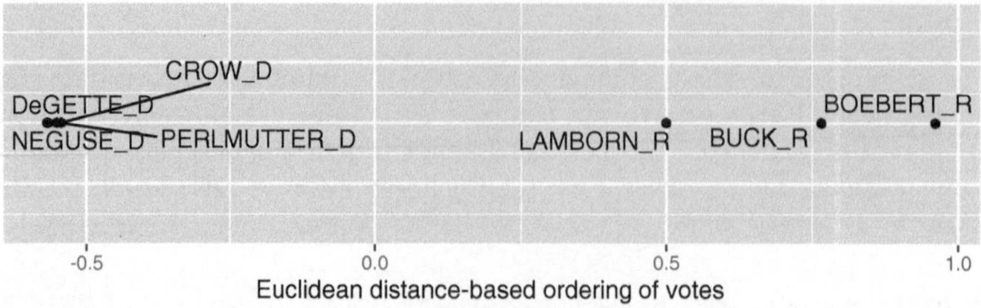

FIGURE 7.5 One-dimensional solution from multidimensional scaling of U.S. House roll call votes, Colorado delegation, 117th U.S. Congress, Euclidean distances.

```
co_h_117_man_mds<-mds(co_117_man_comp_distances, ndim=1, type="ordinal")

co_mds_df_man <- as.data.frame(co_h_117_man_mds$conf) %>%
  rownames_to_column(var = "name") %>%
  mutate(name = str_remove_all(name, "_CO"))
```

Usually the dimensional graphics are visibly sensitive to the choice of distance metrics, although the general conclusions likely remain the same. And with only seven objects, a single dimension fits the data well given the low Stress value. Next, however, we will investigate votes in the U.S. Senate — all 100 Senators — and compare a one- and two-dimensional model fit. We'll go back to the Senate data, and repeat the steps on the U.S. House data.

Scaling votes in the Senate

We start with the combined House and Senate data and select on the Senate chamber:

```
s_memb_votes_117 <- hs_memb_votes_117 %>%
  filter(chamber=="Senate") %>%
  select(-chamber)
```

With the U.S. Senate votes saved in *s_memb_votes_117*, entering `table(s_memb_votes_117$bioname)`reveals that 949 roll call votes were taken, and that Vice President Harris cast two votes, along with temporary appointee from Georgia, Senator Loeffler. We will remove their two votes to limit the dataset to the elected Senators serving throughout the session of Congress.

```
s_memb_votes_117 <- s_memb_votes_117 %>%
 filter(bioname != "LOEFFLER, Kelly") %>%
 filter(bioname != "HARRIS, Kamala Devi")
```

Then we follow the same steps from the House votes to form the name-party-State variable and encode the *NA* missing values for votes of 'present' and absences. One additional line is to change the *last_name* of each Senator to title case and to code the two independent Senators as "I" for party affiliation:

```
s_memb_votes_117<- s_memb_votes_117 %>%
  separate(bioname, c("last_name", "first"), sep=", " ) %>%
  mutate(last_name = str_to_title(last_name)) %>%
  mutate(party_code = case_when(
    party_code == 100 ~ "D",
    party_code == 200  ~ "R",
    party_code == 328 ~ "I"))  %>%
  unite(name, c("last_name", "state_abbrev", "party_code")) %>%
  select(-first, -icpsr)  %>%
  mutate(cast_code = case_when(
    cast_code %in% c(5, 10) ~ NA,
    TRUE ~ cast_code))
```

Inspect *s_memb_votes_117* to observe it matches the format of the House data. Next we pivot the data to wide format (saving it as *s_memb_votes_117_wide*) and move the *name* column to the row names of the data table to feed into the `mds()` function.

```
s_memb_votes_117_wide<-s_memb_votes_117 %>%
  pivot_wider(values_from = cast_code, names_from = rollnumber)
```

The data is pivoted, then we move the *name* column to the row names:

```
s_memb_votes_117_wide<-s_memb_votes_117_wide %>%
  column_to_rownames(var="name")
```

Inspect the data table *s_memb_votes_117_wide* to verify it matches the structure of the House votes data to create the distances. We will create a matrix of Euclidean distances, stored as *s_117_euc_distances*.

```
s_117_euc_distances<-dist(s_memb_votes_117_wide, method="euclidean")
```

Like the House distance matrix, this matrix is calculated from each roll call vote without any 'present' or 'absences'. The distance object is too large to view as a whole, but to observe part of it, we can convert it to a matrix and view the first few columns and rows:

```
temp_senate_matrix<-as.matrix(s_117_euc_distances)
temp_senate_matrix[1:5, 1:5]
```

```
##                 Tuberville_AL_R Shelby_AL_R
## Tuberville_AL_R           0.000       8.811
## Shelby_AL_R               8.811       0.000
## Murkowski_AK_R           22.538      21.216
## Sullivan_AK_R            12.832      12.408
## Sinema_AZ_D              28.611      27.772
##                 Murkowski_AK_R Sullivan_AK_R
## Tuberville_AL_R          22.54         12.83
## Shelby_AL_R              21.22         12.41
## Murkowski_AK_R            0.00         19.75
## Sullivan_AK_R            19.75          0.00
## Sinema_AZ_D              18.01         26.13
##                 Sinema_AZ_D
## Tuberville_AL_R       28.61
## Shelby_AL_R           27.77
## Murkowski_AK_R        18.01
## Sullivan_AK_R         26.13
## Sinema_AZ_D            0.00
```

```
s_117_euc_mds<-mds(s_117_euc_distances, ndim=1, type="ordinal")
s_117_euc_mds$stress
```

```
## [1] 0.0243
```

The one-dimensional solution Stress is .024, suggesting a one-dimensional structure fits the pattern of 'yeas' and 'nays'. Next we convert it to a dataframe, bringing the row names to a column, then graphing it with `ggplot()`:

```
s_117_euc_mds <- as.data.frame(s_117_euc_mds$conf) %>%
  rownames_to_column(var = "name")
```

Figure 7.6 visualizes the structure. Along a single dimension, Democrats are more closely clustered together, apparently voting with greater party discipline, than are Republicans. The `geom_text_repel()` function labels observations where a given label fits, in this case pointing out the Republicans with more moderate voting records based on similarity of roll call votes. Inspect the individual scores for each Senator with `s_117_euc_mds %>% arrange(D1)`.

```
ggplot(s_117_euc_mds) +
  geom_point(aes(x = D1, y = 0)) +
  geom_text_repel(aes(x = D1, y = 0, label = name)) +
  labs(x = "Euclidean distance-based ordering of votes", y = " ") +
    theme(axis.text.y = element_blank(),
          axis.ticks.y = element_blank())
```

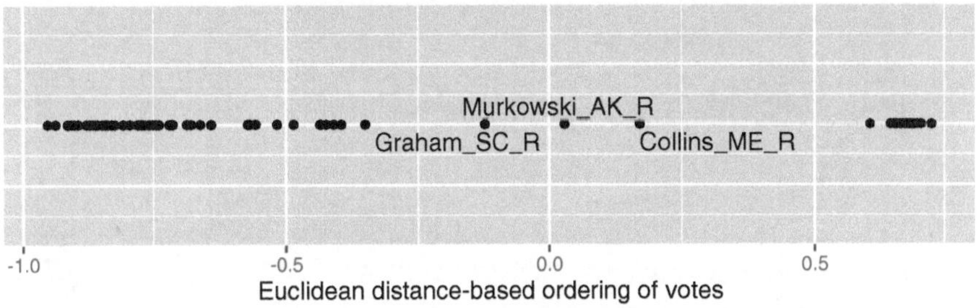

FIGURE 7.6 One-dimensional solution from multidimensional scaling of U.S. Senate roll call votes, 117th U.S. Congress, Euclidean distances.

We will investigate whether a two-dimensional representation provides any insight into voting patterns. We calculate the scores for each Senator on the two dimensions:

```
s_117_euc_mds2<-mds(s_117_euc_distances, ndim=2, type="ordinal")
s_117_euc_mds2$stress
```

```
## [1] 0.02038
```

The Stress improves only a negligible amount, suggesting little improvement in fit over two dimensions. We create the dataframe, and then feed it into `ggplot()` for the visualization.

```
s_117_euc_mds2<- as.data.frame(s_117_euc_mds2$conf) %>%
  rownames_to_column(var = "name")
```

Figure 7.7 depicts the two-dimensional MDS structure of roll call votes in the Senate. As in the last figure, the tick marks and numerical scaling are excluded with the code within `theme()`, to avoid giving the impression that the numeric scores have an underlying meaning tied to the points.

```
ggplot(s_117_euc_mds2) +
  geom_point(aes(x = D1, y = D2)) +
```

```
geom_text_repel(aes(x = D1, y = D2, label = name)) +
labs(x = "Dimension 1", y = "Dimension 2") +
  theme(axis.text.x = element_blank(),
        axis.ticks.x = element_blank())
```

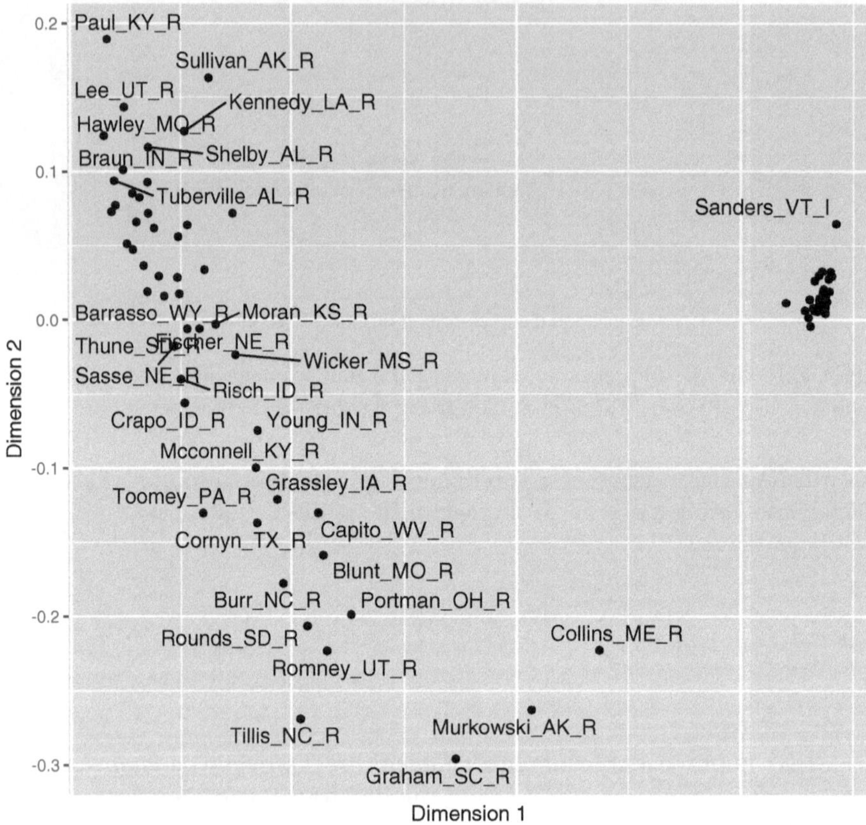

FIGURE 7.7 Two-dimensional solution from multidimensional scaling of U.S. Senate roll call votes, 117th U.S. Congress, Euclidean distances.

```
  theme(axis.text.y = element_blank(),
        axis.ticks.y = element_blank())
```

```
## List of 2
##  $ axis.text.y : list()
##   ..- attr(*, "class")= chr [1:2] "element_blank" "element"
##  $ axis.ticks.y: list()
##   ..- attr(*, "class")= chr [1:2] "element_blank" "element"
```

```
##  - attr(*, "class")= chr [1:2] "theme" "gg"
##  - attr(*, "complete")= logi FALSE
##  - attr(*, "validate")= logi TRUE
```

The two-dimensional results clearly show the same left to right difference observable on Dimension 1, with Collins, Murkowski, and Graham having the more centrist voting record along it. The two Senators Hawley and Sanders anchor both ends of the dimension. And of course the close clustering of Democrats reveals a high degree of party unity in voting, perhaps to be expected given their slim majority. And like the first, the second dimension shows distinct differences among Republicans. On this second dimension, Graham anchors the bottom of the figure, followed by Tillis and Murkowski. Perhaps this dimension reflects a willingness to vote against the party leadership. Of course, the interpretation of the dimensions is necessarily impressionistic and somewhat subjective, and the arrangement of the points is dependent on the votes and the chosen distance metric. In presenting MDS results, researchers typically investigate the sensitivity of the results to the chosen metric.

7.5 Clustering feelings and votes

The term "Cluster analysis" refers to a wide range of analytical methods. Like MDS, it relies on distance metrics as a data input. Unlike MDS, however, rather than uncover a dimensional structure of the data, the purpose is to identify clusters or groups of similar political phenomena. A common approach to cluster analysis is hierarchical agglomerative clustering (HAC) (Bartholomew et al., 2008), which is reviewed conceptually below.

Cluster analysis, like MDS, starts with a matrix of distances to begin identifying clusters. The typical end product of a cluster analysis is a specific visualization, a dendrogram, which displays the formation of clusters in a hierarchy of similarity.

The 'agglomerative' term means that the clustering process is bottom-up, starting with the maximal number of clusters, then ending up with all observations combined into one cluster. The HAC algorithm proceeds in a series of steps: (1) each observation begins as its own cluster; (2) the algorithm merges the two closest clusters; (3) clusters become larger and are merged until all observations belong to a single cluster. The process of forming clusters yields overarching, 'hierarchical' clusters as the name implies.

The algorithm's search for clusters requires specific rules for how it should join clusters together. At each step of the clustering process, the input distance matrix is reduced and recalculated among the clusters. The rules are often one of the following: "Single linkage": the distance between the two closest points in different clusters, "Complete linkage": the distance between the two farthest points in different clusters, or "Average linkage": the average distance between all pairs of points in two clusters. The rule for combining clusters determines how clusters are formed, so the results of any cluster analysis are sensitive to the type of linkage used; for an illustration of different linkage effects, see Gareth *et al* (2013).

Hierachical AC is implemented with the `hclust()` function from the base **R**. The default linkage method is the complete linkage. To illustrate the interpretation of the analysis output, the dendrogram, we will start with the Colorado congressional votes. Given the Colorado matrix of Euclidean distances *co_mds_df_euc*, a HAC is:

```
hac<-hclust(co_117_euc_distances)

hac
```

```
##
## Call:
## hclust(d = co_117_euc_distances)
##
## Cluster method   : complete
## Distance         : euclidean
## Number of objects: 7
```

The dendrogram

The results in *hac* show that the default complete linkage method was used; alternatives are specified with the argument `method=`, such as `method="single"` or `"average"`. The output is the dendrogram, accessible with the `plot()` function:

```
plot(hac)
```

Figure 7.8 displays the resulting dendrogram. The diagram resembles a tree, with branches showing the sequence of cluster formations from pairs to larger group formations. The vertical axis labeled "Height" refers to the distances at which the clusters merge; clusters joined together in branches at higher levels of height are less similar. At the bottom of the figure, the labels or 'leaves', show that at first, the algorithm identified pairs such as Neguse and DeGette, and given the odd number of representatives, one singleton, Lamborn. Then in the next stage the algorithm clustered the four Democrats, then the three Republicans. The dendrogram provides a visual representation of similarity, while forming hierarchically larger (and less similar) groups.

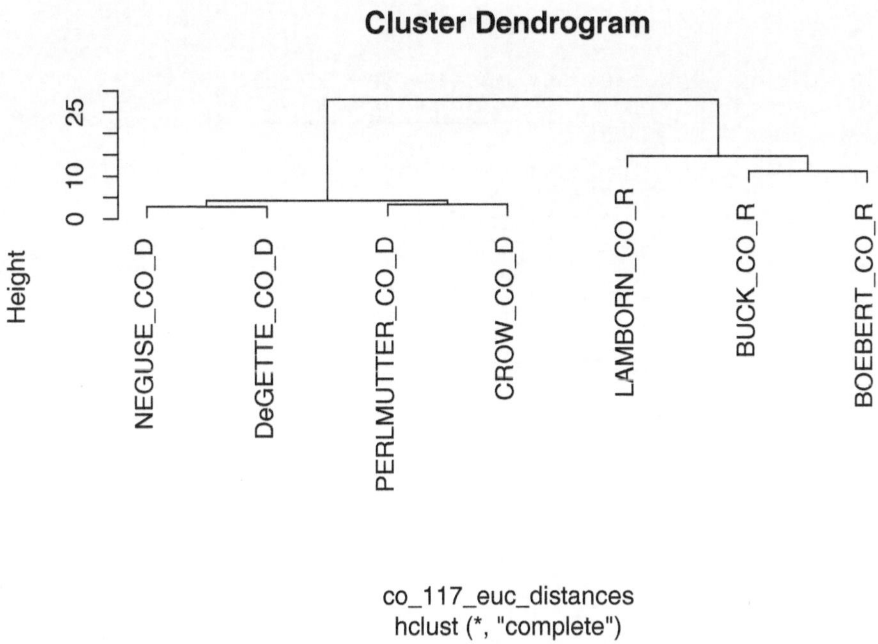

FIGURE 7.8 Dendrogram based on the Euclidean distance matrix of roll call votes cast by the Colorado delegation.

To illustrate a more complex dendrogram, apply the `hclust()` function to the set of full Senate distances, *s_117_euc_distances.*

```
hac_s<-hclust(s_117_euc_distances)
hac_s
```

```
##
## Call:
## hclust(d = s_117_euc_distances)
##
## Cluster method   : complete
## Distance         : euclidean
## Number of objects: 100
```

Plotting a legible dendrogram with 100 objects is not feasible as a small figure on a printed page. Try plotting it and adjust the size to make the labels visible, `plot(hac_s)`. Like any other data visualization in **R**, there are endless ways to customize the appearance of a dendrogram; packages to do so are reviewed at the end of the chapter. Figure 7.9 displays a dendrogram with smaller text labels for the leaves, a suppressed title, and manually rotated, `plot(hac_s, cex=.5, main=NULL, xlab=" ")`.

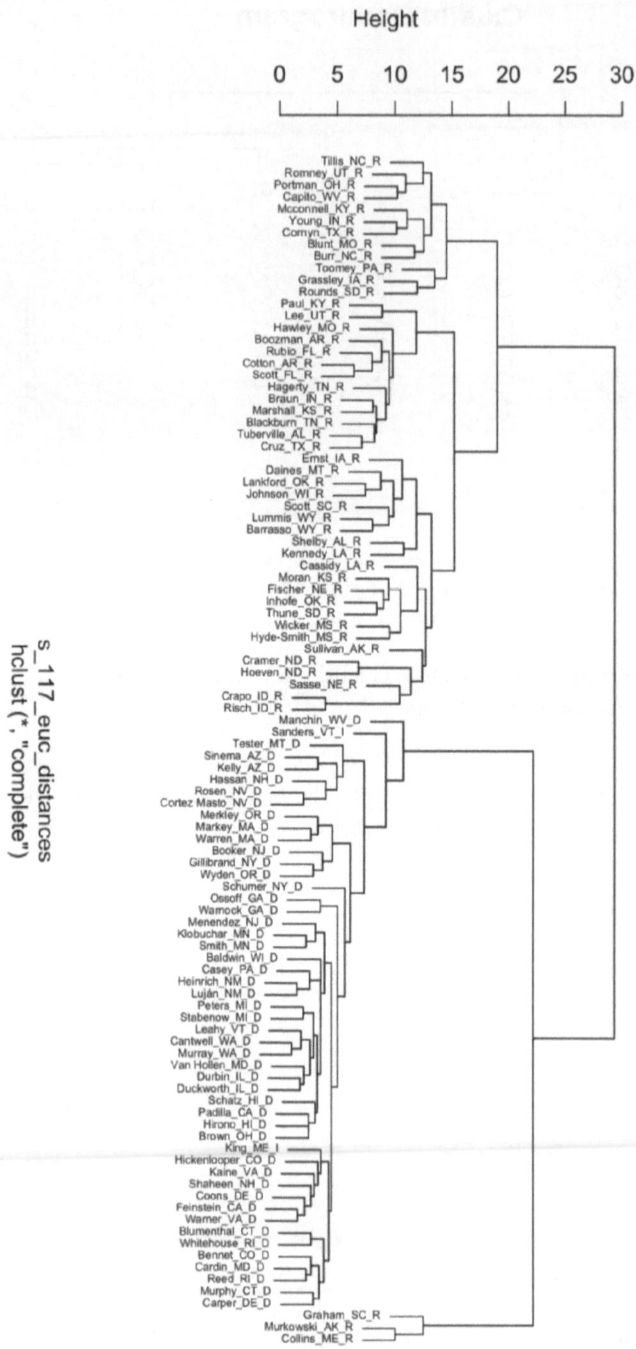

FIGURE 7.9 Dendrogram displaying the clusters of U.S. Senators in the 117th Congress, based on Euclidean distances of roll call votes.

The Senate dendrogram provides a contrasting view of the patterns of roll call votes. Unlike the MDS output, there is no underlying order, beyond similarity within pairs and the increasing hierarchy of larger clusters. Senators Murkowski and Collins pair together, and have a voting record dissimilar enough from other Senators that they are joined by Graham only in later state of clustering after other Senators have been paired into groups of a dozen or more. The party caucuses are divided in two, joined together only in the last stage of the clustering algorithm. Take some time to reflect on other aspects of the dendrogram that interest you.

General steps in scaling and clustering

A scaling and clustering analysis starts with a chosen dissimilarity matrix. While there is no one correct distance metric, different input metrics lead to different outputs. In the examples reviewed in this chapter, each of the measures forming the input matrix are on the same scale (0 - 100 degrees; 0 or 1), but in other cases where measures differ widely, standardization may be necessary. And some consideration should be given to the tradeoffs between the characteristics of different distance metrics given the research question (James et al., 2013). Next, decide on the number of dimensions for the MDS solution. Since the goal of the analysis is a visualization, usually researchers find arrangements for one or two dimensions, while checking the Stress statistic to observe how well the configuration represents the data. Once you have run the MDS algorithm, interpret the resulting configuration by focusing on the relative distances between points: look for spaces, clusters, and directions of points. Are there readily apparent clusters; which observations comprise the clusters, and what do the clustered observations have in common? Keep in mind that the axes in an MDS plot do not have an inherent meaning, and neither do the exact scores of the points on each axis. Assess whether the MDS solution aligns with theoretical expectations or reveals meaningful patterns; in writing up results, include fit statistics and a sensitivity analysis. In a cluster analysis the linkage method affects the formation of clusters, so researchers often try multiple linkage methods to observe how clusters form differently in a dendrogram.

Resources

The package **unvotes** (Robinson, 2021) organizes votes of member states in the United Nations; the same analyses presented in this chapter could be applied to study patterns of country-level voting.

The **wnominate** (Poole et al., 2024) package estimates the ideological scores from roll call votes described in this chapter from Voteview.com. It could be applied to votes from **unvotes** or any legislative body.

As explanatory visualizations, dendrograms are frequently 'pruned' (such as excluding one or two branches) or customized. Popular packages for customizing dendrograms include **dendextend** (Galili, 2015) and **ggdendro** (de Vries and Ripley, 2021).

7.6 Exercises

Knit your R code into a document to answer the following questions:

(1) Using the *polfigures.csv* data, re-create the two-dimensional representation of the figures (and social groups) using Manhattan distances. How does the different metric change observations about the arrangement of figures and groups in the two-dimensional space?

(2) Using *poll_ft.csv*, subset the data to just the feeling thermometer scores on social groups, with a command such as `socialgroups<-poll[,c(21:39)]`, limiting the data table to only social group feeling thermometers. Using this subsetted data, `socialgroups`, use `mds()` to represent the differences in group feelings within a one-dimensional space. How would you interpret the resulting arrangement of the figures?

(3) Import the *polfigures_ft.csv* dataset, and create a one-dimensional figure based on Manhattan distances with MDS. Explain the code, each step in the analysis, and interpret the resulting figure.

(4) Pick a U.S. State with a relatively large delegation to the U.S. House of Representatives. Subset the 117th US Congress roll call voting data to this State and the House delegation. (If you have not already, you will need to download the original raw data from Voteview.com and follow the steps to tidy and subset the data.) Using cluster analysis, create a dendrogram visualizing similarity in voting across the delegation. Try an alternative linkage method to the default. Explain the steps to create the dendrogram, and interpret the results.

(5) Create a distance matrix on characteristics of the four country-level components of the Human Development Index in the file *hdi_2021.csv* and visualize it. To construct the distance matrix, use the `scale()` function to create standardized z scores for each variable, prior to applying `dist()`. How do you interpret the figure?

8

Patterns in Text as Data

Chapter 8 introduces methods for quantifying relationships between texts and the thematic content of texts.

Learning objectives and chapter resources

By the end of this chapter, you should be able to (1) import unstructured text data; (2) prepare a structured corpus and 'tokenized' text; (3) apply scaling and clustering methods to classify text documents by meta characteristic; (4) create word clouds to summarize thematic content; and (5) apply topic modeling to identify thematic content. The material in the chapter requires the **quanteda** (Benoit et al., 2023, 2018; Watanabe and Müller, 2023) package and ancillary **quanteda.textstats**, along with **readtext** (Benoit and Obeng, 2023), **stopwords** (Benoit et al., 2021), and **seededlda** (Watanabe and Xuan-Hieu, 2023), all of which will need to be installed individually. We will apply the **smacof** package (Mair et al., 2022) to scaling text data and rely on **tidyverse** (Wickham, 2023b)packages including _stringr___ (Wickham, 2022) for data import, management, and visualization. The material uses the following datasets *federalist.zip* and *reg_comments_sampled.csv* from https://faculty.gvsu.edu/kilburnw/inpolr.html.

8.1 Text as data

In 1897 Wincenty Lutosławski, a Polish scholar of literary texts, published *The Origin and Growth of Plato's Logic: With an Account of Plato's Style and the Chronology of His Writings.*[1] As implied by the title, Lutosławski found evidence to order by date Plato's dialogues, through the distinctive elements of

[1] I first learned of his work and the term "stylometry" from a lecture by Maciej Eder on his **stylo** package (Eder et al., 2023) at the Digital Humanities Summer Institute, University of Victoria, British Columbia, on June 10, 2016.

Plato's writing style as it changed over time throughout his life (Lutosławski, 1897). Lutosławski measured Plato's writing style through counting variation in verb forms and various other parts of speech, coining the term "stylometry" to describe the systematic measurement of writing style. His theory — that an individual author's distinctive writing style could be measured in part through word frequencies — was remarkably prescient.

While Lutosławski counted frequencies by hand, modern researchers use algorithms for the tedious work. Beyond simple counting, these algorithms classify texts based on stylistic characteristics, just as Lutosławski classified Plato's works into time periods. One enduring example of this is the analysis of the disputed authorship of *The Federalist* papers (Mosteller and Wallace, 1963), which we will investigate in this chapter. By leveraging computational tools, we can measure stylistic markers such as word frequencies to uncover patterns and relationships in texts. We will apply the scaling tools from Chapter 7 on texts to find evidence of authorship.

The field of "text as data" has rapidly grown in political science and other disciplines (Benoit, 2020). The term refers to the process of treating written or spoken language—such as floor speeches in Congress, country statements at the United Nations, or local news stories—as quantifiable data. Researchers analyze patterns in texts to uncover insights into political behavior, public opinion, or social processes. A common approach involves analyzing word frequencies within and across documents, using these patterns to test hypotheses or discover relationships.

For example, political scientists might quantify the emotional sentiment of candidate statements by political party or assess the authorship of texts. These analyses often rely on computational methods to preprocess texts, breaking them into smaller units of analysis, such as individual words or phrases, a process known as "tokenization". Tokens, which may represent words, phrases, or even parts of speech, form the foundation for further analysis.

Two key distinctions in analyzing text as data shape how researchers approach their work. First, researchers study patterns in "function" words or "content" words. Function words, such as pronouns, articles, and prepositions (e.g., "they," "the," "and," "in"), serve as the structural glue that holds sentences together. Content words, on the other hand, carry substantive meaning and include nouns, verbs, adjectives, and adverbs (e.g., "faction," "run," "beautiful," "quickly"). Function words are often used to explore stylistic features of texts, such as authorship or linguistic style, while content words are more useful for analyzing thematic content or topics.

Second, researchers approach these analyses from a "bag of words" or a "strings as words" approach. In a bag of words approach, the order of words is disregarded. Instead the occurrences of words, or co-occurrences across texts, are measured. In the strings as words approach, the order of words is considered to assess patterns in their arrangement. The bag of words approach is foundational in text analysis and forms the basis of the methods we will explore in this chapter.

The analysis within this chapter begins with *The Federalist* papers, applying a bag of words approach to investigate the authorship of the texts. We will tokenize the documents, construct matrices of term frequencies and distances between papers, then apply the multidimensional scaling analysis from Chapter 7 to uncover patterns in authorship. We will see how Lutosławski's insights into writing style remain relevant in the modern era of computational text analysis. Then we will shift to methods for uncovering thematic content of *The Federalist* and next the content of a bureaucratic regulation proposal affecting the interpretation of Title IX of US civil rights law.

8.2 Organizing text as data

In Figure 1.1 from Chapter 1, the process of data analysis involving "Transform" is especially important in text as data. Unstructured text data must be transformed into a structured format for analysis. A text document is imported into **RStudio** and organized into different data structures for further analysis. Typically, this process means importing raw text data (plain text files ending with *.txt*) and then storing it within the **RStudio** environment in a particular data structure.

One commonly used data structure is a corpus object. A corpus is a structured collection of text files that not only preserves metadata (such as author, date, or source) but also facilitates access to the text for further processing. The next step is breaking the text into smaller analytical units known as "tokens". Tokens are typically words or word phrases, but they can also represent sentences or even parts of speech. A more precise term for tokens is n-grams, where "n" refers to the number of combined words. For example, a "1-gram" represents a single word, while a "2-gram" refers to a two-word phrase, such as "of the" or "it is." Tokenizing a corpus enables the measurement of term frequencies.

Once tokenized, the text can be organized into a document features matrix (DFM), a foundational data structure for text analysis. In a DFM (also known as 'document term matrix'), rows correspond to n-grams (or tokens), columns represent documents in the corpus, and each cell contains the frequency of a term's occurrence within a specific document. This matrix provides a numerical representation of the text, which can be used for analyses based on the bag of words approach. In the examples below, we import text files, tokenize them, and construct term document matrices as a basis for further exploration and analysis.

Importing text documents

Text as data may often be organized as individual files, such as news media stories, debate transcripts, or court decisions — one file per story, speech, or opinion. For example, Project Gutenberg (2024) or The Hathi Trust (2024) are two sources for public domain text-based primary sources. We will import *The Federalist* papers in individual text file format, then review more familiar methods of working with text data in CSV format.

The *Federalist* papers were a set of 85 essays written by John Jay, Alexander Hamilton, and James Madison, published under the pen name 'publius' in New York newspapers from the years 1787 to 1788. In defense of the institutional arrangements of the new *Constitution* proposed in the summer 1787 Philadelphia convention, Jay wrote numbers 2-5 and 64, while most were written by Hamilton and Madison individually, and as co-authors numbers 18-20. The set of papers with disputed authorship — either Hamilton or Madison — are numbers 49-57, 62, and 63.

Download the set of *Federalist Papers.* The *.zip* file contains individual text files for each one. We will use the function `readtext()` from the **readtext** package to import the documents, which will organize the text documents into a dataframe. Unzip the folder of *Federalist* papers and observe how each one is saved as a separate text file. The *Federalist Papers* filenames are organized with the number of the essay first, separated by an underscore, then the known or disputed author. For example, *Federalist* 1 is titled *01_Hamilton.txt.*

To read in all text documents within a working directory, the function would be `readtext("*.txt")`. The character `*` in `readtext()` is a "wildcard character", meaning that it represents any and all file character strings or names. In this context, it means `readtext()` will attempt to read in every file of any name in the directory, as long as it is a text file (".txt"). For the function to work without a file path `("your file path/*.txt")`, first set the working directory to the location of the files (Session...Set Working Directory). We read in each text file with the `readtext()` from the **readtext** package:

```
library(tidyverse)
library(readtext)
library(quanteda)
```

```
fed_papers<-readtext("*.txt")

fed_papers
```

```
## readtext object consisting of 85 documents and 0 docvars.
## # A data frame: 85 x 2
##   doc_id          text
##   <chr>           <chr>
## 1 01_Hamilton.txt "\"General In\"..."
## 2 02_Jay.txt      "\"Concerning\"..."
## 3 03_Jay.txt      "\"The Same S\"..."
## 4 04_Jay.txt      "\"The Same S\"..."
## 5 05_Jay.txt      "\"The Same S\"..."
## 6 06_Hamilton.txt "\"Concerning\"..."
## # ... with 79 more rows
```

The papers are imported as 85 separate rows of a dataframe with a column for the *doc_id* (the text file name), and *text* the full text of each document. Notice that the function reports `0 docvars`, which are meta-level characteristics. We can modify the `readtext()` function to include meta characteristics such as the author (or purported author) and include the characteristics when it is read, by specifying two additional arguments, `docvarsfrom` for the source of the metadata and `docvarnames` for the intended name of the metadata. The metadata is pulled from two parts of the filename based on the underscore character that separates name and author:

```
fed_papers<-readtext("*.txt", docvarsfrom="filenames",
                     docvarnames=c("number", "author"), dvsep="_")
```

Now the *number* and *author* variables for each paper specify these two characteristics. The `docvars()` function with *fed_papers* as an argument returns each document variable. The selection of meta characteristics should be related to theoretical considerations, a feature related to the analysis task at hand. In this case, the meta characteristics are the basis for investigating authorship of the disputed series of *The Federalist* papers.

8.3 Corpus, tokens, and the document features matrix

Though the text is imported, it is not yet a corpus. To create one we apply `corpus()` on the imported *readtext* object *fed_papers.*

```
corp_fed_papers<-corpus(fed_papers)

corp_fed_papers
```

```
## Corpus consisting of 85 documents and 2 docvars.
```

```
## 01_Hamilton.txt :
## "General Introduction For the Independent Journal. HAMILTON T..."
##
## 02_Jay.txt :
## "Concerning Dangers from Foreign Force and Influence  For the..."
##
## 03_Jay.txt :
## "The Same Subject Continued: Concerning Dangers From Foreign ..."
##
## 04_Jay.txt :
## "The Same Subject Continued: Concerning Dangers From Foreign ..."
##
## 05_Jay.txt :
## "The Same Subject Continued: Concerning Dangers From Foreign ..."
##
## 06_Hamilton.txt :
## "Concerning Dangers from Dissensions Between the States For t..."
##
## [ reached max_ndoc ... 79 more documents ]
```

The result shows the corpus as the text for each of the 85 essays, along with the two meta characteristics for author and number of essay.

Tokenizing a text

To create the tokens, the `tokens()` function tokenizes the corpus:

```
tokens_fed_papers <- tokens(corp_fed_papers)

tokens_fed_papers
```

```
## Tokens consisting of 85 documents and 2 docvars.
## 01_Hamilton.txt :
##  [1] "General"      "Introduction" "For"
##  [4] "the"          "Independent"  "Journal"
##  [7] "."            "HAMILTON"     "To"
## [10] "the"          "People"       "of"
## [ ... and 1,778 more ]
##
## 02_Jay.txt :
##  [1] "Concerning"  "Dangers"     "from"
##  [4] "Foreign"     "Force"       "and"
##  [7] "Influence"   "For"         "the"
## [10] "Independent" "Journal"     "."
## [ ... and 1,857 more ]
```

```
##
## 03_Jay.txt :
##  [1] "The"          "Same"         "Subject"
##  [4] "Continued"    ":"            "Concerning"
##  [7] "Dangers"      "From"         "Foreign"
## [10] "Force"        "and"          "Influence"
## [ ... and 1,625 more ]
##
## 04_Jay.txt :
##  [1] "The"          "Same"         "Subject"
##  [4] "Continued"    ":"            "Concerning"
##  [7] "Dangers"      "From"         "Foreign"
## [10] "Force"        "and"          "Influence"
## [ ... and 1,825 more ]
##
## 05_Jay.txt :
##  [1] "The"          "Same"         "Subject"
##  [4] "Continued"    ":"            "Concerning"
##  [7] "Dangers"      "From"         "Foreign"
## [10] "Force"        "and"          "Influence"
## [ ... and 1,502 more ]
##
## 06_Hamilton.txt :
##  [1] "Concerning"  "Dangers"      "from"
##  [4] "Dissensions" "Between"      "the"
##  [7] "States"      "For"          "the"
## [10] "Independent" "Journal"      "."
## [ ... and 2,324 more ]
##
## [ reached max_ndoc ... 79 more documents ]
```

The function applies to the corpus object, *corp_fed_papers*, as an argument. The result is tokens organized by the corpus of 85 papers. With the text for each essay represented by individual tokens, a tokens object consisting of 85 documents, each representing one text file in the corpus and accompanied by two document-level variables (*docvars*). For example, the first document (*01_Hamilton.txt*) is tokenized into units such as "General," "Introduction," and "For," while preserving punctuation marks like "." as separate tokens. Each document contains several hundred to thousands of tokens.

The tokenized punctuation relates to another aspect of text as data, which is whether and why to preprocess texts. Tokenizing at the word level, we are probably not interested in including punctuation marks and numbers. By default, the `tokens()` function will remove whitespace separators (`remove_separators=TRUE` is the default argument) – the space between words. Removing numbers and punctuation is not default and requires the arguments `remove_numbers=TRUE` and `remove_punct=TRUE`. (In addition, a `tokens_remove()` function will remove tokens as needed, named in double quotes in a `c()` function.) We will replace the current tokens object.

```
tokens_fed_papers <- tokens(corp_fed_papers, remove_punct=TRUE,
                            remove_numbers=TRUE)
```

Enter the name of the tokens object *tokens_fed_papers* to observe it now excludes numbers and punctuation. To generate n-grams of any size, use the `tokens_ngrams()` function piped in as an additional line. For 2-grams, the two lines together would be `tokens_fed_papers <- tokens(corp_fed_papers, remove_punct = TRUE, remove_numbers = TRUE) %>% tokens_ngrams(n = 2)`.

Constructing a document-term matrix

From tokens we can construct the document-term matrix. The `dfm()` function creates the document term (features) matrix, the only required input argument being a tokenized text. Without a prior reason to count as different words terms such as "The" and "the", with the argument `tolower=TRUE`, `dfm()` will convert to lower case all of the tokenized terms before constructing the matrix:

```
dfm_fed_papers<-dfm(tokens_fed_papers, tolower=TRUE)
```

Enter the name of the new DFM object , *dfm_fed_papers*, to observe the organization of the tokens with lower case letters. The matrix consists of the 85 papers on rows, while the terms (features) appear on columns. The 8,861 features refers to the unique (preprocessed) words across the documents. Sparsity (91.99%) refers to the proportion of zeroes appearing in the cells. Envisioning the matrix, it would consist of over 90% 0 entries; for instance, the vast majority of words used perhaps in one or two documents but not any others.

Describing the DFM with token frequency

One of the extension packages for **quanteda**, **quanteda.textstats**, provides a range of different functions for describing the result of `dfm()` . It will need to be installed.

```
library(quanteda.textstats)
```

The `textstat_frequency()` function calculates frequency statistics in a document-feature matrix (DFM), returning a dataframe of the statistics. By default it will print out statistics for each feature; we will limit it to the top ten terms, `textstat_frequency(dfm_fed_papers, n = 10)` and store the results in dataframe *freq_stats_fed_papers*.

```
freq_stats_fed_papers <- textstat_frequency(dfm_fed_papers, n = 10)
```

The dataframe includes columns for the feature, frequency (the total count of the feature across all documents), rank (the rank of the feature when ordered by frequency), and document frequency (the number of documents in which the feature appears).

```
freq_stats_fed_papers
```

```
##    feature frequency rank docfreq group
## 1      the     17993    1      85   all
## 2       of     11869    2      85   all
## 3       to      7082    3      85   all
## 4      and      5102    4      85   all
## 5       in      4450    5      85   all
## 6        a      3991    6      85   all
## 7       be      3835    7      85   all
## 8     that      2788    8      85   all
## 9       it      2544    9      85   all
## 10      is      2189   10      85   all
```

Of course, the top ten terms are all function ('stop') words. It is not until the 28th most frequent word that the list includes a thematic, content term. (Try `n=30` to verify.) Notice the last column "group". The meta characteristics of the text can serve as a grouping variable (`group=author`). Amending the line to read `textstat_frequency(dfm_fed_papers, n = 10, group=author)` will return the statistics divided by author. The top ten are function words, of course.

Even within the top ten, the frequency declines quickly from "the", "of", and "to" or "and". The pattern relates to an empirical rule, Zipf's law, that describes the relationship between the rank and frequency of words in a corpus of text. It states that the frequency of a word is inversely proportional to its rank: the most frequent word occurs approximately twice as often as the second most frequent word, three times as often as the third most frequent word, and so on (Ignatow and Mihalcea, 2017). The law is often presented graphically. Try adjusting the `n=` in `textstat_frequency()` and graph frequency by rank to observe the relationship, `ggplot(freq_stats_fed_papers) + geom_point(aes(y = frequency, x = rank))`. Frequency descends rapidly, so that relatively few words (mostly function) are common, but most words (content) are rare.

Subsetting a dfm by function and content words

Our goal is to create a subset `dfm()` consisting only of function words, which we will use in the scaling analysis. The `dfm_trim()` function will edit a document features matrix based on term frequency and other characteristics. While there is no one authoritative list of function words, various sources identify different sets of terms as function words. We use a list, or dictionary, of content words from the **stopwords** package (Benoit et al., 2021), which contains multiple lists of function words. (The term "stop words" is another term for "function words".) The package is automatically installed and attached with **quanteda**.

The default list of function words for the English language is the output from the function `stopwords()`. The first 20 words in the list are

```
head(stopwords(), 20)
```

```
##  [1] "i"          "me"         "my"
##  [4] "myself"     "we"         "our"
##  [7] "ours"       "ourselves"  "you"
## [10] "your"       "yours"      "yourself"
## [13] "yourselves" "he"         "him"
## [16] "his"        "himself"    "she"
## [19] "her"        "hers"
```

The total list consists of 175 different terms; enter `stopwords()` to view it. We will create a subset of the matrix consisting solely of the words in the default list, by applying the **quanteda** function `dfm_select()` to the *dfm_fed_papers* object. Within the function, we specify the features matrix object and the pattern of words to select, in this case `pattern = stopwords()`.

```
stop_dfm_fed<- dfm_select(dfm_fed_papers, pattern = stopwords() )
```

Check the DFM object *stop_dfm_fed* to observe it now consists of 123 features, the frequency counts of the words in each document. The subset matrix is saved as *stop_dfm_fed*. Out of the 175 stopwords from the list, 123 were present at least once in *The Federalist* papers.

8.4 Scaling function words in *The Federalist* papers

To create the dissimilarity matrix of distances we use the `dist()` function. The matrix *stop_dfm_fed* is organized the same as the input roll call votes data for the `dist()` function in Chapter 7. Like members of Congress on rows and roll call votes on columns, each *Federalist* paper appears on a row and function words on columns. The entries in the matrix are a frequency count of each word. Across the first row, *01_Hamilton.txt*, the word *for* appears 13 times, while the word *after* appears twice. In *03_Jay.txt* the word *after* appears 0 times. Check the dimensions of the matrix (`dim(stop_dfm_fed)`) and observe that it consists of 85 rows (one for each paper) and 123 columns for the number of function words appearing in it.

We can input it directly into the `dist()` function to create a matrix of pairwise dissimilarities in word usage.

```
dissimat_fed<-dist(stop_dfm_fed)
```

The results stored in *dissimat_fed* is a distance matrix of the pairwise distances between documents, like the distance matrices in Chapter 7. We can convert the distance object to a full matrix using `as.matrix()` . The first four rows and columns:

```
as.matrix(dissimat_fed)[1:4, 1:4]
```

```
##                  01_Hamilton.txt 02_Jay.txt 03_Jay.txt
## 01_Hamilton.txt             0.00      81.19      79.84
## 02_Jay.txt                 81.19       0.00      72.25
## 03_Jay.txt                 79.84      72.25       0.00
## 04_Jay.txt                 95.98      61.74      53.24
##                  04_Jay.txt
## 01_Hamilton.txt       95.98
## 02_Jay.txt            61.74
## 03_Jay.txt            53.24
## 04_Jay.txt             0.00
```

As expected, of course the diagonal entries are 0 representing the distance between each text and itself, while the off diagonals display the Euclidean distance between each paper based on the 123 function words. Larger Euclidean distances measure greater differences between papers in the frequency of usage of the function words. To input the distances and output the scaling solution, we use the **smacof** package from Chapter 7.

```
library(smacof)
```

Before applying the `mds()` function to *dissimat_fed*, we first convert the matrix into an acceptable format for the function. This step is necessary for technical reasons having to do with how the `dist()` function creates the distance matrix from the input document-term matrix. We will store it as a full matrix with `as.matrix()`, then find the scaling solution for two dimensions:

```
dissimat_fed<-as.matrix(dissimat_fed)

nmds.results<-mds(dissimat_fed, ndim=2)
```

The object `nmds.results` contains the results, with the `nmds.results$conf` containing the output, the location of each paper in a two-dimensional space. We save the results as a dataframe, *coord_fed*. Variables containing the coordinates are D1 and D2, for use in a **ggplot2** scatterplot.

```
coord_fed<-as.data.frame(nmds.results$conf)
```

First several lines show structure of the dataframe.

```
head(coord_fed)
```

```
##                      D1      D2
## 01_Hamilton.txt -0.6238 -0.1455
## 02_Jay.txt      -0.7700 -0.3579
## 03_Jay.txt      -0.9532 -0.1824
## 04_Jay.txt      -0.9227 -0.3895
## 05_Jay.txt      -1.1195 -0.2842
## 06_Hamilton.txt -0.2598 -0.3799
```

For labeling points, we will add columns from the row names, and extract parts of the text of each *Federalist* paper column.

```
coord_fed <- coord_fed %>%
  mutate(authors=row.names(coord_fed))

head(coord_fed)
```

```
##                      D1      D2         authors
## 01_Hamilton.txt -0.6238 -0.1455 01_Hamilton.txt
## 02_Jay.txt      -0.7700 -0.3579      02_Jay.txt
## 03_Jay.txt      -0.9532 -0.1824      03_Jay.txt
## 04_Jay.txt      -0.9227 -0.3895      04_Jay.txt
## 05_Jay.txt      -1.1195 -0.2842      05_Jay.txt
## 06_Hamilton.txt -0.2598 -0.3799 06_Hamilton.txt
```

Regular expressions are patterns used to search for and extract specific text. Within the columns of *coord_fed*, we use regular expressions to extract the number of the paper, and the name associated with it. Two `mutate()` functions create a *number* and *author* column:

```
coord_fed <- coord_fed %>%
  mutate(number=str_extract(authors,"[0-9]+")) %>%
  mutate(author=str_extract(authors,
                ("HamiltonMadison|Hamilton|Madison|Jay|Disputed")))

head(coord_fed)
```

```
##                          D1      D2          authors number
## 01_Hamilton.txt -0.6238 -0.1455 01_Hamilton.txt     01
## 02_Jay.txt      -0.7700 -0.3579      02_Jay.txt     02
## 03_Jay.txt      -0.9532 -0.1824      03_Jay.txt     03
## 04_Jay.txt      -0.9227 -0.3895      04_Jay.txt     04
## 05_Jay.txt      -1.1195 -0.2842      05_Jay.txt     05
## 06_Hamilton.txt -0.2598 -0.3799 06_Hamilton.txt     06
##                   author
## 01_Hamilton.txt Hamilton
## 02_Jay.txt           Jay
## 03_Jay.txt           Jay
## 04_Jay.txt           Jay
## 05_Jay.txt           Jay
## 06_Hamilton.txt Hamilton
```

In the first `mutate()` line, the `str_extract()` function from the **stringr** package attached with `library(tidyverse)` uses the pattern `[0-9]+` to search for any sequence of digits in the authors column and to extract it. The expression `[0-9]` matches any single digit, and `+` ensures it captures one or more consecutive digits. See Wickham and Grolemund (2016) to learn more about the use of regular expressions in data analysis. With the *number* and *author* columns for labels, we can graph the twodimensional scaling solution in a scatterplot.

```
ggplot(data=coord_fed) +
 geom_point(aes(x=D1, y=D2, color=author)) +
 geom_text(aes(x=D1, y=D2, label=number, color=author),
           check_overlap = TRUE,
           show.legend=FALSE, hjust=1.3)+
 theme(legend.position="bottom")+
 scale_color_viridis(discrete=TRUE)
```

Figure 8.1 displays the MDS solution visualizing the relationship between papers based on the similarity in usage between the 123 different function words. Each of the 85 papers is a point on the two dimensions, the points closer together are more similar in usage compared to those further away. There are distinct clusters, suggesting that the authors had fairly consistent and distinct writing styles as measured by the frequency of function words. Jay's papers

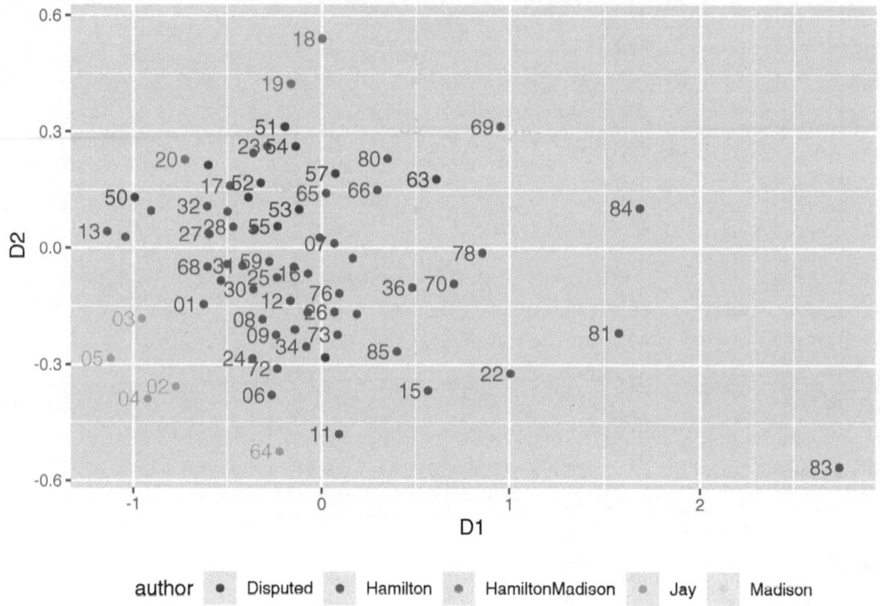

FIGURE 8.1 Two-dimensional scaling of distances between *The Federalist* papers, measured on 123 function words. The papers of each individual author tend to cluster together, with the disputed authorship papers situated between Hamilton and Madison's clusters, although overlapping slightly more with Madison than Hamilton.

form a distinct cluster in the lower-left quadrant, suggesting a distinctive style. Hamilton's papers and Madison's papers divide in clusters, but also overlap through the middle, perhaps reflecting some shared stylistic characteristics or perhaps common variability in writing styles. The disputed papers overlap between both authors, although tending to overlap more with Madison than Hamilton. A working hypothesis for further investigation is that Madison, rather than Hamilton, was the primary author of the disputed works. Further investigating it based on the most frequent words used by the authors, not just the most frequent function word 1-grams, is left for the Exercises at the end of the chapter.

Normalizing frequencies across different text lengths

The Federalist papers are not each the same length. A common method of accounting for differences in text lengths is to measure token frequencies in z scores. For a document-features matrix, the standardized scores are accounted for row-wise: for each cell, subtract the row mean and divide by the row standard deviation. In place of frequency counts in a DFM, each value is expressed as a z score, indicating how many standard deviations the count is above or below the mean for each respective paper. The terms, or features, in the DFM are standardized relative to the counts within each document, removing the effect of paper-specific variability in length.

To see how it works, take the first five rows and columns from *dfm_fed_papers*, save it as matrix *temp_dfm*

```
temp_dfm<-as.matrix(dfm_fed_papers[1:5, 1:5])
```

Because calculations are usually performed columnwise in **R**, making the calculations row-wise requires some additional adjustments. The `rowMeans()` function will calculate row-wise means. The `apply()` function applies a function to a row or column of data; finding the row-wise standard deviation is `apply(temp_dfm, 1, sd)`, where `1` specifies applying the `sd` function to rows instead of columns.

Row-wise means:

```
rowMeans(temp_dfm)
```

```
## 01_Hamilton.txt      02_Jay.txt      03_Jay.txt
##            30.6            25.4            22.6
##      04_Jay.txt      05_Jay.txt
##            21.8            16.4
```

Row-wise standard deviations:

```
apply(temp_dfm, 1, sd)
```

```
## 01_Hamilton.txt      02_Jay.txt      03_Jay.txt
##           57.44           46.50           40.72
##      04_Jay.txt      05_Jay.txt
##           37.34           29.00
```

To subtract each word frequency from the document mean number of words, `(temp_dfm) - rowMeans(temp_dfm)`. Then we divide by the standard deviation, `apply(temp_dfm, 1, sd)`:

```
z_temp_dfm<-(temp_dfm - rowMeans(temp_dfm)) / apply(temp_dfm, 1, sd)
```

Saving it as *z_temp_dfm*, the matrix contains z score transformations:

```
z_temp_dfm
```

```
##                   features
## docs               general introduction    for   the
##   01_Hamilton.txt -0.4631       -0.4979 -0.3064 1.783
##   02_Jay.txt      -0.4818       -0.5463 -0.2452 1.777
##   03_Jay.txt      -0.4813       -0.5550 -0.2603 1.778
##   04_Jay.txt      -0.5303       -0.5838 -0.2357 1.773
##   05_Jay.txt      -0.4966       -0.5656 -0.2897 1.780
##                   features
## docs               independent
##   01_Hamilton.txt      -0.5153
##   02_Jay.txt           -0.5033
##   03_Jay.txt           -0.4813
##   04_Jay.txt           -0.4232
##   05_Jay.txt           -0.4276
```

To save it as a DFM, apply `as.dfm()` : `z_temp_dfm <- as.dfm(z_temp_dfm)`. The same procedures could be applied to the full DFM *dfm_fed_papers* as an input to calculating scaling distances. In the `mds()` function, we could observe how much the standardized scores change the output compared to the raw frequencies. Calculating Euclidean distances on the z scores is equivalent to the Pearson's r, the linear correlation coefficient (Bartholomew et al., 2008); subtracting the correlation matrix from 1 converts it into a matrix of dissimilarities, `1-dist(z_temp_dfm)`. We could save this matrix of dissimilarities as an object and apply it within the `mds()` function.

8.5 Thematic content words in *The Federalist* papers

Often researchers are specifically interested in patterns of content words, as the thematic evidence of texts. In this section we review one familiar graphical approach, word clouds, followed by two others: a commonly used statistic, term frequency inverse document frequency (TFIDF), and the use of topic models. These methods have in common a text pre processing step of removing the function words from a tokenized text prior to any further analysis.

We remove the stopwords from the DFM *dfm_fed_papers* by filtering with `dfm_remove()`, using a more comprehensive list of function words. After attaching the *stopwords* package, we access the more comprehensive list of function words identified in the 1960s, referred to as the 'SMART' list. (See the **stopwords** package helpfile for more information.) We will save the edited DFM to a new object:

```
library(stopwords)
dfm_fed_papers_nostop <- dfm_remove(dfm_fed_papers,
                                    pattern = stopwords())
```

Enter the name of the new dfm *dfm_fed_papers_nostop* and observe the characteristics of the new DFM. The new DFM with the function words removed contains 8,685 features and is 92.83 percentage sparse, meaning that over 90 percentage of the matrix entries are 0 where a particular content word does not appear in a paper.

Content word clouds

Word clouds are widely recognized visual representations of the thematic content of text. A cloud visualizes words in varying sizes, with the size of each word proportional to its frequency in a text. While relatively simple, word clouds are an intuitive way to communicate the subject matter or themes of a text. Creating a word cloud requires pre-processing of the text to remove function words. Otherwise, a word cloud would consist mostly of high frequency function words, obscuring the thematic content words. In the **quanteda.textplots** extension package to **quanteda**, the `textplot_wordcloud()` creates word clouds from document feature matrices. The function will need to be installed: `install.packages("quanteda.textplots")`.

```
library(quanteda.textplots)
```

For *The Federalist* papers, the input document features matrix is *dfm_fed_papers_nostop*. Without any additional arguments, an all-encompassing word cloud can be created with `textplot_wordcloud(dfm_fed_papers_nostop)`. Typically, however, word clouds require some adjustments, such as whether to plot words with a particular minimum frequency and what proportion of the words to draw in a vertical orientation, a common visual practice in word clouds. We will restrict the words that appear to those with a frequency greater than 5 (`min.count=5`) and set a proportion (a quarter) of the words to be drawn at a right angle (`rotation=.25`):

```
textplot_wordcloud(dfm_fed_papers_nostop, min_count=5, rotation = .25)
```

Figure 8.2 displays a word cloud of *The Federalist* papers. The arrangement of horizontal versus vertical words is unrelated to frequency. The results reflect the familiar subject matter of the papers as a whole.

In **quanteda**, a "comparison cloud" divides word clouds by text groups, and within each group visualizes the distinctive words within each – the

FIGURE 8.2 Word cloud of the full text of *The Federalist* papers, function words removed.

common words within a group that are uncommon across other groups. Linking together lines with a pipe operator `%>%` or `|>`, we identify the DFM, then set groups with `dfm_group()`, then plot the comparison word cloud by including `comparison=TRUE` in the `textplot_wordcloud()` function:

```
dfm_fed_papers_nostop %>%
  dfm_group(groups = author) %>%
  textplot_wordcloud(min_count=5,comparison = TRUE)
```

Figure 8.3 displays the comparison cloud. Notice, for example, for Madison's essays the distinction for words "executive," "congress," and "state", reflecting his arguments throughout his essays about the structure of the federal government controlling tyranny by checks and balances. Or for Jay, his particular focus on foreign policy and diplomacy reflected in the words "nations", "treaties", and "war".

FIGURE 8.3 A comparison word cloud of the full text of *The Federalist* papers, function words removed, divided by known author.

8.6 Term frequency × inverse document frequency

A more formal way to assess the subject matter of texts is to identify words (or phrases) from each paper that are unusual across papers, but common enough within one that the terms could illustrate the 'topic' of each document. A common measure is the "Term Frequency Inverse Document Frequency" (TF × IDF) statistic (Manning et al., 2009). The statistic captures this intuition of identifying the distinctiveness of particular words or phrases in a text, to capture its thematic content. Because its calculation can be automated in an **R** function, its usefulness is that it can be applied to large sets of documents to investigate subject matter, rather than requiring a researcher to hand-code, or closely read, each document.

To illustrate how it works, consider three short, illustrative excerpts from the following *The Federalist* papers:

```
Fed10<-"If a faction consists of less than a majority
relief is supplied by the republican principle"  # (16 words total)

Fed51<- "In republican government the legislative authority
necessarily predominates" # (9 words total)
```

```
Fed78 <- "The standard of good behavior for the continuance in office
of the judicial magistracy" # (14 words total)
```

To illustrate, we will calculate TF*IDF for three terms: (1) "republican", (2) "judicial", and "the", on these short excerpts, before automating it across the entirety of *The Federalist* papers.

There are two components: TF and IDF. The TF, term frequency, in simplest form is the number of words in a text that are that term, $TF_{t,d}$, the number of terms t in document d. For the term "republican" in *Fed10*, that frequency is $TF = 1$. The frequency could also be measured in relative or proportional terms: the number of times a term appears in a document divided by the total number of terms in the document. For example, "republican" in Fed10: It is 1 out of 16 (1/16) words, $TF = .0625$. The proportional frequency is often used to normalize over texts of different lengths.

To create a measure of 'distinctiveness', however, we should also consider how common the word is in other texts of a corpus. If "republican" appears frequently in other papers, then it would not be so distinctive of *The Federalist* paper 10. Thus the term is weighted, or discounted, through the inverse document frequency, IDF, to account for how common the term is across all documents.

To calculate the IDF we find the total number of documents in the corpus, N, then divide it by DF_t, the term document frequency — the number of documents where the term appears. In this example comparing 10, 51, and 78, there are three documents forming the corpus, so $N = 3$. For DF_t, the number of documents where the term appears, "republican" appears, is in two documents. So the IDF $= \frac{N}{DF_t} = \frac{3}{2} = 1.5$. Unusual terms — appearing in relatively few documents — have higher IDF scores.

There is an additional adjustment to the IDF. We need to consider how we should weight a term that appears in all documents. For example, the term "the" appears in all documents, so it should be minimally weighted. If we take the logarithm of $\frac{N}{DF_t}$ for $log\frac{3}{3}$, the IDF is 0 (the logarithm of 1 is equal to 0), meaning that terms not distinctive of a particular paper are scored 0, so the more common the word is across documents, the lower the IDF weight. (Usually the base-10 logarithm is used, the IDF for a term t, is $IDF_t = log_{10}\frac{N}{DF_t}$.)

All that is left to do is combine TF and IDF. We apply the IDF weight to TF — multiply the IDF and TF together.

The TF×IDF statistic is $TF_{t,d} * log\frac{N}{DF_t}$.

Compare the different TF×IDF statistics for each term across the documents. For "republican", using a normalized or proportional term frequency, in *Fed10* "republican" TF*IDF = .0625 * 1.5 = 0.09375. For *Fed51* "republican" TF*IDF = .125 * 1.5 = 0.1875. For *Fed78*, the term does not appear, so the TF is 0, thus TF*IDF = 0.

Across the three documents, the TF*IDF term for "republican" could be interpreted as most important or indicative of *Fed78*. The term "the" is found in all three documents (thus the IDF term is 0), making the TF*IDF term 0 for "the" in each document. The term "judicial" does not appear in *Fed10* and *Fed51*, thus scoring a 0 TF*IDF while in *Fed78* the TF (1/14) = .07143.

Compared to "republican" the IDF is higher, since "judicial" appears in just one document, $IDF = log(3/1) = .4771$, resulting in a TF×IDF score of $.07143 * .4771 = 0.03408$. Overall, the term "republican" is more distinctive in *Fed78*, "the" is of course indistinctive, while "judicial" is distinctive and exemplifies the concept of *Fed78*.

Automatically calculating the TF x IDF

With some additional steps, the function `dfm_tfidf()` in the core **quanteda** package calculates tf-idf statistics for each term within a document features matrix. For example, given the input DFM **dfm_fed_papers_nostop**, calculate the TF for each feature with `dfm_weight(dfm_fed_papers_nostop)` . For a relative or proportional frequency: `dfm_weight(dfm_fed_papers_nostop, scheme = "prop")`, which is useful to normalize across texts of different lengths.

By default, given an input DFM, the function `dfm_tfidf()` calculates the IDF term with a base-10 logarithm; for the *Federalist* DFM, the function would be `dfm_tfidf(dfm_fed_papers_nostop)`. For the product of the two terms, two arguments `scheme_tf="prop", scheme_df="inverse"` result in the DFM weighted by the IDF, the product TF×IDF:

```
tfidf_dfm_fed<-dfm_tfidf(dfm_fed_papers_nostop, scheme_tf="prop",
                              scheme_df="inverse")
```

The TF×IDf statistics stored in `tfidf_dfm_fed` appear in place of the frequency counts. Typing the object name will return that it now consists of a `Document-feature matrix of: 85 documents, 8,685 features (92.83% sparse) and 2 docvars`. Applying the `as.matrix()` function and indexing the DFM for the first three rows and columns:

```
as.matrix(tfidf_dfm_fed)[1:3,1:3]
```

```
##                  features
## docs                 general introduction independent
##    01_Hamilton.txt 0.0002624     0.003107   0.0002994
```

```
## 02_Jay.txt       0.0001823   0.000000   0.0005546
## 03_Jay.txt       0.0002045   0.000000   0.0009334
```

The frequency counts in the DFM are replaced by the TF×IDF statistics. Note that in DFM function words are excluded, but the statistics could have been just as easily calculated with function words included; this matrix could be used as an input to the `dist()` function for a scaling or cluster analysis. The statistics are often compared across documents to summarize the contents of documents. In this case, what interests us is the set of highest scoring TF×IDF terms for each *Federalist* paper.

The `topfeatures()` function returns a set of rank-ordered features for the whole DFM or optionally for a DFM by document variables. For example, `topfeatures(tfidf_dfm_fed, groups=author)` returns the highest scoring TF×IDF terms for each author.

We have previously observed the inspection of a DFM as a matrix. We convert it to a matrix with `as.matrix()` and save it as a matrix object, `mat.tfidf`:

```
mat.tfidf <- as.matrix(tfidf_dfm_fed)
```

An interesting characteristic of the TF×IDF statistic is that when ranked within texts, the highest scoring terms can potentially be interpreted as thematic indicators of a text's contents. We can arrange the terms within the matrix with two functions, `head()` to limit the terms to the top ten or so, and `sort()` to arrange the terms from highest to lowest TF*IDF scoring. With an index, we select the row corresponding to a *Federalist* paper. (*Federalist* number 10 by Madison is on row 10 `mat.tfidf[10,]`.)

For *Federalist* 10:

```
head(sort(mat.tfidf[10, ], decreasing=TRUE), n=10)
```

```
##     faction    majority   democracy controlling
##    0.007862    0.004388    0.004129    0.004114
##    republic     parties        cure   factious
##    0.003833    0.003618    0.003273    0.003273
##    property   faculties
##    0.003123    0.003097
```

And *Federalist* 78:

```
head(sort(mat.tfidf[78, ], decreasing=TRUE), n=10)
```

```
##    courts     judges judiciary       void   statutes
## 0.008903   0.005284  0.004998   0.004514   0.004247
```

```
## judicial    tenure preferred  superior  statute
## 0.004103  0.003012  0.003012  0.002584  0.002534
```

Notice the three most distinctive terms for *Federalist* 10 are "faction", "majority", and "democracy", while for *Federalist* 78 it is "courts", "judges", and "judiciary". For anyone familiar with the subject matter of the essays, the highest scoring terms reflect the thematic content of the essays.

Topic models

Topic models are algorithms designed to classify text themes that represent the core subject matter of the text. Like the use of TF×IDF statistics with content words, topic models identify distinctive clusters of words as illustrative of a text's subject matter or that of a meta characteristic. There are many different refinements and uses of topic models, but generally a topic model groups terms from a document into coherent topics to identify the topic of a document — and among other capabilities can assign topic proportions to documents. For example, we could apply a topic model to *The Federalist* papers and investigate the topics of each known author, comparing those to the disputed set of papers — perhaps providing additional clues about their authorship.

In this section we will introduce and apply a particular type of topic model, a 'Latent Dirichlet Allocation (LDA)' model (Ignatow and Mihalcea, 2017) to *The Federalist* papers and in a second example, comments on a proposed U.S. federal bureaucratic rule. There are many different approaches to topic modeling algorithms. Similar to the interpretation of a small set of TF×IDF terms, we will use the model to identify sets of words identified as particularly characteristic of a document — or an author, or any other meta characteristic. Like multidimensional scaling, topic models techniques are highly inductive (Ignatow and Mihalcea, 2017); in scaling, a researcher picks a number of dimensions, and in topic models a number of topics. Given a number of topics, the algorithm then finds the terms in the document most likely comprising those topics. While not reviewed in this chapter, with a "seeded" or guided topic model, the researcher provides a starting point of terms to guide the algorithm in selecting topics.

The **seededlda** package (Watanabe and Baturo, 2023) provides topic modeling functions. Given an input DFM of raw frequencies, the function `textmodel_lda()` will find `k` number of topics specified in advance.

```
library(seededlda)
```

The raw frequency DFM from *The Federalist Papers* is *dfm_fed_papers_nostop*. The simplest use of the algorithm is to identify a set of topics from it. The selection of the number of topics `k` is essentially exploratory if not guided by theory. Here we select three topics to begin exploring the algorithm's identification of the topics:

```
fed_topics_3 <- textmodel_lda(dfm_fed_papers_nostop, k =3)
```

The results of the topic model are stored in _fed_topics_3. Various functions help to unpack the model.

The `terms()` function lists the top terms associated with each topic. The function returns the top ten words by default:

```
terms(fed_topics_3)
```

```
##         topic1     topic2         topic3
##  [1,] "union"    "power"        "government"
##  [2,] "us"       "states"       "people"
##  [3,] "nations"  "constitution" "may"
##  [4,] "upon"     "state"        "state"
##  [5,] "war"      "upon"         "must"
##  [6,] "states"   "executive"    "can"
##  [7,] "foreign"  "authority"    "states"
##  [8,] "peace"    "powers"       "one"
##  [9,] "power"    "shall"        "public"
## [10,] "national" "legislative"  "every"
```

Since the purpose is to interpret the topic, usually the default number of terms is sufficient, although a modification like `terms(fed_topics_3, n=15)` would return the top 15 terms. The selection of these terms by the model illustrates an important aspect of it; unlike the TF×IDF terms per *The Federalist* paper, these are the terms most probable given the topic. Interpreting the meaning of the topics is, of course, subjective.

One way to interpret the topics is to form sentences of phrases from each. An interpretation could be that *topic1* refers to the theme of the authority and power of the U.S. States and federal government under the constitution (e.g., "power," "states," "constitution"). The *topic2* set refers to the role of government and its relationship with the public or people (e.g., "government," "people," "public"), while *topic3* refers to national unity and foreign affairs (e.g., "union", "war", "foreign"). Of course *The Federalist* number 85 in total and are mostly written on various discrete subjects, although the three higher-order topics could perhaps be interpreted as a set of unifying themes across the essays.

The algorithm also assigns each of the texts the most likely topic to be found within it. The `topics()` function returns the most likely topic (out of the three) for each essay, as in `topics(fed_topics_3)`. Enter it, and the list of topics will scroll past.

To see the topics for the first few papers:

```
head(topics(fed_topics_3))
```

```
## 01_Hamilton.txt        02_Jay.txt       03_Jay.txt
##          topic3           topic3           topic3
##      04_Jay.txt       05_Jay.txt 06_Hamilton.txt
##          topic1           topic3           topic1
## Levels: topic1 topic2 topic3
```

For example, the topic model assigned *01_Hamilton.txt* to *topic2*, meaning the model determined that *topic2* has the highest proportion of words in this essay. Across each one, a few tentative patterns emerge.

The majority of *The Federalist* papers, including most authored by Hamilton, Madison, Jay, and the disputed papers, are assigned *topic2*, suggesting that the document encompasses a central or recurring theme across the essays. Hamilton's essays are in a sense more diverse than the others, being assigned to all three topics. Madison's essays are nearly all assigned to *topic2*, but a few are associated with *topic1*. Jay's three essays are assigned to *topic2*, *topic3*, and *topic3*, respectively. The Disputed papers are mostly assigned to *topic2*, suggesting thematic similarity with many of Hamilton and Madison's known works. Unfortunately, the distribution of the three main topics do not provide much insight into the authorship of the Disputed papers.

Within each paper, we can inspect the proportion of the papers assigned to each topic by tabulating the "theta matrix", which is the distribution of each topic across the papers. The matrix is stored as a variable within the model results, `fed_topics_3$theta`. Tabulating the first few lines:

```
head(fed_topics_3$theta)
```

```
##                 topic1  topic2 topic3
## 01_Hamilton.txt 0.2229 0.19192 0.5852
## 02_Jay.txt      0.2676 0.15908 0.5733
## 03_Jay.txt      0.2666 0.15325 0.5801
## 04_Jay.txt      0.4958 0.03564 0.4686
## 05_Jay.txt      0.4691 0.03585 0.4950
## 06_Hamilton.txt 0.5276 0.04815 0.4242
```

The document *01_Hamilton.txt* was assigned to *topic2*, because the largest share of terms were assigned to *topic2* (.57) or 57 percent. The algorithm assigned about 19 percent to *topic1* and 24 percent to *topic3*.

These topics form over the entire corpus of individual *Federalist* papers. Another approach is to find topics within authors, grouping the DFM by author and then applying the topic model to each group. A useful application of topic models is to organize topics by known author, to compare subject matter from named to the disputed authorship.

First group the DFM by author, using the `dfm_group()` function from **quanteda**.

```
dfm_fed_grouped <- dfm_group(dfm_fed_papers_nostop, groups =
                             docvars(dfm_fed_papers_nostop, "author"))
```

We identify the authors by the `docvars()` function. Next fit the model, selecting three topics:

```
fed_group_lda <- textmodel_lda(dfm_fed_grouped, k = 3)
```

The model is fit by grouping all the papers together by author, ignoring the sequence of papers or even the differences between, for example, essay *01_Hamilton.txt* and *06_Hamilton.txt*. Both of those documents would simply be lumped together as Hamilton's tests. Next we view topic proportions for each known author, which will be contained within *theta*:

```
fed_group_lda$theta
```

```
##                 topic1 topic2  topic3
## Disputed        0.7879 0.1680 0.04412
## Hamilton        0.4057 0.4710 0.12336
## HamiltonMadison 0.2915 0.1253 0.58320
## Jay             0.4196 0.2599 0.32055
## Madison         0.6724 0.2101 0.11750
```

We would review the terms within each topic through `terms(fed_group_lda)`. The "theta matrix" shows, intriguingly, that viewed from the author, *topic1* is assigned as the majority for only the Disputed and Madison authored texts. Hamilton and Jay tend to write on a more varied subject matter across the three, as did Jay.

Selecting the appropriate number of `k=` topics to cluster together from texts is subjective and highly interpretative. Of course, the number of topics chosen will influence the interpretability of the results. With no objective standard, the 'best' number of topics depends on the particular corpus of documents at hand, the subject matter and research question addressed with it. In this case of *The Federalist*, a researcher would want to compare these results with a larger number of topics. Within the study of topic models there are additional statistics to assess results for a particular k value. (See the **quanteda** help files, Benoit and Stefan Müller (2024), and Silge and Robinson (2017).) A common strategy is to fit several models with a range of 'k' values, then compare the results based on the metrics and interpret ability. A model with too few topics risks clumping together discrete subjects of interest, while a model with too many topics risks splitting coherent topics into repetitively incoherent smaller topics (Ignatow and Mihalcea, 2017).

8.7 Corpus themes: comments on a proposed Title IX Rule

We will review one additional application of topic models. On April 13, 2023, the US Department of Education published a proposed rule in the *Federal Register* , titled "Nondiscrimination on the Basis of Sex in Education Programs or Activities Receiving Federal Financial Assistance: Sex-Related Eligibility Criteria for Male and Female Athletic Teams"[2]. The proposed regulation addressed how the department planned to interpret the application of Title IX of the United States Code. Specifically, the proposed rule stated that any school district's outright, categorical ban against transgender athlete participation would be interpreted as a violation of the Title IX non-discrimination clause (Mervosh et al., 2023). The proposed rule generated over 130,000 comments from the public. We will analyze a simple random sample of 1,000 separate comments, to use text as data tools for getting a sense of what the comments are about and written in support or opposition to the proposed rule. Rather than being organized as text documents, the comments are stored within a CSV file:

```
reg_comments_sampled<-read_csv(file="reg_comments_sampled.csv")
```

The `comments` field contains the full text of the submitted comment. We will work through the same steps as before with text data – constructing a corpus of 1-grams, tokenizing it, and constructing and filtering the DFM, all starting with the *comment* field. We start by reading in the corpus and creating the tokens:

```
comments_sampled_corp<-corpus(reg_comments_sampled$comment,
                              docvars=reg_comments_sampled$id)
```

[2]See https://www.regulations.gov/document/ED-2022-OCR-0143-0001/

```
comments_sampled_tok<-tokens(comments_sampled_corp,
                        remove_punct=TRUE, remove_numbers=TRUE)
```

The set of tokenized comments are stored in *comments_sampled_tok*. The first comment, for example:

```
comments_sampled_tok[1]
```

```
## Tokens consisting of 1 document and 1 docvar.
## text1 :
##  [1] "I"          "am"         "writing"
##  [4] "to"         "express"    "my"
##  [7] "deep"       "concern"    "over"
## [10] "the"        "Department" "of"
## [ ... and 148 more ]
```

For getting a sense of the subject matter of the comments, the idea of "keywords in context" is useful. The `kwic()` function will search for particular key terms in a tokens object and return matches, along with the contextual words within which the key term appears. To use the function, specify the tokens object and pattern:

```
kw_support<-kwic(comments_sampled_tok, pattern="support")
```

We save the results in an object *kw_support*. Inspecting the tail end of the results:

```
tail(kw_support)
```

```
## Keyword-in-context with 6 matches.
##    [text950, 8] of Education Miguel Cardona I |
##    [text959, 8] of Education Miguel Cardona I |
##  [text961, 173] and nonbinary youth because I |
##  [text963, 492]       gender identity I do not |
##    [text976, 8] of Education Miguel Cardona I |
##    [text980, 8] of Education Miguel Cardona I |
##
##  support | the Department of Education Department
##  support | the Department of Education Department
##  support | the full inclusion of all
##  support | any regulation that would codify
##  support | the Department of Education Department
##  support | the Department of Education Department
```

The results show the key term surrounded by the five words before and after each occurrence. (To display fewer words, add the argument `window = 3` to change the window to three words, for example.) The results provide a sense of how the comments could be further studied to observe the proportions supporting or opposing the proposed rule. Each line includes the comment text row (e.g., `[text950, 8]`) and the position of the word in the comment (e.g., the eighth word). Of course, the lines display the ambiguity of simply relying on a single word such as "support", since it could be negated with "I do not..."

Altering the `kwic()` search to include the phrase:

```
kw_support_phrase<-kwic(comments_sampled_tok,
                        pattern="I support the Department of Education")
```

```
tail(kw_support)
```

```
## Keyword-in-context with 6 matches.
##    [text950, 8] of Education Miguel Cardona I |
##    [text959, 8] of Education Miguel Cardona I |
##  [text961, 173] and nonbinary youth because I |
##  [text963, 492]      gender identity I do not |
##    [text976, 8] of Education Miguel Cardona I |
##    [text980, 8] of Education Miguel Cardona I |
##
##  support | the Department of Education Department
##  support | the Department of Education Department
##  support | the full inclusion of all
##  support | any regulation that would codify
##  support | the Department of Education Department
##  support | the Department of Education Department
```

The *docname* variable in *kw_support* identifies the comment. Typing `kw_support$docname` would display the comment identifiers, such as "text980". The length of the column of variables would be the total number of comments appearing as form letters of support. Since there are 1000 total comments, the proportion of comments from `kw_support` is

```
length(kw_support$docname) /1000
```

```
## [1] 0.119
```

Or about 12 percent of the submitted comments appear to be verbatim form letters in support of the proposed rule.

To create a topic model, we use the tokens as input for `dfm()`:

```
dfm_comments_sampled<-dfm(comments_sampled_tok, tolower=TRUE)
```

Then remove function words, along with the word we expect to be in nearly every comment, such as "education".

```
dfm_comments_nostop <- dfm_remove(dfm_comments_sampled,
    c(stopwords(), "proposed", "rule", "title",
      "ix", "department", "education"))
```

Given the DFM, we apply the `textmodel_lda()` function. We specify the number of subjects, `k`. We'll set `k=4` to observe the sorts of terms the model tends to cluster together:

```
comments_lda <- textmodel_lda(dfm_comments_nostop, k=4)

comments_lda
```

```
##
## Call:
## textmodel_lda(x = dfm_comments_nostop, k = 4)
##
## 4 topics; 1,000 documents; 3,074 features.
```

```
terms(comments_lda)
```

```
##       topic1     topic2        topic3
##  [1,] "female"   "sports"      "sports"
##  [2,] "males"    "students"    "biological"
##  [3,] "sports"   "women"       "young"
##  [4,] "women"    "intersex"    "students"
##  [5,] "girls"    "nonbinary"   "girls"
##  [6,] "compete"  "transgender" "female"
##  [7,] "athletes" "school"      "males"
##  [8,] "females"  "women's"     "locker"
##  [9,] "men"      "right"       "allowing"
## [10,] "sex"      "play"        "policy"
##       topic4
##  [1,] "girls"
##  [2,] "women"
##  [3,] "sports"
##  [4,] "will"
```

```
##  [5,] "boys"
##  [6,] "schools"
##  [7,] "biological"
##  [8,] "sex"
##  [9,] "female"
## [10,] "women's"
```

The distinction between four topics appears slight. Perhaps there are some differences worth investigating further, such as *topic3* referring to general principles in gender and athletics, while *topic1* and *topic2* focus more on the childrens' experiences in athletics. This example points toward some of the limitations of topic models — the topics may not necessarily make sense *a priori.*

To incorporate word context, we could apply the topic model to a DFM built from multigrams , such as four-word phrases. Doing so would require going back to the `tokens()` function and adding a line `%>% tokens_ngrams(n = 4)` to create the tokens as input for the DFM. Yet each additional n-gram adds features to the DFM and substantially requires more computing power to apply functions such as `textmodel_lda()`. With the greater time and computing power required, these models are not investigated further, but could be explored given a willingness to let the `textmodel_lda()` work through the algorithm.

Resources

The Federalist papers analyzed here are available from the Library of Congress along with other primary sources of text in American government https://guides.loc.gov/federalist-papers/full-text. The Hathi Trust https://www.hathitrust.org/ and Project Gutenberg https://www.gutenberg.org/ contains the full text of millions of books, freely available.

The Party Manifesto Project https://manifesto-project.wzb.eu/ is a hand-coded text as data research resource on the contents of political party platforms.

Google N-gram viewer: https://books.google.com/ngrams/

The **quanteda** package (Benoit et al., 2023) homepage presents reference manuals and tutorials, see https://tutorials.quanteda.io/ for further study.

The **stylo** (Eder et al., 2023) package provides a point-and-click menu for a range of different methods of stylometric analyses in a "bag of words" approach, see https://github.com/computationalstylistics/stylo.

The **tidytext** (Silge and Robinson, 2017) package is a set of text as data functions organized for use in **tidyverse** style R coding, see https://www.tidytextmining.com/index.html.

8.8 Exercises

Knit **R** code into a document to answer the following questions:

(1) Construct additional n-gram analyses of *The Federalist* papers. Describe how the apparent separation of the authors within the papers appears to differ, depending on whether a greater number of most frequent words are used, and whether two- or three-word combinations (rather than single word) n-grams are analyzed. For example, how does the inclusion of content words improve the separation (in the two-dimensional scaling diagram) of the texts by author?

(2) Create a one-dimensional representation of the scaling solution of *The Federalist* papers; start by editing `nmds.results<-smacofSym(dissimat_fed, ndim=2)` from the chapter text. Contrast the one-dimensional representation with the two-dimensional. How would it change any interpretations of the evidence?

(3) Construct a word cloud from the sampled Title IX regulatory comments in the CSV *reg_comments_sampled.csv*. Interpret the results. What additional terms would you trim from the DFM, and why? Remove the terms and redo the word cloud. Does the revised cloud make any substantial difference in interpreting the comments? How or how not?

(4) Using a distance metric of your choice, create and interpret a cluster dendrogram of *The Federalist Papers* similarity. How does the clustering of papers compare to the two-dimensional scaling solution in differentiating between the authors?

(5) Access one of the example datasets within **quanteda**, a collection of speeches by U.S. presidents after being sworn into office ("inaugural addresses") from 1789 to 2017, *data_corpus_inaugural*. (Copy the data to a corpus object, `inaugural_corpus <- data_corpus_inaugural` to begin.) Investigate the subject matter with any of the methods reviewed within this chapter.

9

Analyzing Election Studies

Chapter 9 introduces methods for analyzing election polls, such as the 2020 American National Election Studies (ANES) survey.

Learning objectives and chapter resources

By the end of this chapter, you should be able to: (1) import the American National Election Studies (ANES) survey and explain the structure of the data; (2) adjust the ANES data for the survey sample design; (3) create and interpret tables, descriptive statistics, and cross-tabulations (contingency tables); (4) interpret inferential tests of independence and confidence intervals on means and proportions; (5) visualize tabulations and cross-tabulations through bar charts with error bars. The material in the chapter requires the **survey** and (Lumley, 2021) **kableExtra** (Zhu, 2024) packages, which will need to be installed, as well as packages from the **tidyverse** (Wickham, 2023b) along with a pre-installed package, **knitr** (Xie, 2023b). The dataset required for chapter exercises is the **anes_20** dataframe in the **R** Workspace *anes 2020 survey data.rdata* available from https://faculty.gvsu.edu/kilburnw/inpolr.html.

9.1 Election polling

One of the defining moments in modern American presidential campaigns occurred in 1948, on election day at the end of a campaign that would send incumbent President Truman back to the White House.In that autumn, Truman's campaign was declared all but dead, both by pundits and especially pollsters, based on the relatively new polls conducted by George Gallup and Elmo Roper, who both predicted confidently that Truman would lose (Erikson and Tedin, 2015). So confident were the pundits that on November 4, the *Chicago Daily Tribune* ran an early edition with a banner headline declaring "DEWEY DEFEATS TRUMAN". Famously, Truman was photographed holding it aloft in a victory celebration that day (Jones, 2020). Defying the polls,

Truman won a majority of Electoral College votes. In the popular vote, it was a relatively close 49.5 percent for Democrat Truman, and 45.1 percent for Republican Dewey (Woolley and Peters, 2024).

Yet if pundits had access to a groundbreaking polling initiative launched that year, they might have been much more cautious in predicting a Dewey victory. In 1948, the ANES conducted its inaugural survey using an innovating probability based sampling method, known today as 'area probability cluster sampling'. The ANES developed a more representative sample than the quota sampling methods used at the time by Gallup and Roper. The two principal investigators, social scientists Angus Campbell and Robert L. Kahn fielded a survey in which 662 individuals, representative of the adult US population, were interviewed both before and after the election; the survey respondents were asked a series of questions about the upcoming election, policy issues, and their own demographic background (Campbell and Kahn, 1948).

In the pre-election survey, respondents were asked whether they intended to vote in the presidential election and the party of their vote choice. Among likely voters, 47.09 percent chose the Democratic party and 49.24 percent Republican. The survey showed Dewey with more support than Truman, well within the survey's 3.8 percent margin of error, making the election too close to call. Given the candidate centered focus of American politics, even as early as 1948 a survey question asking about party support would have been a fuzzy indicator of vote choice. In the post-election survey, respondents were asked for which candidate they voted back in November. Among likely voters, 50.2 percent of respondents reported voting for Truman and 42.2 percent for Dewey. While under-counting Dewey's support, support for Truman differed by less than a percentage point, just barely exceeding the margin of error. The ANES could have called the election correctly.

Conducted in each presidential election since 1948, the ANES has grown to become the preeminent election survey in the United States. In Chapter 9 we learn to analyze data from the 2020 ANES. The chapter does assume some prior familiarity with introductory statistics. We will (1) tabulate survey responses; (2) construct cross-tabulations (contingency tables), a fundamental tool for exploring associations between two categorical measures; and (3) calculate sample means and proportions. For each of these, we will account for likely ranges of population-level characteristics given the ANES sample, and test the sample against hypothetical population values. Finally we will create data visualizations commonly used in the study of public opinion. Prior to analyzing the 2020 ANES data, however, we first need to become familiar with basic aspects of national election poll sampling.

Sampling a national electorate

A fundamental distinction in survey sampling is the difference between a simple versus a complex random sample. In a simple random sample of the U.S. electorate, every member of the population (e.g., registered voters or voting age citizens) would have a known and equal probability of selection. For national election polling, implementing a true simple random sample is nearly impossible in practice.

To make national samples feasible, cost-effective, and representative, researchers rely on complex sampling methods. These methods usually involve three key components: (1) sampling within strata, (2) clustering interviews, and (3) applying observation weights. The process begins by dividing the country into strata — geographic or demographic regions — ensuring broad national coverage. Within each stratum, interviews are often clustered in specific areas, which reduces costs but increases sampling uncertainty because individuals within clusters tend to be more similar to one another. This clustering effect necessitates adjustments to standard errors during analysis to account for the inflated uncertainty. Finally, observation weights are applied to correct for unequal probabilities of selection and to address over or under representation of certain groups in the sample (Groves et al., 2011; Weisberg et al., 1996).

Analyzing election survey data almost always requires incorporating strata, clustering, and individual weights into the analysis in order to accurately tabulate responses from the sample and correctly infer population level characteristics. After loading the 2020 ANES data, we will explore its structure, key distinctions in data storage types, and introduce functions from the **survey** package to account for the complex sample characteristics and analyze the survey responses.

The 2020 ANES survey

The 2020 ANES survey dataset is stored within an R Workspace, *anes_2020_survey_data.rdata.*

```
load("anes 2020 survey data.rdata")
```

After loading, the dataframe *anes_20* appears in the Environment pane. Rows are survey respondents, and the columns record the different alpha-numeric variables. The A presents a brief ANES codebook explaining the data. Nearly all of the columns refer to different survey questions. In the Console, enter `names(anes_20)` to list the names of each column. For example, the sixth variable *V201005*. In the A codebook following the name is a variable label, which explains that the subject matter of the question is "*How often does [respondent] pay attention to politics and elections*". In the second section of the codebook the exact wording of the survey question is "How often do you pay attention to what's going on in government and politics?"

While not accounting for the survey design, we can tabulate the contents of this variable with a simple `table()` function, specifying the name of the dataframe and the column following the `$`sign:

```
table(anes_20$V201005)
```

```
##
##              1. Always  2. Most of the time
##                   1158                 1863
## 3. About half the time  4. Some of the time
##                    920                  795
##               5. Never
##                     47
```

The results show the (unweighted) frequencies. For example, 1158 people out of the total sample selecting "Always", which is labeled in the response scale as point 1 on a five-point scale ranging from "1. Always" to "5. Never".

The variables in the survey are either *factor* or *numeric* storage types. Since the columns mostly record individual survey questions, the storage type reflects the survey response scale — whether responses are recorded as numeric scores or verbal response scales. Variable `V201005` is a factor; enter `str(anes_20$V201005)` to observe the variable's storage type. In the factor, numeric scores are place-holders for the verbally labeled responses. The `summary()` or `table()` responses will display the response categories and unweighted frequencies, such as `summary(anes_20$V201005)`.

Variables stored in the numeric storage type are purely numeric scores without a verbal label attached, such as age or a feeling thermometer score. Variable `V202162` records feelings toward labor unions. Entering `str(anes_20$V202162)` shows that it is stored as a numeric score `num`. Because the variable is recorded as numbers, `summary()` will report summary statistics:

```
summary(anes_20$V202162)
```

```
##    Min. 1st Qu.  Median    Mean 3rd Qu.    Max.
##     0.0    50.0    60.0    58.7    70.0   100.0
##    NA's
##      88
```

As with any other numeric variable, the `summary()` function reports the five-number summary (min, max, median and 25th and 75th percentiles), along with the arithmetic mean and number of missing values (`NA's`). Because the factor storage type associates numeric scores with each verbal label, we could instruct R to treat the variable for political interest (`V201005`) as numeric, with the `as.numeric()` function wrapped around the name of the variable, `as.numeric(anes_20$V201005)`:

```
summary(as.numeric(anes_20$V201005))
```

```
##    Min. 1st Qu.  Median    Mean 3rd Qu.    Max.
##    1.00    2.00    2.00    2.31    3.00    5.00
```

Treating the variable as numeric returns quantitative summary statistics, where the first ordered category in the response scale ("1. Always") is assigned a value of 1 and the last category ("5. Never") is assigned a value of 5. These numeric assignments are based on the order of the factor levels, not the labels themselves. The "Always" response is assigned a score of 1 because it is the first factor label, not because it is explicitly labeled as "1." In survey data analysis, raw survey responses and labels are often not immediately usable and typically require transformation. We will see examples of these transformations further below, after establishing the complex survey design.

9.2 Specifying a survey sample design

The standard package for analyzing complex sample data in **R** is **survey** (Lumley, 2021). If needed, install the package `install.packages("survey")` and attach it:

```
library(survey)
```

To account for the complex survey design, analyses must adjust for the three key elements: strata, clusters, and weights. These elements are specified using the `svydesign()` function from the **survey** package. . Once the survey design is established, subsequent analysis functions automatically incorporate these adjustments, modifying observation frequencies and ensuring accurate measures of uncertainty, such as standard errors. For the 2020 ANES survey data, the sample design is specified as follows:

```
anes2020<-svydesign(data = anes_20, strata=~V200015d,
                    ids = ~V200015c, weights=~V200015b, nest=TRUE)
```

The function has four main arguments to identify different parts of the design. The first is the identification of the dataset stored in the **R** environment, `data=`. The remaining arguments specify columns within the dataframe, each associated with a distinct component of the survey design.

The package uses a 'tilde' ~ symbol to indicate a specific variable (usually found next to the number 1 on a keyboard). The second argument, `strata=~`, identifies the strata in the dataset. Third, `ids=~` specifies the interview clusters within the strata, and fourth, `weights=~` assigns the appropriate survey weights. In some cases, such as this example, the argument `nest=TRUE` is also required to indicate that the clusters are nested.

The object anes2020 represents the survey design. When performing analysis with functions from the **survey** package, the survey design object (assigned the contents of `svydesign()`) is specified as an argument instead of the dataset itself. We will explore how to use these functions for analyzing nominal and ordinal variables (saved as factor variables), as well as interval/ratio variables (numeric), focusing on tabulations, means, and proportions.

9.3 Tabulating and cross-tabulating survey measures

The `svytable()` function tabulates the weighted frequencies for a variable, such as *V201005*. Two arguments are required: the variable to tabulate and the survey design object, `svytable(~V201005, design = anes2020)`. The ~ symbol is essential, as it indicates that *V201005* is a column within the *anes2020* survey dataset. Without it, the function will not interpret the variable correctly.

The results display weighted tabulations, where each observation's contribution is scaled by its survey weight. Basically, weights adjust the influence of each observation by multiplying its frequency by its weight, ensuring the tabulations reflect population estimates rather than the sample distribution. To round the tabulations to the nearest integer, add the argument `round=TRUE`; without it, frequencies will include decimals:

```
svytable(~V201005, anes2020, round=TRUE)
```

```
## V201005
##              1. Always   2. Most of the time
##                   1004                  1713
## 3. About half the time   4. Some of the time
##                    986                  1027
##               5. Never
##                     54
```

The `design =` description can be omitted. Compare these weighted tabulations to the previously unweighted tabulations from `table()`. The difference is not trivial, substantially affecting proportion or percentage calculations. Placing the `prop.table()` function around the tabulation will yield proportions:

```
prop.table(svytable(~V201005, anes2020, round=TRUE))
```

```
## V201005
##              1. Always    2. Most of the time
##                0.20987                0.35807
## 3. About half the time    4. Some of the time
##                0.20610                0.21467
##               5. Never
##                0.01129
```

Then multiplying the proportions by 100, percentages:

```
prop.table(svytable(~V201005, anes2020, round=TRUE))*100
```

```
## V201005
##              1. Always    2. Most of the time
##                 20.987                 35.807
## 3. About half the time    4. Some of the time
##                 20.610                 21.467
##               5. Never
##                  1.129
```

For example, from the tabulation, we observe that a narrow majority of about 57 percent of US citizens 18 or older report paying attention to politics "always" or "most of the time".

Formatting tabulations

For proportions or percentages in a print-ready format, we can use the `prop.table()` function alongside `knitr::kable()` from the **knitr** (Xie, 2023b) package. Tables in this format work especially well for inclusion in RMarkdown documents knitted to Word or PDF format.

To store these percentages for further formatting, we assign the output to an object, such as `table1`:

```
table1<-prop.table(svytable(~V201005, anes2020, round=TRUE))*100
```

Next, we use `knitr::kable()` to format the table with reduced decimal places and an appropriate caption. The package is preinstalled in RStudio; without needing to use `library()`, we use it by identifying the package and function separated by two colons: `knitr::kable()`. The result is a table formatted as in Table 9.1:

TABLE 9.1 Survey respondent attention paid to politics, 2020 ANES.

Attention to Politics	Percentage
1. Always	20.99
2. Most of the time	35.81
3. About half the time	20.61
4. Some of the time	21.47
5. Never	1.13

```
knitr::kable(table1, digits = 2,
        col.names = c("Attention to Politics", "Percentage"),
        caption = "Survey respondent attention
          paid to politics, 2020 ANES.")
```

The `caption=` and number of `digits=` are optional arguments. The argument `col.names=()` changes the default column names from `table1` in Table 9.1.

Cross-tabulation (contingency) tables

A cross-tabulation table displays the frequency distribution of two categorical variables, how frequencies on one categorical variable covary with the other. For example, pollsters often include cross-tabulations as part of a press release accompanying a poll, displaying how vote choice varies over demographic characteristics. Cross-tabulations are fundamental to bivariate hypothesis testing, given the identification of one variable as independent and the other dependent. Researchers compare proportions or percentages within levels of an independent variable, to observe variation in levels of a dependent variable.

For example, tabulating how frequency of attention to politics varies by sex (variable *V201600*) creates a contingency table. From `svytable()` add an additional variable with a plus (+) sign in front of the second variable:

```
svytable(~V201005 + V201600, anes2020, round=TRUE)
```

```
##                              V201600
## V201005                      -9. Refused 1. Male 2. Female
##    1. Always                           6     568       429
##    2. Most of the time                 6     856       851
##    3. About half the time              7     384       594
##    4. Some of the time                 4     454       569
##    5. Never                            0      28        26
```

The result is a tabulation of frequencies of political interest (as the row variable) by sex (on the column). Notice that in *V201600* (`svytable(~ V201600,` anes2020, round=TRUE)), 24 respondents refused to answer the question, and are marked by the standard "Refused" code of *-9* in the ANES. For respondent sex, the reason why 24 people refused is beyond the scope of the ANES, and is too small from which to generalize. So typically in an analysis we would convert the "Refused" respondents to a missing value (NA) in the dataset — then construct a new table of respondent sex by political attention.

Since the frequencies are from a sample, percentages or proportions interest us more than raw frequencies. Before creating a table of percentages, first we will observe how to remove responses such as "Refused" from a table, by creating a transformed survey measure of respondent sex as a new column in the *anes_20* dataframe.

Creating and recoding a new survey measure

This process of creating new variables in the data frame and transforming the response scales in survey data is often referred to as "recoding". There are at least a few reasons why recoding could be useful: (1) dropping "Don't know" or "Refused" responses; (2) reversing scale directions such as "1. Always" to "5. Never"; (3) categorizing a measure such as age into groups; or (4) simply creating new alphanumeric variables with memorable names.

For *V201600*, we will clone a copy of it into a new variable *sex*, then convert to a missing value NA the respondents recorded as *-9. Refused.* Then in a revised table, contrast *sex* with variation in political attention. Various **tidyverse** functions facilitate recoding, which starts with creating the variable *sex* as a copy of *V201600*:

```
library(tidyverse)

anes_20 <- anes_20 %>%
  mutate(sex=V201600)
```

Enter names(anes_20) to observe *sex* now appears at the end of the dataset. To recode factor labels, we use fct_recode(). We specify a new label for each old one we want to replace: the order being 'new label = old label'. Labels are specified in quotes, although for *sex* we want to recode the "Refused" label to NULL or NA, two system level values for missing data.

Below a newly recoded *sex* variable is created and recoded from the existing *V201600*. The second line creates the new variable with mutate(sex =), in which fct_recode(V201600, NULL = "-9. Refused") specifies the recoded categories:

```
anes_20<- anes_20 %>%
  mutate(sex = fct_recode(V201600, NULL = "-9. Refused"))
```

The `mutate()` function could have also been expressed as `mutate(sex = na_if(V201600,"-9. Refused"))`: the `na_if()` function assigns a missing value `NA` for values listed in the second argument. Either creates the newly recoded *sex* variable.

One important step in analyzing data with the *survey* package is that each time a new variable is created, the `svydesign()` function must be run again, so that the names of the new variables are stored within the `svydesign()` object.

```
anes2020<-svydesign(data = anes_20, ids = ~V200015c,
           strata=~V200015d, weights=~V200015b, nest=TRUE)
```

Then `svytable()` displays the cross-tabulation:

```
svytable(~V201005 + sex, anes2020, round=TRUE)
```

```
##                          sex
## V201005                   1. Male 2. Female
##    1. Always                  568       429
##    2. Most of the time        856       851
##    3. About half the time     384       594
##    4. Some of the time        454       569
##    5. Never                    28        26
```

Column and row proportions in a contingency table

The function `prop.table()` wrapped around the `svytable()` creates a table of proportions. Using `prop.table()` with argument `margin=1` calculates proportions summing to one along the rows (row proportions). Setting `margin=2` calculates proportions summing to one along the columns (column proportions).

```
prop.table(svytable(~V201005 + sex, anes2020, round=TRUE), margin=2)
```

```
##                          sex
## V201005                   1. Male 2. Female
##    1. Always              0.24803   0.17375
##    2. Most of the time    0.37380   0.34467
##    3. About half the time 0.16769   0.24058
##    4. Some of the time    0.19825   0.23046
```

TABLE 9.2 Survey respondent sex by attention paid to politics, column percentages 2020 ANES.

	1. Male	2. Female
1. Always	24.80	17.38
2. Most of the time	37.38	34.47
3. About half the time	16.77	24.06
4. Some of the time	19.83	23.05
5. Never	1.22	1.05

```
##   5. Never                 0.01223   0.01053
```

As column proportions, summing down the column of entries under *"1. Male"* and *"2. Female"*, the totals are 1. Column proportions reflect the logic of a particular conditional probability. For the column proportions, that probability is: given someone's sex, what is the probability of reporting a particular level of attentiveness to politics? If the hypothesis is, for example, that females are more likely than males to report paying a great deal of attention to politics, then this is the probability of interest and an analysis table should include column rather than row proportions.

From the column proportions, we can observe, for example, that given male sex the probability of reporting "Always" paying attention is about .25 compared to female .17. About 25 percent of males report "Always" paying attention, compared to 17 percent of Females. With sex as the independent variable x, and attention to politics as the dependent variable y, the column proportions are the quantities needed to assess how categories on y vary given values of x — for example, how does attention vary, given that someone is male? Female? Without specifying a margin, `prop.table()` calculates total proportions — the frequencies within each category of column and row given the total frequencies.

To create percentages, the contents of `prop.table()` can be saved to a table object, multiplied by 100 within `knitr::kable()`. Table 9.2 displays a table formatted in this style:

```
table2<-prop.table(svytable(~V201005 + sex, anes2020, round=TRUE),
                   margin=2)
```

```
knitr::kable(table2*100, digits = 2,caption = "Survey respondent sex by
             attention paid to politics, column percentages 2020 ANES.")
```

From Table 9.2, we could conclude that respondents identifying as male rather than female identifiers are more likely to report paying attention always (24.8 percent compared to 17.38 percent) or most of the time (37.38 percent compared to 34.47 percent).

In cross-tabulations, the rule of thumb to remember is 'always percent-agize on the independent variable' — to calculate proportions (or percentages) that sum to 1 within each category of the independent variable. Typically researchers add marginal totals (row and column total proportions or percentages) along with frequencies in the same table. The frequencies and proportions or percentages from `prop.table()` can be combined into one table within word processing software.

9.4 Test of independence for contingency tables

The cross-tabulation of political interest by sex showed an association, yet a separate question is whether the association observed within the sample table is large enough to reject the null hypothesis of no relationship within the population. The standard Chi-squared test of independence tests this null hypothesis. The test compares the observed frequencies within the table to a hypothetical set of expected frequencies under the assumption of independence (no relationship) between the variables. The greater the difference between the observed frequencies from expected, the greater the Chi-squared test statistic, and the greater the evidence against the null hypothesis.

The adjusted Chi-squared test for the complex survey design of the ANES reports an F-statistic (the ratio of two Chi-squared distributions) and an associated p-value for the probability of observing a test statistics that large (or larger) if the null hypothesis of independence is true. Given a chosen 'alpha level', or threshold of the p-value for determining statistical significance and rejecting the null. The standard alpha level ($\alpha = .05$) of .05, means that if the we observe a $p < .05$ we reject the null hypothesis.

To test the null of no relationship between political attention and sex in the ANES population, the function is `svychisq()` in place of `svytable()`:

```
svychisq(~V201005 + sex, anes2020)
```

```
##
##  Pearson's X^2: Rao & Scott adjustment
##
## data:  svychisq(~V201005 + sex, anes2020)
## F = 8.4, ndf = 3.7, ddf = 187.2, p-value = 5e-06
```

The results show a statistically significant association between interest and sex. The test statistic has an F-value of 8.4 (degrees of freedom $\frac{3.7}{187.2}$), with a p-value of 0.000005, well below a significance threshold of 0.05. From the p-value we reject the null hypothesis of independence between interest and sex. The test result suggests that the observed relationship between interest and sex is much stronger than would be expected on the basis of chance and sampling error alone, so we can conclude that interest and sex are related within the sampled population. Note that the test does not measure the strength or direction of association; the test result would be the same, for example, if the categories in interest were scrambled.

Contingency tables controlling for a third factor

In a contingency table we could observe that two measures are related (a y covarying with x), but the possibility remains that the relationship is spurious — any observed association in the relationship between x and y could be the result of a third variable z that influences both. To test the extent to which the relationship within a contingency table persists when controlling for a third factor, we subset the data on values of the third factor and recalculate the contingency table. To demonstrate, we will cross-tabulate party identification by religious service attendance, then on levels of *sex*, to observe whether there is evidence — controlling for sex — that Democrats report less religiosity than Republicans. To do so, we will work through some useful variable recoding functions.

Recoding for a table of party identification by religiosity

The two relevant variables from the survey are *V201228* for party identification, and for religious service attendance the variable *attend*, constructed from *V201452* and *V201453*. First check the existing response labels in party identification, `table(anes_20$V201228)`. Variable *V201228* is stored as a factor; recoding functions from the **forcats** package (Wickham, 2023a) in the **tidyverse** modify the factor levels. The recoding function below with `fct_recode()` assigns the "Refused", "Don't know", and "No preference" respondents to the Independent category, while excluding the "Other party" respondents.

```
anes_20 <- anes_20 %>%
  mutate(partyid = fct_recode(V201228,
      "3. Independent" = "-9. Refused",
      "3. Independent" = "-8. Don't know",
      "3. Independent" = "0. No preference {VOL - video/phone only}",
      NULL = "5. Other party {SPECIFY}"),
        partyid = fct_relevel(partyid, "1. Democrat", "2. Republican",
                              "3. Independent"))
```

The function `fct_relevel()` sets the order of the labels. Enter `table(anes_20$partyid)` and observe that after reassigning the respondents, only three categories remain tabulated: Democrat, Republican, and Independent.

Creating a measure that summarizes a survey respondent's attendance at religious services requires combining responses from two different survey questions. First, respondents are asked whether they attend religious services at all:

```
table(anes_20$V201452)
```

```
##
##     -9. Refused -8. Don't know            1. Yes
##              30              2              2266
##           2. No
##            2485
```

Then if they respond "Yes", they are asked how frequently:

```
table(anes_20$V201453)
```

```
##
##                -9. Refused              -8. Don't know
##                         11                           1
##              1. Every week        2. Almost every week
##                        727                         528
## 3. Once or twice a month      4. A few times a year
##                        375                         565
##                   5. Never
##                         59
```

The two variables are combined into one through the `case_when()` function inside `mutate()`. The function is similar to `filter()`, in that `case_when()` evaluates whether a condition is true for each observation, and if so, assigns that observation a particular score in the new variable. For example in the third line below, `case_when()` evaluates whether observations `V201452 == "2. No"`; then for all observations where the responses for *V201452* are recorded as "2. No", the responses for *attend* are assigned "1. Never". For each line, along the right-hand side after the tilde symbol ~ appears the resulting scale in *attend*, ranging from "1. Never" attend services to "5. Every week".

```
anes_20<-anes_20 %>%
  mutate(attend = case_when(
    V201452 == "2. No" ~ "1. Never",
    V201453 == "5. Never" ~ "1. Never",
    V201453 == "4. A few times a year" ~ "2. A few times a year",
    V201453 == "3. Once or twice a month" ~ "3. Once or twice a month",
    V201453 == "2. Almost every week" ~ "4. Almost every week",
    V201453 == "1. Every week" ~ "5. Every week"))
```

The first condition matched in `case_when()` is from *V201452*, while the others are the variation in response scales from *V201453*. The result is a five-point attendance scale:

```
table(anes_20$attend)
```

```
##
##                    1. Never    2. A few times a year
##                        2544                      565
## 3. Once or twice a month     4. Almost every week
##                         375                      528
##                5. Every week
##                         727
```

A further change to the scale could be to combine the responses, "A few times a year" with "Once or twice a month", so that respondents are arranged from "Never", to infrequently ("A few times a year" or "Once or twice a month") to near weekly and weekly:

```
anes_20 <-anes_20 %>%
  mutate(attend = fct_recode(attend,
        "2. few times a year, monthly" = "2. A few times a year",
        "2. few times a year, monthly" = "3. Once or twice a month",
        "3. almost every week" = "4. Almost every week",
        "4. every week" = "5. Every week"))
```

Once the measures are recoded, we respecify the survey design object *anes2020* so that it accounts for the new variables *partyid* and *attend*:

```
anes2020<-svydesign(data = anes_20, ids = ~V200015c,
          strata=~V200015d, weights=~V200015b, nest=TRUE)
```

Then we can create the tables. The contingency table with (column) proportions to display the percentage of each attendance category identifying as Republican, Democrat, or Independent is `prop.table(svytable(~partyid + attend, design=anes2020, round=TRUE), margin=2)`. Saving the table as *table3*, converting it to percentages and formatting with `knitr::kable()`, the result appears in Table 9.3.

```
table3<-prop.table(svytable(~partyid + attend, design = anes2020,
                            round=TRUE), margin=2)
```

TABLE 9.3 Contingency table of party identification by religious service attendance, column percentages, 2020 ANES.

	1. Never	2. few times a year, monthly	3. almost every week	4. every week
1. Democrat	36.12	37.55	27.11	25.76
2. Republican	23.11	35.21	47.19	49.70
3. Independent	40.77	27.23	25.70	24.55

```
library(kableExtra)
knitr::kable(table3*100, digits = 2,
        caption = "Contingency table of party identification
        by religious service attendance, column percentages, 2020 ANES.") %>%
  kable_styling(full_width = FALSE, latex_options = c("scale_down"))
```

The code to generate a table similar to Table 9.3 could be as simple as `knitr::kable(table3*100, digits = 2)`; the code above includes additional options to format the table on a printed page. Table 9.3 shows that of those "Never" attending services, 23 percent identify as Republican compared to 36 percent as Democrat, a 13 point difference. At the other end of the scale, about half of those attending services "every week" identify as Republican versus only about 26 percent as Democrat. Enter `svychisq(~partyid + attend, design = anes2020)` to observe a statistically significant F-statistic.

Yet given the possibility that these relationships break down when controlling for sex, we recreate the cross-tabulation on each value of the *sex* variable. In constructing control tables such as these, the goal is to test whether any observed relationship between the independent and dependent variable persists for each level of the control variable. If the relationship between party identification and religious service attendance appears at least somewhat null across levels of *sex* (`1. Male` and `2.Female`), then this pattern would be evidence of spuriousness. Of course, it is unlikely that *sex* would be a source of spuriousness; it is much more likely that we would observe *sex* as an additive effect shifting party identification and religious attendance in a particular direction.

To construct the table, for the value of *sex* `1. Male`, we subset the survey design object with `subset(anes2020, sex== "1. Male")` in place of the survey design object:

```
svytable(~partyid + attend, subset(anes2020, sex== "1. Male"),
        round=TRUE)
```

```
##                  attend
## partyid           1. Never 2. few times a year, monthly
##   1. Democrat          375                          158
##   2. Republican        314                          144
```

TABLE 9.4 Party identification by religious service attendance, column percentages, male respondents only.

	1. Never	2. few times a year, monthly	3. almost every week	4. every week
1. Democrat	30.54	37.44	14.35	25.00
2. Republican	25.57	34.12	57.40	54.49
3. Independent	43.89	28.44	28.25	20.51

```
##    3. Independent       539                          120
##                      attend
## partyid               3. almost every week 4. every week
##    1. Democrat                          32            78
##    2. Republican                       128           170
##    3. Independent                       63            64
```

The result is a "control table", a tabulation of religious service attendance by party identification, controlling for gender. Creating a formatted table from `knitr::kable()` results in Table 9.4:

```
table4<-prop.table(svytable(~partyid + attend,
                   subset(anes2020, sex== "1. Male"), round=TRUE),
                   margin=2)
```

A simple table without a caption would be `knitr::kable(table4*100, digits = 2)`.

```
library(kableExtra)
knitr::kable(table4*100, digits = 2,
   caption = "Party identification by religious
   service attendance, column percentages, male respondents only.") %>%
   kable_styling(full_width = FALSE, latex_options = c("scale_down"))
```

Table 9.4 shows, for example, among males the gap between Democrats and Republicans among weekly service attendees grows to 29 points. About 54 percent of weekly attendees identify as Republican. Among males, party appears dependent on religiosity. Then creating an additional table for the other level of the *sex* variable, "2. Female", results in Table 9.5. Again, the relationship persists between party and religiosity. Among non-attendees, 41.84 percent identify as Democrats, and only 20 percent as Republican, while among weekly attendees, 45.24 percent identify as Republican compared to 26.51 percent as Democrat:

```
table5<-prop.table(svytable(~partyid + attend,
                   subset(anes2020, sex== "2. Female"), round=TRUE),
```

TABLE 9.5 Party identification by religious service attendance, column percentages, female respondents only

	1. Never	2. few times a year, monthly	3. almost every week	4. every week
1. Democrat	41.84	37.79	37.59	26.51
2. Republican	20.72	36.05	39.05	45.24
3. Independent	37.44	26.16	23.36	28.24

```
                  margin=2)
```

The **kableExtra** (Zhu, 2024) package is optional, for formatting the tables on the printed page.

```
library(kableExtra)
knitr::kable(table5*100, digits = 2, caption = "Party
      identification by religious service attendance, column
      percentages, female respondents only") %>%
  kable_styling(full_width = FALSE, latex_options = c("scale_down"))
```

The two tables are constructed for the purpose of evaluating – at each measured level in *sex*, do respondents reporting less-frequent religious service attendance tend to identify as Democrats, compared to Republicans? If sex is a source of spuriousness in the relationship between political party identification and service attendance, then the table would display no relationship between party and service attendance. Yet across both tables, the direction and strength of the relationship between party and service attendance persists. Checking the Chi-squared test on each sub-sample (as in `svychisq(~partyid + attend, subset(anes2020, sex== "2. Female"))`) would show a significant association.

Among males and females, the gap between Democrats and Republicans among those reporting that they "Never" attend services is about 5 and 21 percent, respectively. Among those attending services weekly, the gap between Republicans and Democrats is 29 and 18 percent. Rather than being a source of spuriousness, *sex* is better described as an additive factor — pushing up Republican party identity among males and Democratic party identity among females. Notice that among non-attendees, 41.84 percent of females identify as Democrats, an 11 percent gap compared to males, while only 45.24 percent of females attending services weekly identify as Republican, a similar gap compared to males.

9.5 Confidence intervals and tests on proportions and means

Given the proportions of the respondents within each category of party identification from `svytable(~partyid, anes2020)`, two functions from the **survey** package can be combined to estimate confidence intervals on proportions.

The function `confint()` calculates, by default, 95 percent confidence intervals. For a variable stored as a factor, with ordered or nominal categories, the `svymean()` function calculates the proportions based on a mean (from each category scaled to a value of 1), and wrapped within `confint()` reports the lower and upper bounds of a confidence interval. Because there are missing values recorded for some respondents, the argument `na.rm=TRUE` is necessary:

```
confint(svymean(~partyid, anes2020, na.rm=TRUE))
```

```
##                          2.5 % 97.5 %
## partyid1. Democrat     0.3183 0.3591
## partyid2. Republican  0.3014 0.3372
## partyid3. Independent 0.3245 0.3595
```

The shorthand interpretation of the results would be that the table displays the range of proportions of Independents, Democrats, or Republicans, for which we can be 95 percent confident that the true population-level proportion lies. For example, for Democrats, our best estimate (with 95 percent confidence) is that between 32 to 36 percent of the voting age U.S. citizenry identifies as a Democrat. While the confidence interval for Republicans contains the lowest proportion at 30 percent, the confidence intervals all substantially overlap, indicating the true population proportions of each are indiscernibly different. Of course, in this example it is important to keep in mind the assumptions made in recoding it so that anyone choosing "something else" is treated as independents.

The same functions applied to a numeric variable results in a mean and confidence interval on the underlying scale. For example, the feeling thermometer for the "U.S. Supreme Court" ranges from 0 to 100. The sample mean is

```
svymean(~V202165, anes2020, na.rm=TRUE)
```

```
##         mean  SE
## V202165 59.5 0.5
```

And a 95 percent confidence interval is

```
confint(svymean(~V202165, anes2020, na.rm=TRUE))
```

```
##          2.5 % 97.5 %
## V202165 58.51  60.49
```

The sample mean is 59.5 degrees, while the 95 percent confidence interval for the mean feeling toward the Court ranges from 58.5 to 60.49 degrees.

Tables of means for levels of a categorical variable

The `svyby()` function can be combined with `svymean()` to calculate a table of means on a categorical (factor) variable. For example, we could create age groups based on the survey question *V201507x* recording the survey respondent's age, and compare feeling thermometer scores across age groups. Feelings toward "transgender people" is recorded in variable *V202172*.

We first create an age category variable, with `case_when()` . Then a second `mutate()` function stores it as a factor and sets the factor levels.

```
anes_20<-anes_20 %>%
   mutate(agecat= case_when(
     V201507x %in% 15:29 ~ "18 to 29",
     V201507x %in% 30:49 ~ "30 to 49",
     V201507x %in% 50:59 ~ "50 to 59",
     V201507x %in% 60:100 ~ "60 and above")) %>%
  mutate(agecat = as_factor(agecat),
        agecat = fct_relevel(agecat, "18 to 29",
        "30 to 49", "50 to 59", "60 and above"))
```

Check the categories with `table(anes_20$agecat)`. To create the table we reset the `svydesign()` function.

```
anes2020<-svydesign(data = anes_20, ids = ~V200015c,
          strata=~V200015d, weights=~V200015b, nest=TRUE)

svyby(~V202172 ,~agecat, anes2020, svymean, na.rm=TRUE)
```

```
##                    agecat V202172     se
## 18 to 29         18 to 29   66.24 1.6865
## 30 to 49         30 to 49   59.92 0.8281
## 50 to 59         50 to 59   56.35 1.1294
## 60 and above 60 and above   55.16 0.9076
```

The syntax in `svyby()` references first the numeric variable, followed by the categorical, and `svymean` identifies the function to apply to *V202172* across levels of *agecat*. The results show for each age group the mean feeling score (under `V202172`) and the standard error of the mean (under `se`). Among 18- to 29-year-olds, the sample mean feeling toward transgender people is 66.24 degrees, and among those 60 and above 55.16 degrees. Calculating confidence intervals:

```
confint(svyby(~V202172 ,~agecat, anes2020, svymean, na.rm=TRUE))
```

```
##                 2.5 % 97.5 %
## 18 to 29        62.93  69.54
## 30 to 49        58.30  61.55
## 50 to 59        54.13  58.56
## 60 and above 53.38  56.93
```

The youngest age group feels warmer toward transgender people compared to older age groups. The confidence intervals show, for example, a statistically significant difference between younger respondents (18 to 29 years old) compared to older groups.

Hypothesis tests on means

The `svyttest()` function formally tests a mean against a hypothetical value or tests for a statistically significant difference in means across two levels of a categorical variable. A 'one-sample' t-test compares the mean of a continuous variable in the population to a specific hypothetical value. A 'two-sample' t-test checks for a significant difference in means between two groups. Researchers usually emphasize confidence intervals around sample means, but a hypothesis test is useful for more directly comparing sample means to a specific, hypothetical population value.

For example, we may be interested in testing whether there is evidence of population-based differences, by gender, in feeling thermometer scores. One approach would be to calculate and compare 95 percent confidence intervals. Another is to directly test the hypothesis that the mean feeling differs across groups, a two-sample test. To compare, for example, feelings toward the "Me too movement" (feeling thermometer *V202183*), the `svytest()` function argument is the feeling thermometer *V202183* by *sex*:

```
svyttest(V202183 ~ sex, anes2020, na.rm=TRUE)
```

```
##
##  Design-based t-test
##
## data:  V202183 ~ sex
## t = 7.7, df = 49, p-value = 6e-10
```

```
## alternative hypothesis: true difference in mean is not equal to 0
## 95 percent confidence interval:
##   7.74 13.24
## sample estimates:
## difference in mean
##              10.49
```

The results display the t-distribution score, corresponding degrees of freedom for the test, followed by the p-value. The small degrees of freedom are determined by the design effect of the survey, the complexity of the sampling design, so that the t-test accounts for the additional variability introduced by stratification and clustering. For this and other adjustments to statistical tests, see Lumley (2011). Because the default test is whether the difference in *V202183* means across groups is 0, the alternative is nondirectional, whether the difference is not equal to 0. The sample difference in the mean is 10.49, a difference large enough to reject the null hypothesis of no difference within the population, given the minuscule p-value. The 95 percent confidence interval shows the range for the difference of 7.74 to 13.24 degrees.

9.6 Data visualization for election polls

Bar plots

Perhaps the most common data visualization type for public opinion research is the bar chart, sometimes accompanied by "error bars" to display confidence intervals on a proportion or percentage. Responses to survey questions measured on nominal or ordinal scales, such as vote choice or a Likert type scale (e.g., "Favor strongly" to "Oppose strongly"), are visualized as bar charts. Within the `ggplot()` function, `geom_bar()` will plot bar heights corresponding to a survey response quantity such as a frequency, proportion, or percentage.

To create bar charts, instead of sending the entire *anes_20* dataframe directly to `ggplot()`, we first create a smaller data frame containing only the quantities to be plotted. For this example, we will visualize attitudes toward capital punishment ("Do you favor or oppose the death penalty for persons convicted of murder?") using variable *V201345x*. To enhance the analysis, we will also compare these attitudes by respondent sex in side-by-side bar plots.

To plot the percentages supporting and opposing the death penalty, we start by tabulating and cross-tabulating the measures. Using `svytable()`, we calculate the proportions, convert the results into percentages, and save the tabulation as a miniature dataset in a dataframe:

```
table1<- prop.table(svytable(~V201345x, anes2020, round=TRUE))*100
table1<-as.data.frame(table1)
table1
```

```
##                  V201345x  Freq
## 1       1. Favor strongly 42.07
## 2   2. Favor not strongly 21.10
## 3 3. Oppose not strongly 19.35
## 4      4. Oppose strongly 17.48
```

From `as.data.frame()` (or `as_tibble()`), the dataframe `table1` has two column headings, *V201345x* and *Freq*. We will relabel these columns to "Response" and "Percentage", then send the data to `ggplot()`:

```
colnames(table1) <- c("Response", "Percentage")
```

Enter the table name `table1` and observe how the column names have changed. Then as displayed in Figure 9.1 `geom_bar()` plots the bar heights with an argument `stat="identity"`, which instructs `geom_bar()` to plot the heights 'as is', without changes:

```
ggplot(table1, aes(x = Response, y = Percentage)) +
geom_bar(stat = "identity") +
labs(x = "Death penalty attitude", y = "Percentage")
```

To contrast death penalty support by respondent sex, we construct the cross-tabulation and save it as a dataframe, `table2`:

```
table2<- prop.table(svytable(~V201345x + sex, anes2020, round=TRUE),
                    margin=2)*100
table2<-as.data.frame(table2)
table2
```

```
##                  V201345x       sex  Freq
## 1       1. Favor strongly   1. Male 44.65
## 2   2. Favor not strongly   1. Male 21.15
## 3 3. Oppose not strongly   1. Male 17.88
## 4      4. Oppose strongly   1. Male 16.33
## 5       1. Favor strongly 2. Female 39.75
## 6   2. Favor not strongly 2. Female 21.08
## 7 3. Oppose not strongly 2. Female 20.62
## 8      4. Oppose strongly 2. Female 18.55
```

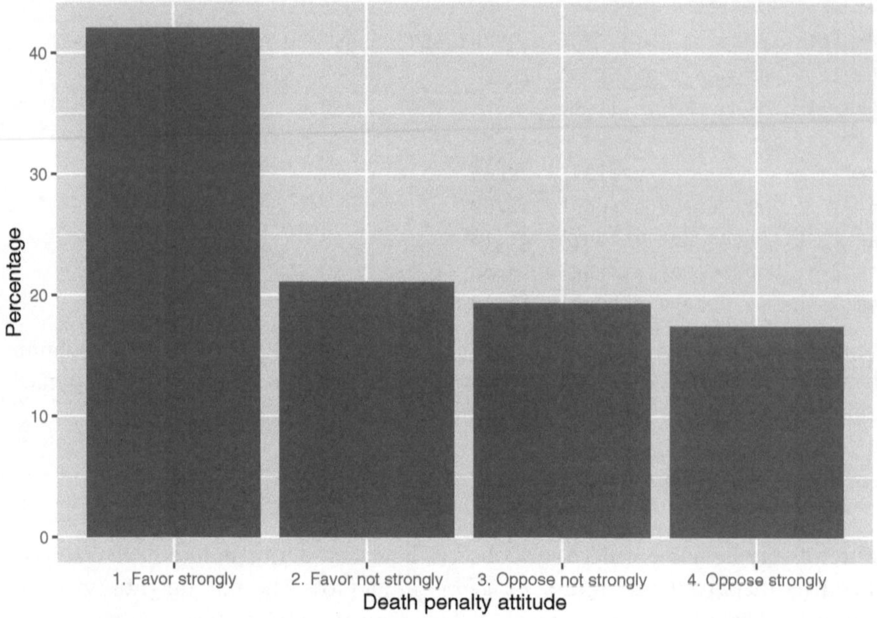

FIGURE 9.1 Attitude toward capital punishment (death penalty) in the 2020 ANES survey, categorized by strength of favor or opposition. Bars represent the proportion of respondents in each category.

Notice there are three columns, which we will rename, then pass on to `ggplot()`.

```
colnames(table2) <- c("Response", "sex", "Percentage")
```

The aesthetic argument `fill=sex` colors the bars by sex, while in `geom_bar()`, `position="dodge"` places the bars side by side rather than stacked. Figure 9.2 displays the bar graph with error bars:

```
ggplot(table2, aes(x = Response, y = Percentage, fill = sex)) +
  geom_bar(stat = "identity", position = "dodge") +
  labs(x = "Response", y = "Percentage") +
  theme_minimal() +
  theme(legend.position = "bottom") +
  scale_fill_grey()
```

The line `theme_minimal()` removes the grey background and `scale_fill_grey()` distinguishes the bars in grey versus black. The single bar plot visualizes the tabulation, the side-by-side bars display the cross-tabulated percentages. Because the percentages in the bar plot are the percentages of males and

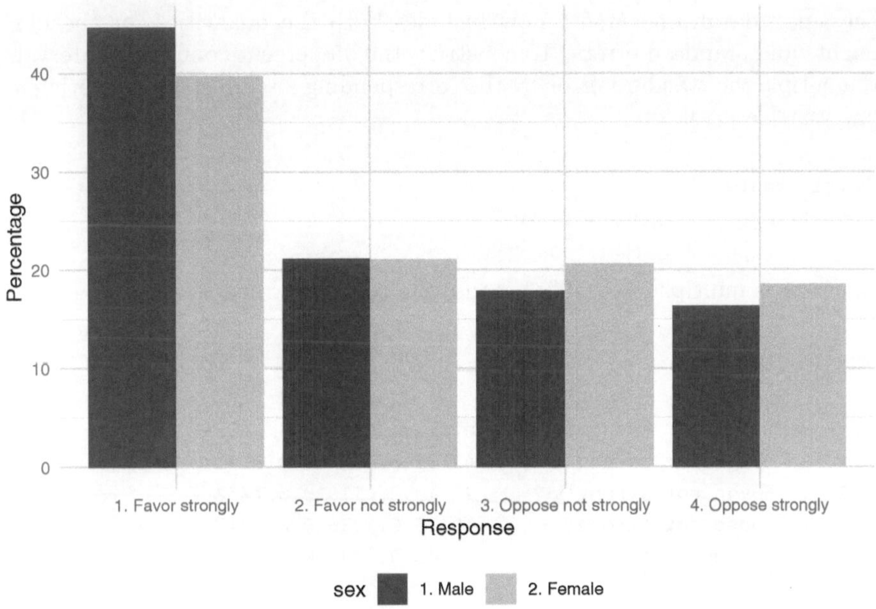

FIGURE 9.2 Attitude toward capital punishment (death penalty) by respondent sex in the 2020 ANES survey, categorized by strength of favor or opposition. Bars represent the proportion of respondents by sex in each category.

females holding each position, these quantities in the bars match the column percentages from the cross-tabulation.

Bar graph with error bars

For both bar graphs, we can plot the 95 percent confidence intervals as error bars on each one. We calculate the error bar quantities for each subset of sex, and then combine these quantities with the dataframe for `ggplot()`.

The first step is to calculate the standard error of level of support, relying on the `svymean()` function.

```
error_table1<-svymean(~V201345x, anes2020, na.rm=TRUE)
```

It is converted to a dataframe and attached as additional columns onto *table1* with `cbind()`:

```
error_table1<-as.data.frame(error_table1)
table1<-cbind(table1, error_table1)
```

The small dataframe *table1* now includes both the quantities for the bar heights and standard errors. To construct the 95 percent confidence interval, we multiple the standard error by the corresponding z score of 1.96, creating a new variable *error*:

```
table1<-table1 %>%
  mutate(error = (1.96*SE)*100)
```

The error is multiplied by 100, because the bars are scaled as percentages.

```
table1
```

```
##                 Response Percentage   mean       SE
## 1       1. Favor strongly      42.07 0.4206 0.010141
## 2   2. Favor not strongly      21.10 0.2110 0.007772
## 3 3. Oppose not strongly      19.35 0.1936 0.007045
## 4      4. Oppose strongly      17.48 0.1747 0.008413
##   error
## 1 1.988
## 2 1.523
## 3 1.381
## 4 1.649
```

The dataframe includes row names, which can be deleted with `row.names(table1)<-NULL`. The dataframe *table1* is fed into `ggplot()`. The geometry `geom_errorbar()`, specifying the minimum and maximum values of the error bars, specified as `aes(ymin = Percentage - error, ymax = Percentage + error)`. Figure 9.3 displays the bar plot with error bars marked in black.

```
ggplot(table1, aes(x = Response, y = Percentage)) +
geom_bar(stat = "identity") +
labs(x = "Death penalty attitude", y = "Percentage") +
geom_errorbar(aes(ymin = Percentage - error, ymax = Percentage + error),
      width = 0.25) +
  theme(legend.position = "bottom") +
  theme_minimal()
```

The length of the black lines along the bars marks the 95 percent confidence intervals, with the width (`width = 0.25`) a commonly used adornment to make the bars more noticeable.

For a bar plot displaying the cross-tabulation of attitude by sex, the steps are essentially the same. We already have *table2* with a label *Response* for attitudes

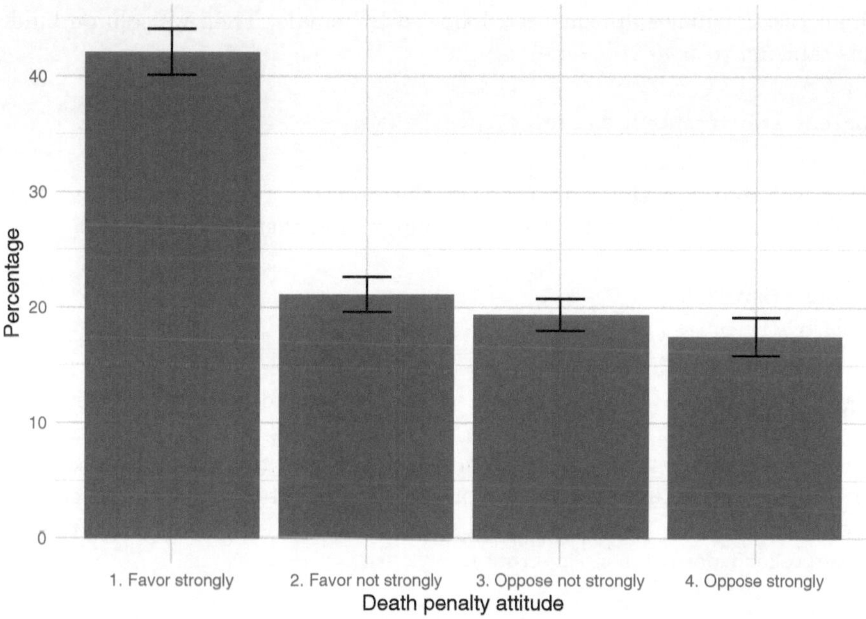

FIGURE 9.3 Attitude toward capital punishment (death penalty) in the 2020 ANES survey, categorized by strength of favor or opposition. Error bars measure the margin of error (95 percent confidence intervals) on each category.

toward the death penalty, *sex* and *Percentage*. To calculate the standard errors, we rely on `svymean()` with `subset()` for each level of sex.

```
table2_m<-svymean(~V201345x , subset(anes2020, sex=="1. Male"),
                  round=TRUE, na.rm=TRUE)*100

table2_f<-svymean(~V201345x , subset(anes2020, sex=="2. Female"),
                  round=TRUE, na.rm=TRUE)*100
```

We convert the two tables to dataframes.

```
table2_m<-as.data.frame(table2_m)
table2_f<-as.data.frame(table2_f)
```

The data table `table2` is arranged in long form; we row bind the `table2_m` and `table2_f` data tables and then column bind the result back to `table2`:

```
table2_m<- rbind(table2_m, table2_f)
```

Quantities for males appear first, followed by female. Then we column bind this table on to *table2*:

```
table2<-cbind(table2, table2_m)
```

Enter the name of the data table at the prompt (`table2`) to observe the combined tables. We create the *error* column, and then plot the result:

```
table2<-table2 %>%
  mutate(error = (1.96*SE)*100)
```

Figure 9.4 displays the two versions of the graphic, one reoriented to a horizontal bar chart, a popular variation on the presentation of bar charts, often used to make verbal labels more legible or to simply change the orientation so that the graph is wider and shorter vertically. The horizontal bars are arranged by adding `coord_flip()` and the legend is moved to the bottom of each graphic `theme(legend.position = "bottom")`.

```
ggplot(table2, aes(x = Response, y = Percentage, fill=sex)) +
  geom_bar(stat = "identity", position="dodge") +
  labs(x = "Death penalty attitude", y = "Percentage") +
  geom_errorbar(aes(ymin = Percentage - error,
                    ymax = Percentage + error),
      position = position_dodge(width = 0.8), width = 0.25) +
  coord_flip()+
  theme_minimal() +
  theme(legend.position = "bottom") +
  scale_fill_grey()
```

Note that for the error bars grouped by sex, the additional argument in `geom_errorbar()`, `position = position_dodge(width = 0.8)`, lines up the error bars with the 'dodged' or side-by-side bars.

Weighted histograms

In Chapter 4 visualizations of feeling thermometers ignored the survey design. We can apply a weight to the visualization as an argument to a geometry in `ggplot()`. For example, a weighted histogram of feelings toward 'Donald Trump' takes the observations and multiplies each one by the weight, adjusting the bar heights accordingly. The weight argument is simply `weight=V200015b`, as in Figure 9.5:

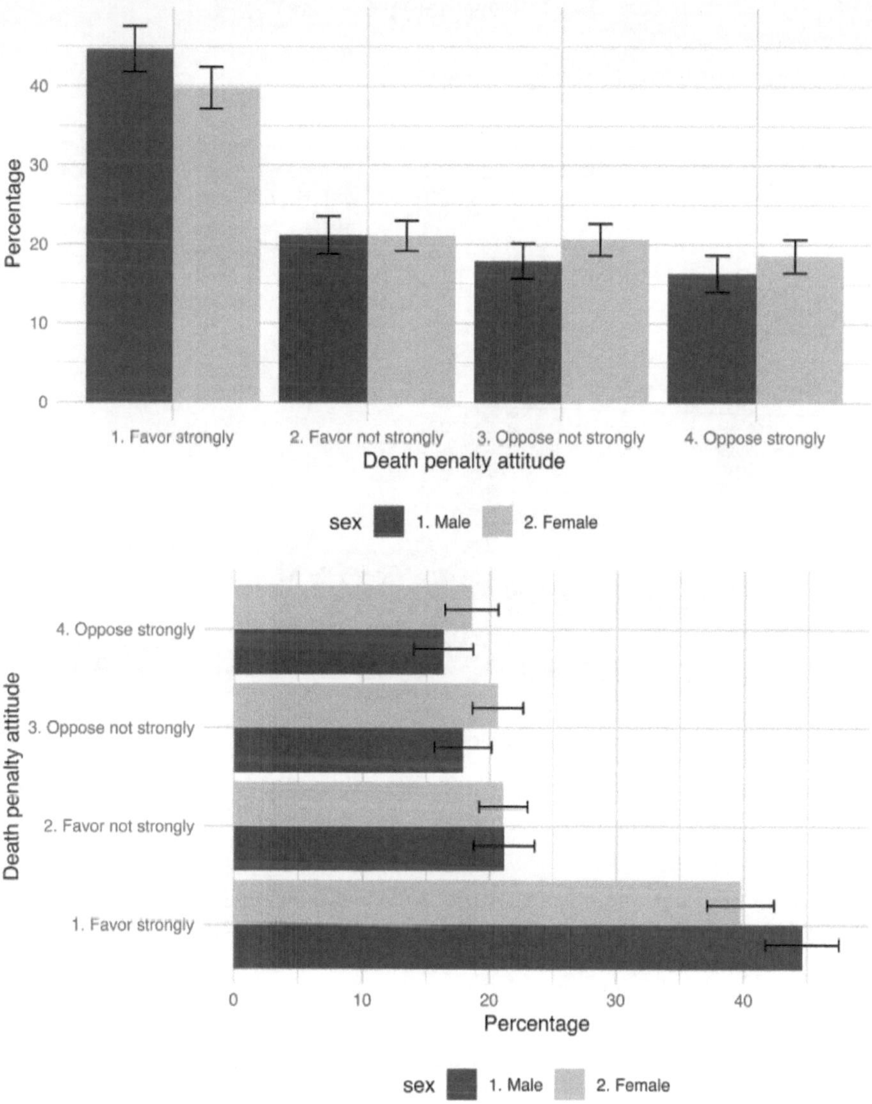

FIGURE 9.4 Attitude toward capital punishment (death penalty) by sex in the 2020 ANES survey, categorized by strength of favor or opposition. Error bars measure the margin of error (95 percent confidence intervals) on each category.

```
ggplot(anes_20) +
   geom_histogram(aes(x=V202179, y=after_stat(width*density)*100,
                      weight=V200015b), bins=10) +
   labs(x="feelings toward 'Donald Trump'",
       y="Percentage of respondents")
```

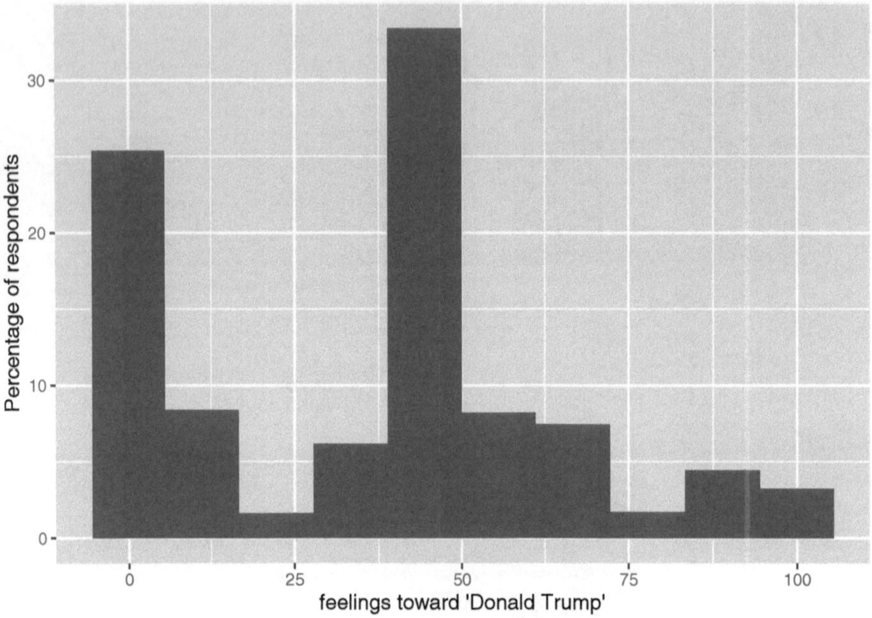

FIGURE 9.5 A weighted histogram of feelings toward Donald Trump, 2020 ANES.

The *survey* package includes a histogram visualization function `svyhist()`, along with other visualization types for survey data described by a `svydesign()` function. (See the help files for the package for details.)

Resources

The American Association for Public Opinion Research https://aapor.org/, the leading professional association for public opinion researchers, maintains primers for topics in survey data analysis and particularly election polling.

The American National Election Studies (ANES) https://electionstudies.org/ maintains the catalog of ANES data and other resources.

9.7 Exercises

Knit **R** code into a document to answer the following questions, based on the 2020 ANES data:

(1) Tabulate percentages and create a bar graph from a nominal or ordinal measure. In a paragraph, identify the question wording and response scale; explain any recoding or transformations of the responses in the variable and describe the results. Be sure to set the `svydesign()` and find the relevant proportions with `svytable()`.

(2) Create a crosstabulation between any nominal or ordinal survey measure and respondent age in groups — if necessary use the example in this chapter to create a variable that categorizes age. Identify the independent and dependent variables, write a hypothesis relating the two variables. In the table include frequencies and appropriate percentages, and describe the extent to which the table supports your hypothesis.

(3) Following the previous question, construct a set of control tables on sex. Controlling for this third measure, does the relationship still hold? How or how not?

(4) Create a histogram for a continuous measure across levels of a categorical variable such as sex or age group. Apply the weight to `ggplot()`, and record percentages rather than frequency counts on the appropriate axis. Interpret the graphic.

(5) Construct a crosstabulation of respondent gun ownership by party identification (Democrat, Republican, Independent), displaying frequencies and appropriate percentages. Locate the gun ownership variable, constructing a categorical measure of 0, 1, and 2 or more guns. In a paragraph answer the question, 'How much more likely are Republicans to own guns than Democrats?' Describe the table and include an inferential test of independence.

(6) Locate the respondent authoritarianism variable, and calculate 95 percent confidence intervals on mean levels of authoritarianism by educational attainment. Describe the results in a paragraph – is authoritarianism related to education? How so?

(7) Find the modern sexism, respondent age group, and sex variables. To what extent is there evidence of an age gap in attitudes toward sexism, and how do attitudes toward sexism vary by age? Does any relationship persist controlling for sex?

10

Linear Models

Chapter 10 introduces the estimation and interpretation of linear regression models.

Learning objectives and chapter resources

By the end of this chapter you should be able to: (1) explain the form of the simple and multiple linear regression model; (2) interpret model estimates; (3) interpret measures of model fit; (4) explain the data generating assumptions necessary for regression estimation and inference; (5) apply simple and multivariate regression to US presidential election fundamentals and mass party identification; and (6) construct tabular and graphical presentation of model estimates. The material in the chapter requires the following packages: **ggplot2** and **readr** from the **tidyverse** (Wickham, 2023b), **ggrepel** (Slowikowski, 2023), **survey** (Lumley, 2021), and two new packages requiring installation, **memisc** (Elff, 2023) and **jtools** (Long, 2023). The material uses the following datasets: *prez.csv* and *anes 2020 survey.rdata* from https://faculty.gvsu.edu/kilburnw/inpolr.html.

Introduction

Linear models are perhaps the most common methodological form of quantitative social science. We have already encountered a linear model in Chapter 1, an election forecast. The phrase 'linear models' often refers to a particular type of linear model, an "Ordinary Least Squares" (OLS) regression model , which is a fundamental statistical technique used to estimate the relationship between a Y (or dependent variable) and one or more X (or independent variables). The primary purpose of OLS regression is to identify the best-fitting linear model that explains how the independent variables relates to the dependent variable.

10.1 Simple linear model of election fundamentals

In Chapter 10 we review the fundamentals-based model of US presidential elections, starting with the incumbent president's share of the Democratic plus Republican popular vote as related to presidential approval. Then from there we examine the model in full.

Read in the **prez.csv** dataset:

```
# library(tidyverse)
prez<-read_csv(file="prez.csv")
```

This election fundamentals dataset consists of 19 observations of presidential elections from 1948 to 2020, displayed in Table 10.1. The table consists of 19 rows, one for each presidential election, and seven variables, characteristics of each election year. The column *year* records the election year. The column *incprez* records the name of the incumbent president. The column *vote* records the incumbent candidate's share of November popular vote for two major party candidates. The last three columns are measures of the three fundamentals: (1) *approval*: incumbent Gallup approval in late June or July; (2) *qgdprate*: quarterly growth rate in GDP for the second quarter of the election *year*; and (3) a binary indicator of incumbency status: 0, and 1 for a candidate running for their party's third term in the White House.

Let's look further at the example from the presidential election forecast by visualizing the scatterplot of the incumbent party's share of the two-party vote and presidential approval, presented in Figure 10.1. The code for the scatterplot includes a linear regression line ('line of of best fit') that relates the vote share to approval.

```
library(ggrepel)
ggplot(data=prez) +
  stat_smooth(method="lm", se=FALSE, aes(x=approval, y=vote),
              color="red") +
  labs(y="incumbent percentage share of two-party vote",
       x="Presidential approval, summer of election year") +
  geom_text_repel(aes(x=approval, y=vote, label=year), alpha=.6)  +
  geom_point(aes(x=approval, y=vote), alpha=.6, show.legend = FALSE) +
  scale_x_continuous(limits=c(20, 80)) +
  theme(legend.position="bottom")
```

At the center of OLS regression lies the concept of the "line of best fit," which is the linear equation that best describes the relationship between the

TABLE 10.1 Data for the model of presidential elections, 1948-2020. The column *vote* records the incumbent's percentage of the Democratic plus Republican candidate popular vote, *approval* the incumbent Gallup approval rating in late June or July of year, *qgdprate* the gross domestic product growth rate for the second quarter (Q2) of year, and *incstatus* an indicator 0 or 1 for incumbents running for a third party term in the White House.

year	incprez	vote	approval	qgdprate	incstatus
1948	truman	52.31	40	1.7	0
1952	truman	44.71	32	0.2	1
1956	eisenhower	57.40	73	0.8	0
1960	eisenhower	49.50	61	-0.5	1
1964	johnson	61.10	74	1.1	0
1968	johnson	43.56	40	1.7	1
1972	nixon	60.70	56	2.3	0
1976	ford	48.00	45	0.7	1
1980	carter	44.70	31	-2.1	0
1984	reagan	58.80	54	1.7	0
1988	reagan	53.40	48	1.3	1
1992	hwbush	46.54	38	1.1	1
1996	bclinton	54.74	52	1.7	0
2000	bclinton	50.26	55	1.8	1
2004	wbush	50.70	48	0.8	0
2008	wbush	45.70	28	0.5	1
2012	obama	51.10	46	0.4	0
2016	obama	48.20	50	0.3	1
2020	trump	46.86	38	-9.1	0

observations for independent and dependent variables. This line is drawn through the scatterplot in Figure 10.1. The equation of a line in slope-intercept form is $Y = a + bX$, where Y is the dependent variable, X is the independent variable, a is the intercept of the line on the Y-axis, and b is the slope of the line. The line equation in the figure is $Y = 34.78 + 0.338X$.

The slope (b) quantifies the change in the dependent variable for a one-unit change in the independent variable. The y-intercept, 34.78, is the linear prediction of the incumbent share of the vote when presidential approval is at 0, a level of approval not observed within the dataset and hypothetical. The slope .338 estimates how a one-unit (1 percent) change in approval changes the incumbent share of the vote, by .338 percent. For each one-unit change in approval percentage, on average, the incumbent's share of the two-party vote changes by about a third of a percentage point. So it takes about a three-point change in approval to move the incumbent vote share by one point.

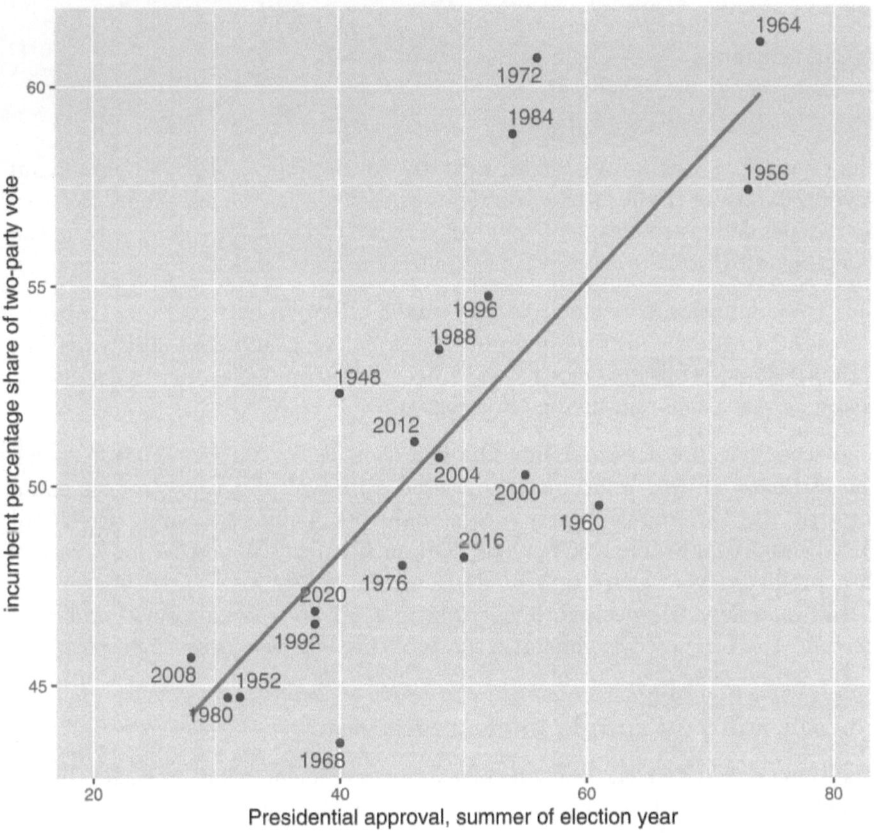

FIGURE 10.1 Linear trend relating incumbent party percent of the two-party popular vote (on the Y-axis) to presidential approval from June or July of each election year (on the X-axis).

In Figure 10.1, the OLS trend line is estimated and displayed through the `stat_smooth()` function, specifying a linear model (“`method="lm"`”) and the name of the variables on the X and Y axes. This function appears in the code as the line just below the `ggplot(data=prez)` line.

Given the data, apart from the figure the linear model estimates are estimated with the `lm()` function, specifying Y and X separated by a “tilde” character: `lm(vote ~ qgdprate, data=prez)`, the data source.

```
lm(vote ~ approval, data=prez)
```

```
##
## Call:
## lm(formula = vote ~ approval, data = prez)
```

```
##
## Coefficients:
## (Intercept)     approval
##      34.781        0.338
```

(The ~ symbol is usually located next to the number 1 key on USA layout keyboards). Here the `lm()` function requires that we list the dependent followed by independent variables, and an identification of the data source with `data=`. There are two broad purposes for estimating a linear model.

One is explanation. For example, the model offers an explanation – through the slope – of how close the relationship is between approval and vote. The estimate .338 tells us how much, on average, the incumbent vote percent should change given a one-unit change in approval.

The second is prediction. Using the equation, $Y = 34.78 + 0.338X$, given the estimated linear relationship between approval and the vote, we could forecast the Democratic party's expected vote share. Assuming President Biden's approval rating is 39 percent, then substituting 39 percent for X results in a predicted vote share of $Y = 34.781 + 0.338(39)$, or $Y = 47.96$ percent. Based on solely the average linear relationship between approval and vote percent, the predicted Democratic share of the vote would be 47.96 percent.

Correlation and simple linear regression

The slope and the linear Pearson correlation coefficient relate to each other in a specific way. The linear correlation, r, is a measure of the linear relationship between two variables. The correlation quantifies the strength and direction of the linear relationship between two interval or ratio scaled (numeric) measures, ranging from -1 to 1: a value of -1 indicates a perfect negative linear relationship (as one variable increases, the other decreases proportionally); a value of 0 means no linear relationship; a value of 1 signifies a perfect positive linear relationship (both variables increase or decrease together proportionally). Because variables can have different units or scales (income measured in dollars vs. age in years), the correlation standardizes these differences by converting the variables to z scores. Since the z scores expresses how far a value is from the mean in terms of standard deviations, it converts both variables to the same scale. Pearson's r correlation is defined in these terms

$$r = \frac{\sum z_x z_y}{N}$$

Here z_x and z_y are z scores for the two variables, and N is the number of paired observations. The correlation is the average product of the standardized values, providing a numerical summary of how the two variables move together. With sample data, the formula denominator is $N - 1$ to account for sample variability. The `cor()` function in R automatically assumes sample data:

```
sum(scale(prez$vote) * scale(prez$approval)) / (length(prez$vote) - 1)
```

```
## [1] 0.788
```

The result is equivalent to

```
cor(prez$vote, prez$approval)
```

```
## [1] 0.788
```

From the equation of the line $Y = a + bX$, the slope is related to Pearson's r (the linear correlation) through the formula:

$$b = r\frac{s_Y}{s_X}$$

in which s_Y and s_X are the standard deviations of Y and X. Since the standard deviation is always a positive quantity, b and r will always share the same sign — if Y tends to increase with X, b (and r) will be positive; if Y decreases as X increases, b (and r) will be negative.

$$b = r\frac{s_Y}{s_X} = .788 * \frac{5.471}{12.75} = .3381$$

This result matches the slope reported from the `lm()` function.

We can rearrange these terms to see that $r = b\frac{S_X}{S_Y}$. Pearson's r correlation is the slope multiplied by the ratio of the spread (standard deviation) in X (approval) relative to spread in the Y (vote percent). Substituting in terms,

$$r = .3381 * (12.75/5.471) = .788$$

Line of best fit

The OLS line is estimated by finding the line (out of all possible lines) that minimizes the sum of the squared differences between the observed (Y_i) and predicted ($\hat{Y}_i$) values of the dependent variable for each election year, $\Sigma(Y_i - \hat{Y}_i)^2$. Minimizing this sum of squared differences is known as the 'least squares criterion'. For any observation on Y, the difference between the observed and expected value on Y is known as a residual.

In the context of the model of election fundamentals, these residuals represent the error in the prediction of the model for each year's incumbent vote share. Figure 10.2 annotates the residuals for four observations. The two in the upper half of the figure are positive (for 1972 and 1984), while in the lower half they are negative (for 1960 and 1968). The residual is the distance along the dashed vertical line.

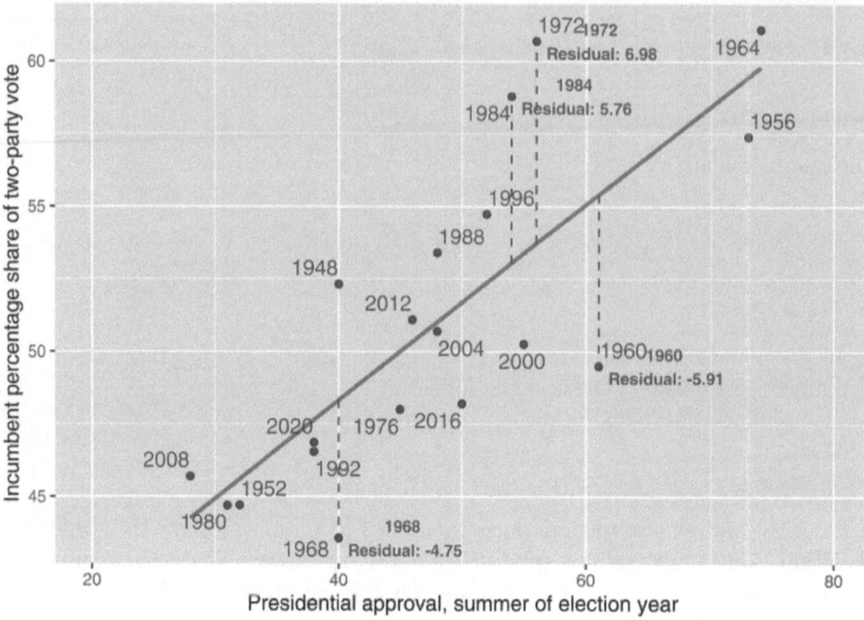

FIGURE 10.2 The dashed vertical lines mark the residuals along the Y-axis, the difference between the observed and expected. The residuals for 1984 and 1972 are positive, because the incumbent candidate's vote share over performed model expectations based on presidential approval. The residuals for 1968 and 1960 are negative, because the incumbent under-performed.

For Figure 10.2, finding the best fitting line means for each observation, calculating the difference between each observed value of the dependent variable and its corresponding predicted value (based on the regression line), squaring these differences to ensure that negative and positive deviations do not cancel each other out, and then summing these squared differences. This summed difference is the "sum of squared residuals", arithmetically $\Sigma(y_i - \hat{y}_i)^2$. Through OLS, the y-intercept a and slope b of the line are the values minimizing this sum. By doing so, OLS regression ensures that the resulting line of best fit is positioned in such a way that it minimizes the prediction error across all observations.

10.2 Understanding and evaluating model fit

After estimating the line, we want to evaluate how well it fits the data: How close are the predicted values to the actual observations on incumbent party support and GDP growth? The size of the residuals — the differences between the observed and predicted values for — provides a starting point to answer this question.

Root mean square error

Root Mean Square Error (RMSE) is a common measure of model fit. It quantifies the average magnitude of residuals, capturing how well the model's predictions match the observed data. RMSE is calculated as:

$$Root\ mean\ square\ error\ (RMSE) = \sqrt{\frac{\Sigma(y_i - \hat{y}_i)^2}{N}}$$

Here, y_i represents the observed values, while $\hat{y}_i$ the predicted values, and N the number of observations.

As a measure of fit, it reflects how well the model's predictions match the actual data. The RMSE finds the square root of the average squared residuals, capturing a measure of the size of the errors in the same units as the dependent variable. Lower RMSE values indicate better model fit, as the predictions are closer to the actual observations. However, RMSE is sensitive to outliers because squaring the residuals amplifies larger errors, which can disproportionately influence the result.

Within an **R** object that stores model results, there are additional pieces of information about the model, beyond the slope and y-intercept. We can take the `lm(vote ~ approval, data=prez)` and store the results as *model1*:

```
model1<-lm(vote ~ approval, data=prez)
```

The model residuals are stored within the object as `model1$residuals`. (Enter `names(model1)` to observe the other stored variables.) We could apply the `mean()` and `sqrt()` functions directly to `model1$residuals` or extract and save the residuals to a separate object and create an `rmse` object to save the RMSE:

```
residuals <- model1$residuals

rmse <- sqrt(mean(residuals^2))

rmse
```

```
## [1] 3.279
```

The result, 3.279, measures the average magnitude of the errors between the predicted and actual values of the incumbent percentage share of the vote. On average, the predictions made by the model are off by about 3.279 percentage points from the actual values.

Because the dependent variable is measured in percentages, the RMSE is relatively straightforward to interpret: each unit corresponds directly to a percentage point. Still, there is no definitive "cutoff" for what constitutes a good or bad fit. The quality of the fit is relative to the specific context of the data, the purpose of the analysis, and expectations about how well a model should perform, often in comparison to other models.

R-squared

Another metric for assessing model fit is based on the idea of reducing error. Imagine that we are interested in predicting a future vote based on the model's slope and y-intercept — basically forecasting a future vote based on the historical relationship between approval and vote. To evaluate the model's usefulness, we compare it to a baseline of using the mean vote percentage to predict the incumbent vote share in a future election. The extent to which the linear model improves upon this baseline, reducing error, forms the basis of a model fit statistic.

To construct the metric, we define the following quantities:

1. ESS (Error Sum of Squares), the sum of the squared residual errors (residuals), measures the total error in the mode's predictions. Arithmetically: ESS $= \Sigma(y_i - \hat{y}_i)^2$, where y_i are the observed y values, and $\hat{y}_i$ are the predicted values.
2. TSS (Total Sum of Squares), the total variation in the observed data, is calculated as the summed squared difference between each observed y and the mean. Arithmetically: TSS $= \Sigma(y_i - \bar{y})^2$.
3. RSS (Regression Model Sum of Squares) is the variation in the observed data explained by the regression model. It is calculated as the summed squared difference between the predicted value $\hat{y}_i$ and the mean $\bar{y}$. Arithmetically: RSS $= \Sigma(\hat{y}_i - \bar{y})^2$.

The RSS quantifies how much better the regression model's predictions are compared to simply using the mean of Y as a predictor. It measures how much additional information X provides about predicting Y beyond the mean of Y itself. The sum of ESS and RSS equals TSS, $ESS + RSS = TSS$, the total variation in Y splits into two parts: the explained variation RSS and unexplained variation ESS. The ESS is the error in this model prediction — the difference between the observed Y values and the predicted Y values. The ratio $\frac{RSS}{TSS}$ is the proportion of the total variation in Y accounted for by the predictor X.

This quantity, the $\frac{RSS}{TSS}$ is known as R^2 ("r-squared"), a measure of model fit.

$$R^2 = \frac{RSS}{TSS} = 1 - \frac{ESS}{TSS}$$

We can rewrite R^2 as $1 - ESS/TSS$. It is equal to the square of the Pearson's r correlation. Because it represents the proportion of the total variation in Y explained by the model (RSS) relative to the total variation in Y (TSS), R^2 ranges from 0 to 1. Values closer to 1 indicate that the model explains more of the variation in the outcome, reflecting a better fit, while values closer to 0 indicate a poorer fit. While we have already calculated Pearson's r, we will calculate the R^2 term by calculating the RSS and TSS terms, starting with the TSS.

The `lm()` function automatically records the fitted values, the $\hat{y}_i$, in the model object *model1*. While we have already extracted the residuals, the predicted values are stored as the `fitted.values` in *model1*.

The mean vote is `mean(prez$vote)`, or 50.96. To calculate the TSS, we first subtract this mean from the vote:`prez$vote-mean(prez$vote)`. Then we square this difference, `(prez$vote-mean(prez$vote))^2`. Finally, we sum up these squared differences.

```
total_squares<-(prez$vote-mean(prez$vote))^2
TSS<-sum(total_squares)

TSS
```

```
## [1] 538.8
```

Similarly, the RSS, regression sum of squares is equal to the `fitted.values` from *model1* minus the `mean(prez$vote)`, squared, then summed up:

```
RSS <- sum((model1$fitted.values - mean(prez$vote))^2)

RSS
```

```
## [1] 334.6
```

Since we already have the residuals calculated in the dataset, stored as a variable `residuals`, we can simply sum these squared differences:

```
ESS<-sum(prez$residuals^2)

ESS
```

```
## [1] 204.2
```

According to the formula, $R^2 = \frac{RSS}{TSS}T$.

```
RSS/TSS
```

```
## [1] 0.6209
```

The result, .6209, is the R^2 statistic. The statistic is stored within the `summary()` of the model object, and you can verify it directly by typing `summary(reg1)$r.sq` to see the computed value of RSS/TSS.

The R^2 can be interpreted from a few different perspectives. In simple linear regression R^2 is equivalent to the squared Pearson correlation (r^2) between the observed y_i values and the predicted $\hat{y}_i$ values. Both R^2 and r^2 measure the proportion of the variation in the dependent variable that can be explained by the independent variable(s).

The key interpretation is that R^2 represents the "proportion of variance explained": About 62 percent of the variance in the dependent variable (*vote*) is explained by the independent variable (*approval*). Since R^2 is a proportion between 0 and 1, an R^2 of .62 suggests that the model does a reasonably good job of accounting for variability in incumbent party vote share. Nonetheless, about 38 percent of the variance is not explained by the model.

Next, of course, the R^2 can be interpreted as a measure of model fit. Various thresholds are suggested for differences between a 'good', 'fair' or 'poor' fit. These thresholds should generally be used with caution, because R^2 is sensitive to the specification of the model and the context of the data. Still, from this perspective, the R^2 of .62 reflects a good fit of the model. This model's R^2 could be compared with another model of the same dependent variable and linear form, but different independent variable, to compare which model better explains the variance. Typically researchers rely on multiple measures to assess model fit (such as RMSE).

Residuals and fitted values

The residuals and fitted values are often plotted together. Beyond analyzing residuals for assessing the quality of model fit, residuals are important in assessing whether the regression model itself may be problematic — that is, whether it even makes sense to propose a linear relationship, and whether any of the various assumptions regarding how the underlying data were generated can be met by the model.

To plot residuals versus fitted values the quantities can be extracted and merged with the data for *ggplot2*, or plotted separately, appearing in Figure 10.3.

```
residuals <- model1$residuals
fitted_values <- model1$fitted.values

ggplot(data = prez, aes(x = fitted_values, y = residuals)) +
  geom_point(alpha = 0.6) +
  geom_hline(yintercept = 0, linetype = "dashed",
             color = "red") +
  labs(x = "Fitted Values", y = "Residuals") +
  theme_minimal()
```

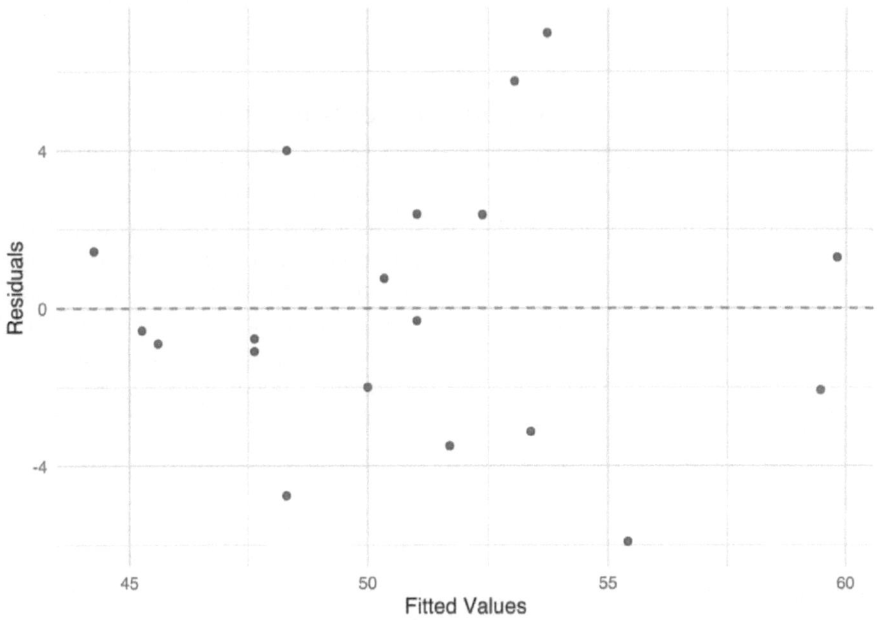

FIGURE 10.3 A plot of residuals versus fitted values from the simple linear regression of incumbent part vote share on presidential approval. While there is variation in residual size, there is no discernible pattern in the residuals suggesting nonlinearity, as was evident from the scatterplot of vote by approval.

For a simple linear regression model, the plot of residuals versus fitted values does not provide a great deal more information than the scatterplot of the original X and Y. However, inspecting residuals is an important part of regression analysis. As the number of variables and observations in the model increase, residuals are often useful in assessing potential problems. In any plot of residuals versus fitted values, there are two general trends to look for along the fitted values:

1) the mean of residuals (marked in this graphic by the horizontal line) should be roughly 0. If it is, then the assumption underlying the model that the relationship between X and Y is linear, holds.
2) the spread of the residuals should be roughly the same. This pattern is important, because if the spread differs, such as there being higher spread at higher ends of the fitted values, then the model fit is problematic.

Within Figure 10.3, both of these conditions appear to be met; what to do when the canditions are not is covered in Chapter 11.

10.3 Multiple linear model of election fundamentals

Through multiple linear regression , where "multiple" refers to the inclusion of more than one independent variable, we can address the research question of how the popular U.S. presidential vote varies with the "fundamentals" of elections. Previously, we considered only presidential approval, but the theory of fundamentals suggests that we must also account for economic growth and party term in office. By expanding the number of independent variables X from one to three: X_1 election year summer presidential approval; X_2 Q2 election year GDP growth; and X_3 whether the incumbent president's party has held the White House for one or more than one term — we assess the simultaneous influence of these factors on vote share.

A key advantage of multiple linear regression is that it allows us to isolate the effect of each variable while "controlling" for the influence of the others. In other words, this approach helps disentangle the unique contribution of each independent variable to the dependent variable (vote share), providing a clearer picture of how each factor relates to election outcomes in the presence of alternative explanations.

In a multiple regression model, the linear relationship is extended to account for how Y relates not just to a single X (such as presidential approval), but also to other factors simultaneously. With three independent variables, the form of the model becomes:

$$Y = a + b_1X_1 + b_2X_2 + b_3X_3$$

In the equation, a is still the y-intercept, and there there are additional slopes b_1, b_2, and b_3 for each independent variable X_1, X_2, and X_3. These slopes quantify how each X relates to Y holding the other X variables constant.

The OLS estimator still proceeds as usual, finding the minimized sum of squared residuals to estimate the y-intercept a and the slopes (b_1, b_2, and b_3).

By doing so, the model provides the best linear fit to the observed data while accounting for the simultaneous effects of all predictors.

```
lm(vote ~ approval + qgdprate + incstatus, data=prez)
```

```
##
## Call:
## lm(formula = vote ~ approval + qgdprate + incstatus, data = prez)
##
## Coefficients:
## (Intercept)     approval     qgdprate    incstatus
##      41.172        0.248        0.612       -4.843
```

The model estimates look familiar. There is an estimated y-intercept, followed by three slopes. The intercept a now measures the predicted vote share (41.17 percent) when all three of the X independent variables are equal to 0. It is a hypothetical vote share for an incumbent president with 0 percent approval, 0 percent growth in GDP, and not running for a third term for their party. Each slope b measures the expected change in Y (vote share) given a one-unit change in the corresponding X, assuming all other predictors remain constant. In other words, the slopes capture the unique contribution of each predictor while holding the others constant.

The slopes in a multiple regression are often called "partial regression coefficients," emphasizing that each coefficient represents the partial effect of its predictor on Y. This idea is often described using the Latin phrase *ceteris paribus* ("all else equal"). For example, all else equal, a 1 percentage point increase in approval is associated with a 0.24 percentage point increase in expected vote share. Another essential phrase to include in interpretations is "on average." Thus, all else equal, on average, a 1 percentage point increase in approval is associated with a 0.24 percentage point increase in vote share, and a 1 percentage point increase in GDP growth is associated with a 0.61 percentage point increase in vote share. Notice that the slope for approval is smaller in the multiple regression compared to the simple regression. Once we control for GDP growth and incumbency, the relationship between approval and vote share is attenuated, reflecting shared influence among the predictors.

There is a slight difference in interpretation between the slopes for *approval* and *gdprate* compared to *incstatus*. The variable *incstatus* is dichotomous, taking values of 1 (for third-term attempts) or 0 (for fewer than three terms). Slopes for dichotomous variables, like *incstatus*, represent changes in the y-intercept rather than incremental changes per unit. Specifically, the slope for *incstatus* reflects the difference in the predicted baseline vote share (y-intercept) between third-term and non-third-term incumbents. Here's how it works:

The *incstatus* variable creates two separate regression prediction equations, illustrating the shift in the y-intercept based on whether an incumbent party is running for a third term or not.

For non-third-term candidates, while the equation is

$$Y = a + b_1X_1 + b_2X_2 + b_3 * X_3$$

Since X_3=0 for these candidates, the term $b_3 * 0$ drops out, leaving:

$$Y = a + b_1X_1 + b_2X_2$$

Substituting in the y-intercept and slope for non-third term incumbents, the equation becomes:

$$Y = 41.172 + .248X_1 + .612X_2$$

For third term candidates, X_3 is 1, so the equation includes the term $b_3 * 1$:

$$Y = a + b_1X_1 + b_2X_2 + b_3 * 1$$

Substituting the values for the intercept and slopes gives:

$$Y = 41.172 + .248X_1 + .612X_2 - 4.843$$

This can be rearranged to emphasize the "time for a change" penalty applied to third-term candidates:

$$Y = (41.172 - 4.843) + .248X_1 + .612X_2$$

which simplifies to

$$Y = 36.33 + .248X_1 + .612X_2$$

In this equation for third-term candidates, the base vote share (y-intercept) is lower, at 36.33 compared to 41.172 for non-third-term candidates. This reflects the "time for a change" penalty, which results in a drop in predicted support for third-term incumbents.

For dichotomous variables such as *incstatus*, the slope b_3 represents the "bump" or "penalty" applied to the y-intercept depending on the value of the variable. In this case, it quantifies the decrease in baseline support (almost 5 percentage points) for third-term candidates. Tools for generating tables and figures, which are reviewed at the end of this chapter, can help visualize and clarify these interpretations in practical applications.

Overall, keep in mind two key points in interpreting multiple regression estimates:

The magnitude of the slopes in a multiple regression model does not necessarily indicate the "importance" or "strength" of one predictor compared to another. Each slope is measured in the scale of its corresponding independent variable, which often makes direct comparisons across predictors misleading. To make slopes more comparable, consider standardizing the predictors. One common approach is to scale predictors by dividing them by two standard deviations (See Gelman (2008) for an explanation), so that the slope captures movement through the predictor on a common metric. Another is to attempt, where feasible, to place predictors on a common scale, such as 0 to 1 to make comparisons more meaningful.

Second, the R^2 is still interpreted as the proportion of variance in the dependent variable that is explained by the independent variables. For example, in the following model:

```
model1<-lm(vote ~ approval + qgdprate + incstatus, data=prez)
summary(model1)$r.sq
```

```
## [1] 0.8162
```

The R^2 value indicates that 81% of the variation in vote share is explained by presidential approval, GDP growth, and incumbency status. However, keep in mind that R^2 will always increase as more predictors are added to the model, regardless of whether those predictors have a meaningful relationship with the dependent variable. As such, avoid focusing solely on maximizing R^2 or judging the quality of a model by the R^2 alone. Instead, use R^2 in combination with other metrics, such as adjusted R^2, and consider the theoretical relevance of your predictors when assessing model quality.

10.4 Population regression model

Regression modeling is most commonly applied to sample data, such as survey responses from a sample of voters, rather than population-level data like the complete record of U.S. presidential elections. In these applications of it, the goal is not only to estimate model parameters for the sample, but also to infer characteristics of the broader population. Achieving valid population-level inferences requires careful consideration of a set of assumptions about both the model and the underlying sample data.

These assumptions become apparent when we express the regression model with an explicit error term:

$$Y = a + B_1X_1 + ... + B_kX_k + e$$

Here, the error term e represents the unknowns — factors not captured by the model. The error could be due to omitted variables that influence Y, random variation in Y, or even errors in measuring Y. Since we typically lack access to complete population data, e accounts for the uncertainty inherent in modeling sample data.

To estimate the model and make valid inferences from this model's estimates from sample data, a series of assumptions must be satisfied about both the error term and the sample data. We review these assumptions below; this discussion is necessarily brief and conceptual; more in-depth explanations are found in the sources cited at the end of this chapter.

The first and most obvious assumption is simply linearity: the relationship between X and Y is linear throughout X. As a result, the expected value of Y, given X, is expressed as a linear model — a linear function of the predictors $Y = a + B_1X_1 + ... + B_kX_k + e$.

Second, there are no omitted independent variables. This assumption follows from the error term e being independent of and uncorrelated with the X, so any omitted variables (the factors not included in the model) do not bias the estimation of the slopes. If e were correlated with the predictors, then the model could not isolate the individual slopes for each X on Y.

Third, the errors have a mean of zero — whatever omitted sources of influence are pushing the error up, there are opposing forces pushing it down — the errors 'cancel out', so that we can model Y as a function of the y-intercept and slopes.

Fourth, the errors e are independent across observations. One observation's error does not influence another. This assumption is often referred to as the errors being "independently and identically distributed" ('i.i.d.'). Fifth, the variation in the errors is constant across the X, known as the assumption of "homoskedasticity". This assumption ensures that the variability in Y is the same regardless of the value of X.

In addition, we must assume that the errors e follow a normal distribution. This assumption is particularly important for inferring population-level parameters, as it supports the construction of hypothesis tests and confidence intervals. Violations of the constant error variance (homoscedasticity) assumption can also pose challenges for population inference, leading to incorrect standard errors and invalid test results. Fortunately, there are diagnostic tools and statistical tests that can help identify whether these assumptions hold or if adjustments are needed.

To explore these assumptions and the implications of the population regression model, we will analyze data from the 2020 ANES. Specifically, we will estimate a multiple regression model examining the demographic factors that explain an individual's party identification on a seven-point scale ranging from "Strong Democrat" at 1, to "Strong Republican" at 7.

Models and election surveys

In Chapter 9, we estimated survey statistics by accounting for differential response rates and unequal selection probabilities in the 2020 ANES. These statistics were calculated using weighted adjustments, to better reflect the population from which the survey was drawn. Additionally, when estimating population-level characteristics (parameters), the standard errors were adjusted to account for the survey sample design. These adjustments reflect the increased uncertainty in estimates due to the complex sample design compared to a simple random sample. We apply these same principles when estimating linear models.

To illustrate the basic concepts involved in interpreting regression model estimates in light of the population regression model, we develop a simple regression model. Specifically, we examine party identification as the dependent variable, modeled as a function of religious service attendance as the independent variable:

$$Y = a + B * attendance + e$$

When using party identification as a dependent variable in a regression model, we are implicitly assuming that the variable has at least an interval level of measurement. While party identification is measured on a seven-point ordinal scale, it is common in social science research to treat such scales as continuous for regression analysis. This approach aligns with one of the key assumptions of the population regression model: that the error term e is continuous and normally distributed. Treating the scale as numeric reflects this assumption.

Although we cannot directly observe population-level errors, we can inspect the residuals (the differences between observed and predicted values in the sample) for clues about whether this assumption is reasonable. Residual diagnostics will help us evaluate the appropriateness of treating the party identification scale as a continuous variable.

In this analysis, we will go beyond substantively interpreting the y-intercept and slope estimates to also assess whether there is sufficient evidence to reject the null hypothesis. Specifically, we will test whether the slope for religious service attendance is effectively zero in the population, indicating no relationship between attendance and party identification.

The dependent variable is the same seven-point party identification scale created in Chapter 9. The independent variable, religious service attendance, is a measure already included in the 2020 ANES survey data.

Linear models with the 2020 ANES

Load the **R** Workspace *anes 2020 survey data.rdata*:

```
load(file="anes 2020 survey data.rdata")
```

The use of "analytic weights" (also referred to as "post-stratification weights") from the ANES in a linear regression model is conceptually similar to their use in calculating a weighted mean. Without weights, all observations in the regression model are implicitly given equal weight, effectively treating each observation as if it has the same importance (weight = 1). However, with analytic weights, the regression model adjusts to give some observations more or less influence, based on their corresponding weight values. This approach is known as "weighted least squares" (WLS). For a more detailed discussion of WLS in **R**, see Fox and Weisberg (2019).

In **R**, applying a weighted least squares estimator is straightforward: simply include the `weights=` argument in the `lm()` function. In this analysis, we use the same weighting variable (*V200015b*) applied in Chapter 9. This variable accounts for differential probabilities of selection and nonresponse in the ANES survey, ensuring that the regression model better reflects the underlying population.

Before estimating the model, we need to ensure that both the party identification and religious service attendance variables are stored as numeric variables in *anes_20* rather than as factors. The `lm()` function requires the dependent variable to be numeric for a linear regression. Additionally, the attendance variable should be numeric to estimate a single slope reflecting the change in party identification for a one-unit increase in attendance, rather than treating it as a categorical variable. We create these two numeric measures:

```
anes_20 <- anes_20 %>%
  mutate(partyid7pt = fct_recode(V201231x,
          NULL = "-9. Refused",
          NULL = "-8. Don't know"))

anes_20 <- anes_20 %>%
  mutate(partyid_num = as.numeric(partyid7pt)) %>%
  mutate(attend_num = as.numeric(attend))
```

By default, the `lm()` function treats factor variables as categorical measures, converting them into a series of dichotomous variables (0 or 1) for each category. This model setup is similar to how we modeled *incstatus* in the election forecast model. Further details on the use of factors for categorical or ordinal measures in regression models are explained later in this chapter.

We begin by estimating the regression model using the `lm()` function:

```
regmodel<-lm(partyid_num~ attend_num, weight=V200015b, data=anes_20)
```

Here, we include the `weight=` argument to apply the weight variable (*V200015b*), to the observations. Saving the model estimates in the object *regmodel* allows us to store the results and inspect the evidence from the sample toward the population level model. Typing the name of the object reveals the slope and y-intercept:

```
regmodel
```

```
##
## Call:
## lm(formula = partyid_num ~ attend_num, data = anes_20, weights = V200015b)
##
## Coefficients:
## (Intercept)   attend_num
##       3.213        0.388
```

For a more detailed view of the regression results, including model fit and statistical tests for inference about the population regression model, we use the `summary()` function:

```
summary(regmodel)
```

```
##
## Call:
## lm(formula = partyid_num ~ attend_num, data = anes_20, weights = V200015b)
##
## Weighted Residuals:
##    Min     1Q Median     3Q    Max
## -8.772 -1.592  0.008  1.567  8.487
##
## Coefficients:
##             Estimate Std. Error t value Pr(>|t|)
## (Intercept)   3.2127     0.0615    52.2   <2e-16 ***
## attend_num    0.3877     0.0286    13.6   <2e-16 ***
## ---
## Signif. codes:
## 0 '***' 0.001 '**' 0.01 '*' 0.05 '.' 0.1 ' ' 1
##
## Residual standard error: 2.16 on 4723 degrees of freedom
##   (58 observations deleted due to missingness)
## Multiple R-squared:  0.0376, Adjusted R-squared:  0.0374
## F-statistic:  184 on 1 and 4723 DF,  p-value: <2e-16
```

The summary output includes detailed information about the model, such as the estimated coefficients, standard errors, R^2, and significance tests. To compare the impact of weights, rerun the model with `lm()` but without the `weight=` argument and compare the unweighted and weighted results. The `summary()` function provides a detailed overview of the regression model results. Here are the key components:

`Weighted Residuals:` This section presents summary statistics on the residual errors, adjusted for the analysis weights. The residuals indicate how far the observed values deviate from the predicted values. A median close to zero and symmetry in the minimum, quartiles, and maximum values suggest that the model fits the data reasonably well.

`Coefficients:` This section should be read row-wise. Each row provides information about one of the regression coefficients: The first row describes the y-intercept `(Intercept)`, representing the predicted value of the dependent variable when the independent variable is zero. The second row corresponds to the independent variable *attend_num* detailing the estimated slope.

The column labeled `Estimate` display the estimated coefficients: the y-intercept and slope for attendance. Of course, these quantities are based on a survey sample and are subject to sampling variability. The `Std. Error` column quantifies this variability, indicating how much the coefficient estimates are expected to vary across different samples.

`Signif. codes:` This section reports the p-values associated with hypothesis tests for the regression coefficients. The null hypothesis for each test is that the coefficient equals zero (no relationship). The significance codes (e.g., ***) categorize p-values into ranges: For the row `attend_num`, the three asterisks *** indicate that the p-value is less than .001. This p-value means there is less than a 0.1 percent probability of observing a slope at least as large as 0.3877 in magnitude if the null hypothesis of no relationship were true.

Using the conventional threshold of $p < .05$ for statistical significance, we reject the null hypothesis and conclude that religious service attendance is significantly associated with party identification. Keep in mind the p-value represents the probability of obtaining a test statistic at least as extreme as the one observed, assuming the null hypothesis is true. It is not a measure of the effect size or practical significance but rather the strength of evidence against the null hypothesis.

Finally, at the bottom of the `summary()` model fit and significance statistics are presented: The `Residual standard error` (2.16) reports the average size of the residuals, measured in the same units as the dependent variable. The `Multiple R-squared` is the R^2 statistic, 0.0376 indicating that about 3.76 percent of the total variation in party identification is explained by religious service attendance. The `Adjusted R-squared` (0.0374) slightly adjusts this value to account for the number of predictors in the model, although with only one independent variable, the adjustment is minimal. Finally, the `F-statistic` tests whether the model as a whole explains a significant amount of variation in the dependent variable. With a p-value of <2e-16, we reject the null hypothesis that the model provides no explanatory power, concluding that religious service attendance is significantly associated with party identification, though the overall explanatory power is modest.

In the section that follows, we review the concept of the standard error of the regression slope.

Measuring uncertainty of the population parameters

The standard error (SE) of a regression slope estimate quantifies how much the estimated slope would vary if we repeatedly drew different samples from the same population and computed the slope for each sample. The same concept applies to the y-intercept. A smaller standard error indicates greater precision in the estimates, meaning that repeated ANES sampling would yield slope estimates close to the one observed in *regmodel*. Conversely, a larger standard error suggests more variability in the slope estimates across different samples.

In a simple linear regression without clustering, stratification, or analytic weights — such as in a survey unlike the ANES — the standard error of the slope is given by the formula:

$$SE_b = \frac{s}{\sqrt{\sum (x_i - \bar{x})^2}}$$

in which s is the adjusted standard deviation of the residuals, $s = \sqrt{\frac{ESS}{n-2}}$. In the formula, ESS is the sum of squared residuals. It is divided by $n - 2$ to accurately account for the 'degrees of freedom' in the model the number of observations n being penalized by 2 for the two parameters (y intercept and slope) estimated in the model. In the denominator, x_i is the value of the X independent variable for observation i; and $\bar{x}$ is the mean of the X independent variable.

Degrees of freedom represent the number of observations free to vary after accounting for the estimated parameters. Like the concept discussed in the χ^2 test of independence in Chapter 9, the concept here refers to the number of freely varying pieces of information (observations) free to vary. In the regression model, each parameter estimated reduces the degrees of freedom by 1.

From the formula, the size of the standard error depends on three main factors:

1) Smaller residuals s. Less variability around the regression line results in a smaller standard error, improving the precision of the slope estimate.
2) Spread of the independent variable $\sum(x_i - \bar{x})$. A wider spread of values in X increases the denominator, decreasing the standard error.
3) Sample size n. Larger sample sizes reduce the residual variability s reducing the standard error.

While accounting for analytic weights and the complex survey design in the ANES complicates this formula, the basic idea remains the same. Weighted least squares and survey-specific adjustments alter both the numerator and denominator, but the fundamental idea—that greater variability in residuals or smaller sample sizes increases uncertainty—still holds.

The next entry in the `Coefficients:` section of the model summary is the t value from Student's t distribution. This statistic is used to test the null hypothesis (no relationship in the population):

$$H_0 : B = 0$$

against the alternative:

$$H_a : B \neq 0$$

The `t value` is calculated as the ratio of the sample slope b to its standard error:

$$t = \frac{b}{SE}$$

For example, in this model the slope $b = .3877$ divided by the standard error $SE = .0286$ yields:

$$t = \frac{.3877}{.0286} = 13.6$$

The corresponding p-value represents the probability of observing a t-statistic as extreme as 13.6 under the null hypothesis. With $p < .05$ as the standard, we reject the null hypothesis, concluding that there is a significant relationship between religious service attendance and party identification.

For large sample sizes (large degrees of freedom), the t-distribution closely approximates the normal z-distribution. This relationship simplifies hypothesis testing in large samples.

Accounting for the sample design in the ANES survey

While the `weights=` argument in the `lm()` function adjusts the regression model to account for the observation weights, it does not correct the standard errors for the complex survey design. Even with `weights=` applied to the model, the standard errors are calculated as if the data were obtained from a simple random sample, ignoring the strata and clustering present in the ANES sample design. The complex survey design increases uncertainty, and failing to account for it can lead to underestimated standard errors and overly optimistic inference.

To account for the ANES survey design in a linear regression, we rely on the **survey** package and the equivalent function for estimating a regression model, `svyglm()`. As in Chapter 9, we first set the survey design:

```
anes2020_design<-svydesign(data = anes_20, ids = ~V200015c,
                           strata=~V200015d, weights=~V200015b, nest=TRUE)
```

The syntax for specifying the regression model in `svyglm()` is similar to `lm()`: the dependent variable followed by the tilde ~ with independent variables separated by plus signs. Before proceeding to the multiple regression model, we review the output for a simple linear model of party identification as a function of service attendance.

After setting the survey design, in the next chunk of **R** code, we add the `design=anes2020_design` argument to the `svyglm()` function. The model is estimated and saved as `regmodel2`:

```
regmodel2<-svyglm(as.numeric(partyid7pt) ~ attend_num,
                  design=anes2020_design)

summary(regmodel2)
```

```
##
## Call:
## svyglm(formula = as.numeric(partyid7pt) ~ attend_num,
## design = anes2020_design)
##
## Survey design:
## svydesign(data = anes_20, ids = ~V200015c, strata = ~V200015d,
##     weights = ~V200015b, nest = TRUE)
##
## Coefficients:
##             Estimate Std. Error t value Pr(>|t|)
## (Intercept)   3.2127     0.0730    44.0  < 2e-16 ***
## attend_num    0.3877     0.0349    11.1  5.2e-15 ***
## ---
```

```
## Signif. codes:
## 0 '***' 0.001 '**' 0.01 '*' 0.05 '.' 0.1 ' ' 1
##
## (Dispersion parameter for gaussian family taken to be 4.677)
##
## Number of Fisher Scoring iterations: 2
```

The output of `summary(regmodel2)` for *regmodel2* is similar in format to the output from `lm()`. Notice the `Estimate` values for the y-intercept and slope remain the same as in *regmodel*, reflecting the weighted least squares estimates. The standard errors, however, under the `Std.Error` heading are larger, as they now account for the survey design.

While the p values remain approximately 0 in this example — leading to the same substantive conclusions about the relationship between attendance and party identification — the adjustment for the survey design can meaningfully impact standard errors and p-values in other models. In such cases, failing to account for the survey design might lead to incorrect conclusions from hypothesis tests on the slopes.

Numeric versus ordinal or categorical measures

In *regmodel2*, the religious service attendance variable was entered into the model as a numeric measure. This approach assumes that the variable is at least interval in scale, allowing the model to estimate a single slope. The slope captures the expected linear change in party identification with each one-unit increase in service attendance. For instance, the model assumes that the change from `1. Never` to `2. few times a year, month` is equivalent to the change from `3. almost every week` to `4. every week`. This linear assumption is reflected in the model's treatment of the variable as continuous.

Alternatively, we can treat the independent variable as a categorical (factor) variable. This approach is particularly useful when the variable's scaling is ordinal or nominal, as it allows the model to estimate distinct slopes for each category relative to a reference category. When an independent variable is stored as a factor, both `lm()` and `svyglm()` automatically create a series of dichotomous (0 or 1) variables for each level of the factor, leaving one level out as the reference category.

For example, the religious service attendance variable *attend* has four levels. When entered as a factor, the model creates three dichotomous variables, each representing whether or not an observation falls into one of the categories. One dichotomous variable could indicate whether someone attends services `4. Every week`. Those who attend weekly are scored as 1, while everyone else is scored as 0. This process repeats for the other categories, excluding the reference category, `1. Never`, which serves as the baseline for comparison.

To confirm the storage type of the variable, check its structure with `str(anes_20$attend)`, which would confirm that *attend* is a `Factor w/ 4 levels` as expected. When we include this variable as a factor in the regression model:

```
regmodel3<-svyglm(partyid_num ~ attend,
                  design=anes2020_design)

regmodel3$coefficients
```

```
##                         (Intercept)
##                              3.5878
## attend2. few times a year, month
##                              0.4050
##       attend3. almost every week
##                              0.9612
##              attend4. every week
##                              1.0879
```

The coefficient output (type `summary(regmodel3)` for the full output) lists three dichotomous variables corresponding to the levels of attend, excluding the reference category, `1. Never`. The magnitude of each coefficient reflects the relative change in party identification compared to the baseline category. For instance, attending services "A few times a year or monthly" increases party identification by approximately 0.4 points on the scale toward stronger Republican identification compared to "never" attending. The relative association is about twice as large for "almost weekly" attendees, but there is little difference between "almost weekly" and "every week" attendees, as their coefficients are similar in magnitude.

Whether to treat an ordinal or nominal variable as numeric or as a factor depends on the theoretical purpose of the variable in the model. As a control variable, if it is included purely for statistical control and the specific category level differences are not of theoretical interest, then it is sufficient to treat it as numeric. As a focal variable, however, if the goal is to explore the relative differences between categories or if it makes little sense to assume one constant slope across the range of categories, it should be treated as a factor. Consider which category should be the baseline. Use the `relevel()` function to set a different baseline; to set the baseline as `4. evey week`, use `mutate(attend=relevel(attend, ref="4. every week"))`.

10.5 Multiple linear model of party identification

Next, we will build a multivariate regression model to explore the relationship between various demographic characteristics and party identification. Using multiple regression allows us to examine the unique association of each independent variable with party identification while controlling for the influence of other factors. This approach is particularly useful for understanding how specific characteristics, like educational attainment, interact with political preferences.

For example, a researcher might hypothesize that formal educational attainment has a strong relationship with party identification in the United States, even after accounting for other demographic factors. In this model, educational attainment will serve as the focal variable, while other independent variables are included to control for alternative sources of influence.

The model includes the following independent variables:

edu: five-point scale of educational attainment, treated as a factor with "High school credential" as the baseline reference category. *sex*: dichotomous coded as 0 for male and 1 for female. *race_category*: dichotomous coded as 0 for White, non-Hispanic, and 1 for all other racial or ethnic groups. *age*: continuous age in years. *attend_num*: four-point numeric service attendance scale *never_married* : dichotomous coded as 1 for "Never married" and 0 for any other marital status.

The following code prepares the variables for the regression model. For each variable, it removes missing or invalid responses, recodes categorical variables into the appropriate levels, and creates new variables with mnemonic names for clarity:

```
anes_20<- anes_20 %>%
  mutate(sex = fct_recode(V201600,
                          NULL = "-9. Refused"))

anes_20<-anes_20 %>%
    mutate(race_category=fct_recode(V201549x,
              NULL="-9. Refused",
              NULL="-8. Don't know"))
anes_20<-anes_20 %>%
  mutate(race_category = ifelse(race_category ==
                          '1. White, non-Hispanic', 0, 1))

anes_20 <- anes_20 %>%
```

```
  mutate(age =  V201507x)%>%
  mutate(age = na_if(age, -9))

anes_20<-anes_20 %>%
  mutate(edu = V201511x) %>%
  mutate(edu = fct_recode(edu,
         NULL = "-9. Refused",
         NULL = "-8. Don't know"))

anes_20<-anes_20 %>%
  mutate(never_married = ifelse(V201508== '6. Never married', 1, 0))
```

To measure the association between educational attainment and party identification, with slopes expressed relative to obtaining a high school credential, we reorder the levels of the *edu* variable so that `2. High school credential` serves as the reference category:

```
anes_20 <- anes_20 %>%
  mutate(edu = relevel(edu, ref = "2. High school credential"))
```

Next, we shorten the labels for the levels of the *edu* variable to make the output more readable:

```
anes_20 <- anes_20 %>%
 mutate(edu = fct_recode(edu,
  "1. Less than HS" = "1. Less than high school credential",
  "2. HS cred." = "2. High school credential",
  "3. Some post HS" = "3. Some post-high school, no bachelor's degree",
  "4. Bachelor's degree" = "4. Bachelor's degree",
  "5. Graduate degree"  = "5. Graduate degree"))
```

After creating and modifying the variables, we reset the survey design to ensure consistency with the updated dataset:

```
anes2020_design<-svydesign(data = anes_20, ids = ~V200015c,
                    strata=~V200015d, weights=~V200015b, nest=TRUE)
```

Then in the model, we enter the multiple predictors, so the multiple regression estimates the unique association of each variable with party identification, controlling for the others.

```
regmodel4<-svyglm(partyid_num ~  edu + sex +
                  race_category + age + attend_num +
                  never_married, design=anes2020_design)

summary(regmodel4)
```

```
##
## Call:
## svyglm(formula = partyid_num ~ edu + sex + race_category + age +
##     attend_num + never_married, design = anes2020_design)
##
## Survey design:
## svydesign(data = anes_20, ids = ~V200015c, strata = ~V200015d,
##     weights = ~V200015b, nest = TRUE)
##
## Coefficients:
##                         Estimate Std. Error t value
## (Intercept)              4.82090    0.16524   29.18
## edu1. Less than HS      -0.03507    0.17483   -0.20
## edu3. Some post HS      -0.01764    0.11531   -0.15
## edu4. Bachelor's degree -0.42678    0.11077   -3.85
## edu5. Graduate degree   -0.95418    0.12858   -7.42
## sex2. Female            -0.38690    0.06697   -5.78
## race_category           -1.39717    0.09976  -14.01
## age                     -0.01182    0.00261   -4.52
## attend_num               0.39982    0.03046   13.12
## never_married           -0.60429    0.09479   -6.38
##                         Pr(>|t|)
## (Intercept)              < 2e-16 ***
## edu1. Less than HS        0.8420
## edu3. Some post HS        0.8792
## edu4. Bachelor's degree   0.0004 ***
## edu5. Graduate degree    4.2e-09 ***
## sex2. Female             9.0e-07 ***
## race_category            < 2e-16 ***
## age                      5.2e-05 ***
## attend_num               2.9e-16 ***
## never_married            1.3e-07 ***
## ---
## Signif. codes:
## 0 '***' 0.001 '**' 0.01 '*' 0.05 '.' 0.1 ' ' 1
##
## (Dispersion parameter for gaussian family taken to be 4.116)
##
```

```
## Number of Fisher Scoring iterations: 2
```

A basic interpretation of the slopes, from the summary output, starts with recognizing that the slopes are the estimated association or effect of each predictor on party identification, measured on the seven-point scale, while holding all other predictors constant.

Starting with the y intercept `(Intercept)`, the estimate 4.82 represents the predicted party identification score (leaning Republican) for a strictly hypothetical person, male, White non-Hispanic, very low formal education, never attending services, and married or previously married.

The coefficients for education levels are interpreted relative to the baseline category, `2. High school credential`. Holding other variables constant, having a bachelor's degree (-0.43) or a graduate degree (-0.95) is associated with a shift toward more Democratic identification, compared to having only a high school credential. Both effects are statistically significant ($p < .001$). Interestingly, other education levels (`Less than HS`, `Some post HS`) do not show significant differences compared to the baseline.

Sex, race, and marital status show significant associations. Female respondents (–0.39) tend to identify slightly more Democratic compared to male respondents ($p < .001$). Non-white or Hispanic respondents (–1.40) identify significantly more Democratic than White non-Hispanic respondents, as do single, never married respondents (-0/60). There is a slight, statistically significant association between age and stronger Democratic identification.

Increased religious service attendance (-0.40) persists as a statistically significant variable, associated with a stronger Republican party identification.

The `svyglm()` function does not report traditional fit statistics like R^2, because it is designed for a broad range of complex survey designs, where such metrics may not directly apply. A pseudo-R^2 can be approximated using an `lm()` model with the `weights=` argument. The dispersion parameter reported in the svyglm() output (c.g., 4.116) represents the estimated variance of the residuals, and its square root provides the standard error of the residuals.

Confidence intervals on model estimates

The `confint()` function calculates confidence intervals for the slope estimates, defaulting to 95 percent intervals:

```
confint(regmodel4)
```

```
##                           2.5 %    97.5 %
## (Intercept)              4.4872  5.154604
## edu1. Less than HS      -0.3881  0.318016
```

```
## edu3. Some post HS        -0.2505  0.215233
## edu4. Bachelor's degree -0.6505 -0.203067
## edu5. Graduate degree   -1.2139 -0.694509
## sex2. Female            -0.5222 -0.251644
## race_category           -1.5986 -1.195704
## age                     -0.0171 -0.006537
## attend_num               0.3383  0.461339
## never_married           -0.7957 -0.412867
```

These intervals provide a range within which we can be 95% confident that the true value of each coefficient (in the population) lies. Confidence intervals that do not include zero indicate that the corresponding predictor is statistically significant ($p < 0.05$).

Confidence intervals are calculated manually using the estimated slope, its standard error, and the critical value for a 95% confidence level (ś1.96 for a standard normal distribution). For example, the confidence interval for the slope on *never_married* is calculated as, for the lower bound $-0.60429-(1.96*0.09479) = -0.7901$ and $-0.60429 + (1.96 * 0.09479) = -0.4185$ for the upper. Thus, the 95% confidence interval is [-0.7901, -0.4185]. This interval confirms that the "bump" associated with being never married moves the individual toward stronger Democratic identification, as the entire interval lies below zero.

10.6 Model-based predictions

Using the regression model, we can generate predictions of party identification scores (ranging from 1 "Strong Democrat" to 7 "Strong Republican") for individuals with given characteristics. Since the model is based on sample data, there are two types of intervals to account for uncertainty in these predictions:

Confidence Intervals: These estimate the range within which the average party identification score for individuals with specific characteristics is likely to fall. For example, we might say, "We are 95 percent confident that the average party identification score for individuals with these characteristics lies between these values."

Prediction Intervals: These estimate the range within which the party identification score for a single individual with given characteristics is likely to fall. Because this prediction applies to a specific person, the interval is wider to reflect greater uncertainty.

In most cases, we are more interested in confidence intervals, as they provide insight into the typical or average party identification for groups (e.g., "What is the average party identification for women with a college degree?").

Both intervals can be generated in base **R** using the `predict()` function, specifying either `interval="confidence"` for confidence intervals or `interval="prediction"` for prediction intervals.

To generate predictions from the model, we first create a dataset containing the desired values for the independent variables. The `predict()` function requires a data table, one column per independent variable, the first row reserved for variable names.

For this example, we use the `tribble()` function from the **tidyverse** to create a small prediction dataset. The model *regmodel4* includes the variables *edu*, *sex*, *race_category*, *age*, *attend_num*, and *never_married*. Unless specific values are of interest, a common approach is to use the mean for continuous variables and select representative categories for categorical variables.

We estimate the average party identification for a White, non-Hispanic woman who is never married, holds a high school diploma, and attends religious services at the average rate. Using survey-weighted means for continuous variables: The mean age is 47.5 `svymean(~age, design=anes2020_design, na.rm=TRUE)`. The mean religious service attendance is 0.853 (on a scale of 0 to 3) `svymean(~attend_num, design=anes2020_design, na.rm=TRUE)`.

Note in the prediction dataset, the names of each variable in the first row are prefixed with ~.

```
predictiondata <- tribble(
  ~edu, ~sex, ~race_category, ~age, ~attend_num, ~never_married,
  "2. High school credential", "2. Female", 0, 47.5, 0.853, 0)


predictiondata
```

```
## # A tibble: 1 x 6
##   edu                     sex    race_~1   age atten~2 never~3
##   <chr>                   <chr>    <dbl> <dbl>   <dbl>   <dbl>
## 1 2. High school c~ 2. F~        0  47.5   0.853       0
## # ... with abbreviated variable names
## #   1: race_category, 2: attend_num, 3: never_married
```

Additional rows of prediction values can be added to *predictiondata*.

The arguments of `predict()` are the model for the basis of the prediction, the input data, and the type.

```
predict(regmodel4, predictiondata, interval="confidence")
```

The prediction appears under the heading `link`, which refers to the function used in the `svyglm()` function.

The prediction appears under `link`, which refers to the function used in the `svyglm()` model. The average predicted party identification score is 4.21, slightly toward the Republican side of the seven-point scale. The standard error (0.10) quantifies the uncertainty of the predicted mean.

To compute a 95 percent confidence interval for the prediction, store the results in an object and apply the `confint()` function:

```
pred<-predict(regmodel4, predictiondata, interval="confidence")
confint(pred)
```

```
##    2.5 % 97.5 %
## 1 4.011  4.416
```

The confidence interval indicates that, for an average person with the given characteristics, the true population mean is 95 percent likely to lie between 4.011 and 4.416 on the party identification scale, slightly leaning Republican.

10.7 Describing model estimates in tables and graphs

Standard practice for presenting regression results involves formatting tables neatly to display slopes, statistical significance, and relevant fit statistics, while limiting the number of significant digits. The default `summary()` output is functional but not ideal for professional, legible tables. Packages such as **memisc** (Elff, 2023) improve the appearance of tables and export results to a Word or PDF document. Install the package if necessary, `install.packages("memisc")`. And attach it:

```
library(memisc)
```

The `mtable()` function creates a formatted table of model results. For example, to present the output of *regmodel4*:

```
regtable4<-mtable("Model of party identification" = regmodel4,
                  sdigits=2)
```

In an RMarkdown document rendered to Word, print the table directly, `print(regtable4)` or simply enter the name `regtable4`.

```
print(regtable4)
```

```
##
## Calls:
## Model of party identification: [1] ""
##
## ========================================================
##   (Intercept)                              4.821***
##                                           (0.165)
##   edu: 1. Less than HS/2. HS cred.        -0.035
##                                           (0.175)
##   edu: 3. Some post HS/2. HS cred.        -0.018
##                                           (0.115)
##   edu: 4. Bachelor's degree/2. HS cred.   -0.427***
##                                           (0.111)
##   edu: 5. Graduate degree/2. HS cred.     -0.954***
##                                           (0.129)
##   sex: 2. Female/1. Male                  -0.387***
##                                           (0.067)
##   race_category                           -1.397***
##                                           (0.100)
##   age                                     -0.012***
##                                           (0.003)
##   attend_num                               0.400***
##                                           (0.030)
##   never_married                           -0.604***
##                                           (0.095)
## --------------------------------------------------------
##   Log-likelihood                      -9123.06
##   N                                    4455
## ========================================================
##   Significance: *** = p < 0.001; ** = p < 0.01;
##                 * = p < 0.05
```

Or use `write_html()` to save the table as an HTML file, or `show_html()` to preview it in the **R** Console. To make the table more interpretable, variable names can be relabeled using the `relabel()` function. By default, `mtable()` combines the reference category in factors. For instance, to change row labels for education, such as the row label `edu: 1. Less than HS/2. HS cred.` to `Education: 1. Less than HS credential` use:

```
regtable4 <- relabel(regtable4, `edu: 1. Less than HS/2. HS cred.`
                     = "Education: 1. Less than HS credential")
```

After relabeling, print the updated table `print(regtable4)`.

Rather than presenting tables, researchers often visualize model estimates, focusing on the magnitude of slopes and the uncertainty surrounding them (Kastellec and Leoni, 2007). Figures typically display point estimates with error bars representing 95 percent confidence intervals. Various R packages can produce these figures automatically. In the **jtools** package (Long, 2023), the `plot_summs()` function generates horizontal point and error bar plots directly from a regression model object.

Install the package if necessary, `install.packages("memisc")`. And attach it:

```
library(jtools)
```

For a basic plot of regression results from *regmodel4*, enter `plot_summs(regmodel4)`. By default, variable names from the model output are used; specify interpretable names with the coefs() argument. For example:

```
plot_summs(regmodel4,
  coefs=c("1. Less than HS diploma"="edu1. Less than HS",
   "3. Some post HS" ="edu3. Some post HS",
   "4. Bachelor's degree"="edu4. Bachelor's degree",
   "5. Graduate degree"="edu5. Graduate degree",
   "sex (1 female)"="sex2. Female",
   "race (1 minority status)" ="race_category",
   "age in years"="age",
   "religious service attendance"="attend_num",
   "marital status (1 never married)" ="never_married"))
```

The graph in Figure 10.4 is a common way to visualize slope estimates. The figure is read row-wise, like a table of model estimates. The points represent the estimated coefficients. Horizontal bars represent 95 percent confidence intervals, so the bars overlapping zero indicate that the slope is not statistically significant. An increasingly common practice is for researchers to summarize regression models with graphs, while saving the model output tables for an appendix or supplementary technical report.

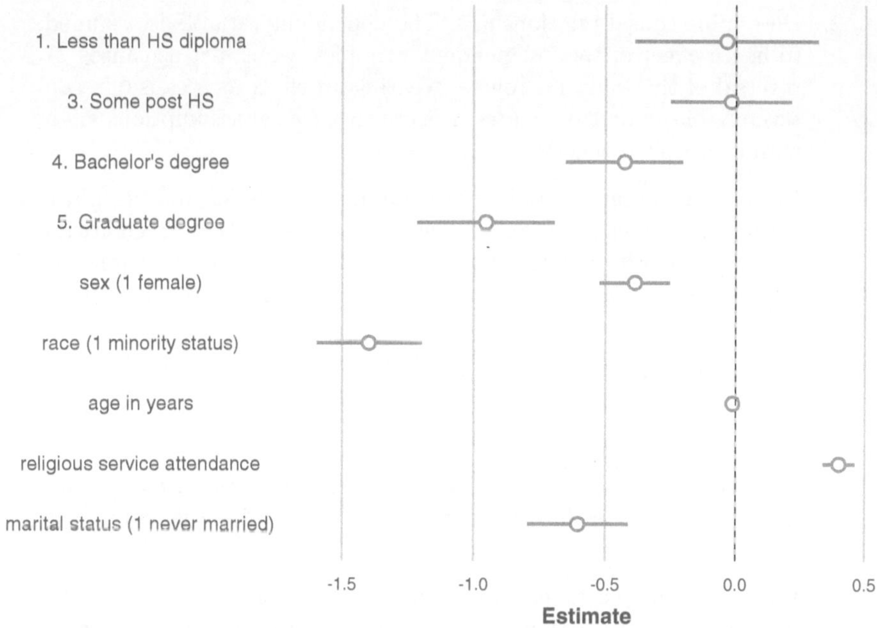

FIGURE 10.4 Coefficient plot created with the **jtools** package showing the magnitude, direction, and statistical significance of slope estimates from a regression model predicting party identification. Each point represents a coefficient estimate, with horizontal lines indicating 95 percent confidence intervals. Estimates to the right of zero suggest a positive association with party identification, while those to the left indicate a negative association. Confidence intervals that do not cross zero are statistically significant.

Key steps in regression analysis

When conducting a regression analysis, follow established practices and draw upon prior literature in the field. Published studies provide examples of how to specify regression models and insights into how to measure theoretical concepts in a model. Your model should replicate, as closely as possible, the approaches in prior studies while addressing the specific nuances of your data. Keep in mind some key steps:

1. Identify the dependent variable: Determine how the dependent variable should be scaled. For continuous measures, interval or ratio scaling is assumed. While ordinal scales like a seven-point party identification scale are often treated as continuous, measures with fewer categories may not meet this assumption and should be treated as categorical.

2. Determine causal relationships: The dependent variable is assumed to be an effect of the independent variables, which act as causes. A reversal of this causal arrow — treating an effect as a cause — can severely bias model estimates. Ensure that causal assumptions align with established theory.

3. Inspect variables: Use univariate summary statistics and visualizations to explore each variable. Identify skewness or other irregularities that may require transformations. Positive skew, for example, often necessitates a log transformation, while other transformations depend on the context and prior studies.

4. Assess linearity: Create scatterplots to assess the relationship between the dependent variable and key independent variables. This helps verify whether the assumption of linearity is reasonable.

5. Estimate and interpret the model: Fit the regression model and interpret the slopes for direction, magnitude, and statistical significance.

6. Inspect residuals: Check residual plots for signs of nonlinearity, nonconstant error variance, or other deviations that might indicate problems with the model.

7. Present results: Prepare at least a well-formatted table summarizing the model estimates. Consider adding a figure to communicate, in a nontechnical way, the results of the model.

In Chapter 11, we review strategies for investigating the plausibility of the assumptions underlying the multiple linear regression model, and we explore further refinements of the model, such as the specification of interactive, moderating relationships between independent variables.

Resources

Numerous textbooks and data analytics websites discuss regression modeling. An accessible textbook covering both fundamentals and more advanced subjects is Fox (2016). For applications with **R** code there is Fox and Weisberg (2019), which includes a popular companion package **car** (Fox et al., 2022a), applied in the following chapter.

10.8 Exercises

Include R code and relevant output in an .Rmd file knitted to Word or PDF.

(1) Construct a presidential election forecast starting with the **prez.csv** dataset. Using the parameter of the model, construct a forecast for the next election given hypothetical values for each of the factors in the forecast. Given the hypothetical values and the resulting forecast, how would you interpret the results?

(2) With the 2020 ANES dataset, estimate a regression of party identification (as dependent variable) on education, gender, and religious service attendance (all as independent X variables). All else equal, were more frequent religious service attendees more likely to identify with the Republican party? How do you know? (Interpret the direction and magnitude, as well as statistical significance of service attendance.) Explain the model and your interpretation in two paragraphs.

(3) Estimate a regression of attitude toward "scientists" ("scientists" feeling thermometer *V202173*), as a function of education and sex. Interpret the marginal relationships and statistical significance — controlling for one demographic characteristic (education), does sex appear to be systematically related to feelings toward scientists? How or how not?

(4) Starting with the same independent variables from the prior model, construct a model of an attitude toward modern sexism (the *sexism_index* variable in the ANES survey). Next, add the religious affiliation category of born again or not, as well as service attendance, as well as the measure of biblical literalism. Do broad religious beliefs, behaviors, and affiliations appear to significantly covary with sexist beliefs given controls for education and gender? How or how not? Show the model results and interpret the results in a paragraph.

(5) Import the US States dataset. Create a Trump or Biden vote percent measure. (a) Use `lm()` to estimate the simple linear relationship between vote percent and the State per capita income. Explain model estimates. (b) Extract from the model summary the R^2 term. Interpret. (c) Construct a scatterplot of each variable, and add the OLS trend line. Is it reasonable to propose a linear relationship between the two measures? And does it appear as if any States might be problematic outliers? How or how not? In the scatterplot, use `ggplot()` along with `geom_point()` and `stat_smooth()` to create the points and line. For example, if the Trump vote

percent measure is named *trump_percent_2020*, then the line for the OLS line could be `stat_smooth(method="lm", se=FALSE, aes(x=pc_income_2020, y=trump_percent_2020), color="red")`. Try labeling States on the figure with **ggrepel** and with a line such as `geom_text_repel(aes(x=pc_income_2020, y=trump_percent_2020, label=state), alpha=.3, label.size = 0.15, max.overlaps=20)`.

6. With the US States dataset, construct a scatterplot on two continuous measures, related to each other on X and Y axes. (For a severly positively skewed measure, consider an appropriate log transformation). Add a linear trend line with `stat_smooth(method="lm", se=FALSE, aes(x=VARNAME, y=VARNAME), color="red")`. Contrast the linear trend line with a "smoother" nonlinear regression line with `stat_smooth(se=FALSE, aes(x=VARNAME, y=VARNAME), color="red", span=WIDTH)`. In this non-linear trend, the `span=` option instructs the smoother to redraw a best fitting line along each width of points along the X axis. For example, if `span=.3`, the smoother would redraw and smooth together a line every .3 points along the X axis. Experiment with different settings to find a nonlinear trend. Contrast the two figures in a paragraph, and be sure to explain why you chose the spanning width.

11

Evaluating and Extending Linear Models

Chapter 11 introduces common problems and remedies to linear models, and how to explore interactive relationships.

Learning objectives and chapter resources

By the end of this chapter, you should be able to (1) construct and interpret diagnostic plots of model residuals and fitted values; (2) identify patterns in residuals to evaluate potential violations of linear model assumptions, such as nonlinearity, error variance, and normality; (3) interpret measures to identify observations as outliers, leverage, and influence; (4) specify and interpret a capturing conditional, interactive relationships between two independent variables. The material in the chapter requires the following packages: **survey** (Lumley, 2021), **dplyr** (Wickham et al., 2023b) and **tibble** (Müller and Wickham, 2023) from the **tidyverse** (Wickham, 2023b), and a new package requiring installation, **car** (Fox et al., 2022a). The material uses the following dataset: *anes 2020 survey.rdata* from https://faculty.gvsu.edu/kilburnw/inpolr.html.

11.1 Examining assumptions underpinning linear models

The assumptions discussed in Chapter 10 about ordinary least squares (OLS) regression form the foundation for evaluating linear models. These assumptions are critical for accurately estimating slopes and y-intercepts and for making valid inferences to population parameters. When these assumptions are violated, the resulting models may be unreliable. Diagnostic tools allow us to detect potential problems, including model misspecification, nonlinearity, nonconstant error variance (heteroskedasticity), or problematic observations like outliers and high leverage points.

The starting point for linear model diagnostics is analyzing residuals. Residuals are a proxy for the unobservable errors e in the population regression model and provide insight into the validity of the model's assumptions. The most fundamental diagnostic plot is the scatterplot of residuals versus fitted values, used to check for linearity, constant error variance, and normality of residuals. Beyond residual analysis, we also evaluate individual observations for leverage, influence, or outlying behavior.

This chapter illustrates common diagnostic techniques using `regmodel4`, the demographic model of party identification introduced in Chapter 10: `svyglm(formula = partyid_num ~ edu + sex + race_category + age + attend_num + never_married, design = anes2020_design)`. We will examine common diagnostic tools to detect and address potential issues in linear models. Some diagnostic tools are not directly compatible with `svyglm()`, so the complex survey design potentially complicates the interpretation of these tools. For example, residuals in survey-weighted models should be weighted to reflect the contribution of each observation to the regression model. The clustering and stratification in the survey can potentially introduce artifacts in residual plots, such as within-cluster correlations. Yet on the upside, a common problem in cross sectional public opinion research is heteroscedasticity, and fortunately the larger standard errors in svyglm() generally address issues of non-constant error variance, reducing the need for additional corrections.

Despite these challenges, diagnostic tools remain essential for evaluating linear models. This chapter reviews a variety of approaches to assess residuals, fitted values, and observations to detect potential problems and ensure model robustness.

Residuals versus fitted values

As discussed in Chapter 10, the most fundamental tool for investigating linear models for problematic features is a plot of residuals versus fitted values. Typically, researchers look for residuals to appear in a random scatter without any clear pattern. Deviations from randomness could indicate a number of potential different problems. First, any clear curvature or systematic trends in the residuals suggests there may be nonlinearities present in the relationship between Y and any X, which would not be adequately modeled by a linear slope parameter. Second, if the residuals clearly increase or decrease in spread as fitted values increase, it indicates that the error variance is not constant – parts of the model fit worse than in other parts. Often it can lead to standard errors that are too small. Third, points that deviate from others within the figure may be evidence of outliers or influential data points, that could potentially influence the model's parameters.

We use the `residuals()` function to extract residuals from the linear model object *regmodel4*, and `fitted()` for the fitted values. We start with the residuals. Typically, researchers inspect residuals for severe departures from normality.

A common starting point is to plot residuals, for example a histogram, `hist(residuals)` inspected for departures from normality. A quantile-quantile plot (QQplot) is often used to more explicitly compare the residuals against a normal distribution. The plot pairs quantiles of the model residuals with residuals from a hypothetical normal distribution, plotting them in pairs; the less the deviation of the observed residuals from the hypothetical, the observations should line up along the 45 degree line. Given the *regmodel4 residuals*, the base R functions `qqnorm(residuals)` will create the QQplot while `qqline(residuals,col="red")` will add a 45 degree line marking the equality of the two quantities.

Figure 11.1 displays the QQ plot of residuals from *regmodel4*, showing party identification as a function of demographic characteristics. The residuals appear to approximate a normal distribution through the central portion of the plot, where the points closely align with the diagonal line. However, at the extremes (both tails), there are clear departures from normality, with points deviating above and below the line. This indicates the presence of outliers or extreme residuals, suggesting that the residuals are not perfectly normally distributed. While mild departures at the tails are not uncommon, they warrant further investigation to assess their potential impact on model assumptions and results.

```
residuals <- residuals(regmodel4)

qqnorm(residuals)
qqline(residuals,col="red")
```

The second plot is residuals paired with fitted values. The residuals and fitted values can be combined into a dataframe for **ggplot2**, or plotted with the base **R** `plot()` function and as a reference `abline()` for a line drawn through the point at 0 on the residuals axis, as in Figure 11.2:

```
fitted_values<-fitted(regmodel4)

plot(fitted_values, residuals, xlab = "Fitted Values",
     ylab = "Residuals", pch=20)
abline(h = 0, col = "red", lwd = 2)
```

For the residuals to meet the assumptions of a linear model, they should be centered around zero and display no discernible pattern across the range of fitted values. In Figure 11.2, the residuals appear relatively random around the horizontal line at zero. However, there are notable outliers and a slight

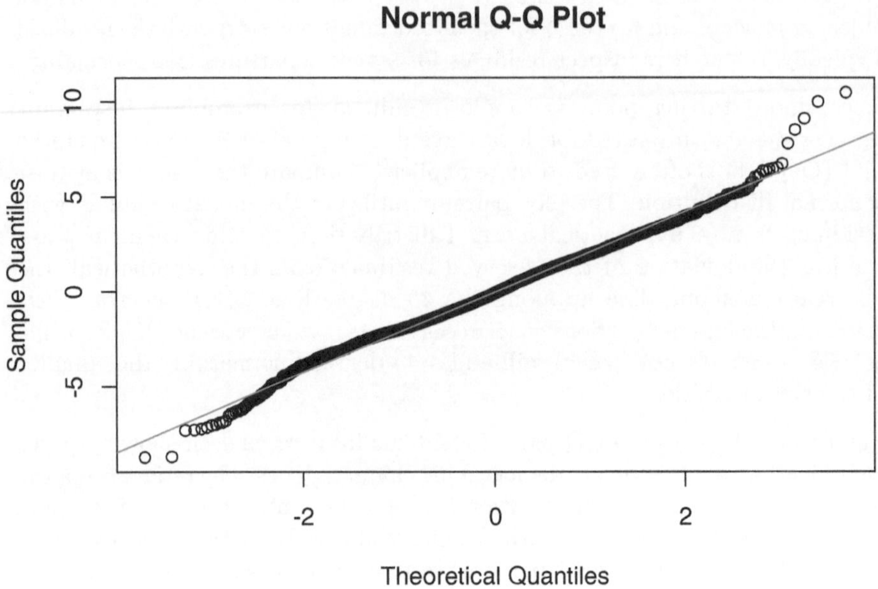

FIGURE 11.1 Normal quantile–quantile (QQ) plot of residuals from the model of party identification as a function of demographic characteristics. The figure displays deviations from normality at the ends of the plot.

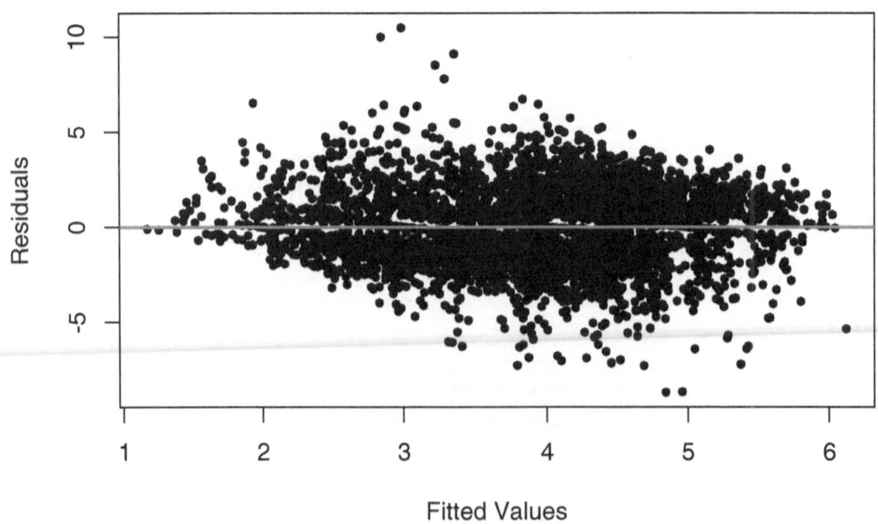

FIGURE 11.2 A plot of model residuals versus fitted values. The line is drawn through the 0 point on the vertical axis.

fanning pattern at both the lower and upper ends of the fitted values, suggesting heteroskedasticity — again, a violation of the constant error variance assumption.

This heteroskedasticity may arise for a few reasons. First, it could stem from the complex survey design of the data, where nonconstant error variance results from differing regional strata or interview clusters. Fortunately, the `svyglm()` estimator adjusts for this design effect by producing robust standard errors, so we need not be overly concerned about this source. Second, the heteroskedasticity may relate to the distribution of the party identification variable, which is concentrated at the extreme points of the scale (Strong Democrat or Strong Republican). The model's demographic predictors may generate larger residuals for these strong partisans.

Addressing this issue could involve respecifying the model. Potential remedies include adding omitted variables to reduce bias, introducing interaction terms to account for conditional relationships, or considering alternative estimators. For instance, the assumption of one slope across all education levels may not hold, suggesting a more flexible specification. Nonetheless, most of these approaches require intermediate to advanced knowledge of linear modeling, which is beyond the scope of this book.

For now, despite the signs of heteroskedasticity, we proceed with the demographic model of party identification, as the tools available within the `svyglm()` framework sufficiently address design-based sources of variability.

While heteroskedasticity can undermine confidence in slope estimates, another potential issue is multicollinearity.

Collinearity of independent variables

Multicollinearity arises when two or more independent variables in a regression model are highly correlated, making it difficult to isolate their individual effects on the dependent variable. In such cases, the regression coefficients may become unstable, inflating standard errors and reducing the statistical significance of the predictors.

In the case of *regmodel4*, the highly significant slopes suggest that multicollinearity is unlikely to be a major concern. However, it is standard practice to assess multicollinearity explicitly using Variance Inflation Factors (VIFs).

VIFs measure how much the variance of a regression coefficient is increased due to correlations with other independent variables in the model. A VIF of 1 indicates no multicollinearity, while higher values signal stronger correlations, making it harder to disentangle a variable's unique contribution to the model.

To calculate VIFs, we use the vif() function from the car package:

```
library(car)
```

```
vif(regmodel4)
```

```
##                   GVIF Df GVIF^(1/(2*Df))
## edu              1.587  4           1.059
## sex              1.373  1           1.172
## race_category    1.461  1           1.209
## age              1.509  1           1.228
## attend_num       1.168  1           1.081
## never_married    1.690  1           1.300
```

The VIFs appear under the heading `GVIF`. As expected, the VIFs for *regmodel4* indicate that multicollinearity is not a concern. All values are close to 1, with the highest being 1.690 for *never_married*, suggesting only a small degree of variance inflation. Generally, VIF values exceeding 5 or 10 are considered a threshold for problematic multicollinearity that may require further attention.

For categorical predictors with multiple levels, the Generalized VIF (GVIF) adapts the standard VIF to account for the number of categories in the variable. By scaling the GVIF, it becomes easier to interpret, similar to the standard VIF used for numeric predictors. This adjustment ensures that multicollinearity is properly assessed in models that include both numeric and categorical predictors (Fox and Weisberg, 2019).

Note that multicollinearity is not inherently problematic. While it can inflate standard errors and reduce the precision of individual slope estimates, it does not affect the overall fit or predictive power of the model. Addressing multicollinearity — for example, by dropping a variable or combining correlated predictors into a single index are some common ways to address it. Yet doing so should be guided by theoretical considerations rather than, for example, dropping a variable from a model simply because of a large VIF.

In this case, the low VIF values suggest that the predictors in *regmodel4* are sufficiently independent, and their effects can be interpreted with confidence.

Unusual observations

Even with a large sample size as in *regmodel4*, it is important to check whether individual observations are disproportionately influencing the regression results. Such observations can distort slope estimates and undermine the reliability of the model. There are three key sources of "troublesome" observations: (1)

outliers, which have excessively large residuals; (2) high leverage observations, which have unusual values on independent variables; and (3) influential observations, which combine both large residuals and high leverage.

To investigate these issues, we can use tools in base **R** to examine specific diagnostics: residuals to detect outliers, "hat" values to identify leverage points, and Cook's distance to measure influence. In addition, the **car** package provides many additional tools for assessing all of these qualities, briefly reviewed in the Resources section of this chapter.

Outliers

Outliers are observations with unusually large residuals, meaning their actual values deviate significantly from the model's predictions. A good starting point for detecting outliers is a residuals versus fitted values plot, which visually highlights observations that do not align with the overall trend. However, numerical diagnostics provide more precise tools for identifying these points.

Studentized residuals are standardized versions of the residuals, adjusted to account for the variability of each residual and the influence of the observation on the model. These residuals are calculated by dividing each residual by its estimated standard deviation. This adjustment makes it easier to compare residuals across observations and detect outliers. Typically, studentized residuals greater than 2 (or greater than 3 in large datasets, such as the ANES survey) signal potential outliers.

The `rstudent()` function calculates studentized residuals. The `summary()` function reports any potential outliers. Storing the residuals in *stand_residuals*:

```
stand_residuals <- rstudent(regmodel4)

summary(stand_residuals)
```

```
##     Min. 1st Qu.  Median    Mean 3rd Qu.    Max.
##  -4.304  -0.719  -0.041  -0.016   0.687   5.181
```

Notably, some residuals exceed 4 in magnitude, indicating observations with extreme deviations. To count the number of residuals greater than 3 in absolute value (a common threshold for identifying severe outliers), we can apply the `abs()` and `sum()` functions:

```
sum(abs(stand_residuals) > 3)
```

```
## [1] 31
```

While 31 outliers may seem notable, they represent a very small fraction of the overall sample size, making their impact on the model likely very limited.

In smaller datasets, however, the presence of even a few outliers can pose a significant challenge. Researchers must carefully decide whether to exclude or retain such observations. And there is a tradeoff: Excluding outliers may reduce the influence of extreme observations but risks introducing selection bias, as the data may no longer fully represent the population. Including outliers ensures all observations are considered but may distort the model estimates, particularly the slopes and standard errors.

To determine whether these outliers meaningfully affect the model, we next investigate their leverage and influence on the regression results.

Leverage and influence

Leverage measures how much an observation influences the regression model's fitted values based on its scores on the independent variables. Observations with high leverage are those that are far from the "center" of the independent variables — meaning they are distant from the mean values across all predictors. High leverage points can have a substantial impact on the regression results because of their unique positions.

High leverage alone does not always indicate a problematic observation. For an observation to disproportionately influence the model, it must combine high leverage with an unusually large residual (an outlier). This combination indicates that the observation significantly alters the slope estimates or fitted values.

A conventional rule of thumb is that leverage values (or "hat values") exceeding 2 to 3 times the ratio of model parameters (number of slopes + intercept) to the sample size are considered unusually large. For instance, in the case of *regmodel4*, with approximately 20 to 30 model parameters and a sample size of 4433, the threshold for high leverage would be around 0.004 to 0.006.

To calculate leverage values, use `hatvalues()` with `summary()` :

```
hat_values <- hatvalues(regmodel4)

summary(hat_values)
```

```
##      Min.  1st Qu.   Median     Mean  3rd Qu.     Max.
## 0.000031 0.000889 0.001557 0.002255 0.002641 0.026908
```

The results for *regmodel4* indicate that some observations exceed the leverage threshold, as the maximum value is notably higher than the third quartile. These points warrant further examination to assess their influence on the model.

Cook's distance combines two key diagnostics — a standardized residual and leverage — into a single statistic to identify influential observations. It measures how much the fitted regression values would change if a particular observation were removed. Observations with large Cook's distance values can disproportionately affect the regression results.

To calculate Cook's distance, use the `cooks.distance()` function and `summary()`:

```
influence <- cooks.distance(regmodel4)
summary(influence)
```

```
##      Min.  1st Qu.   Median     Mean  3rd Qu.     Max.
## 0.000000 0.000027 0.000104 0.000224 0.000265 0.005435
```

A common rule of thumb is that observations with a Cook's distance greater than $\frac{4}{n}$ may be considered influential. Since Cook's distance is always positive, and with $n = 4433$, this threshold is $\frac{4}{4433} \approx 0.0009$. Given that the maximum Cook's distance in the model is only 0.005, none of the observations appear to be especially problematic.

In this section, we explored key diagnostic tools for evaluating the assumptions of linear regression models. By examining residuals for outliers, patterns, and non-constant variance, we checked for potential violations of linearity and homoskedasticity. Measures of leverage, studentized residuals, and Cook's distance allowed us to identify observations that could disproportionately influence model estimates. While a few observations exhibited high leverage or large residuals, their overall impact on the model appears. minimal.

With confidence in the stability of our current model, we now turn to a more advanced topic: interaction terms. Interaction terms allow us to investigate how the relationship between one independent variable and the dependent variable changes depending on the level of another independent variable, a conditional relationship.

11.2 Exploring interactions in linear models

Religious belonging x behavior in party identification

The strength of the relationship between service attendance could vary depending on the religious tradition associated with the individual. For example, consider the evangelical Protestant religious tradition (see Steensland et al, for an explanation (2000)). It is, of course, likely that evangelical Protestants identify more strongly with the Republican party, relative to the rest of the American electorate. And overall, more frequent religious service attendance is associated with stronger Republican party identification. Yet it could also be that among evangelical Protestants, perhaps more frequent religious service attendance does not lead to stronger Republican party identification, relative to the electorate as a whole. This intuition implies an interaction between a religious belonging and behavior, instead of separate linear slopes for attendance and evangelical Protestant belonging.

To illustrate how a linear model accommodates this intuition, we simplify the model of party identification to three independent variables for each of these terms: X_{1evang} for evangelical identity, $X_{2attend}$ for service attendance, and $X_{1evang} * X_{2attend}$ for the interaction between the two. The model:

$$Y = a + b_1 X_{1evang} + b_2 X_{2attend} + b_3 X_{1evang} \times X_{2attend}$$

For non-evangelicals, the value of X_{1evang} is 0, so the slope on the interaction term drops out of the term $b_3 \times 0 \times X_{2attend}$, leaving

$$Y = a + b_1 X_{1evang} + b_2 X_{2attend}$$

as the model.

For evangelicals, however, the value of X_{1evang} is 1, so the interaction term simplifies to $b_3 \times 1 \times X_{2attend} = b_3 \times X_{2attend}$. This slope $b_3 X_{2attend}$ is added to the other slope on $X_{2attend}$

For evangelicals: $Y = a + b_1 X_{1evang} + (b_2 + b_3) X_{2attend}$

For non-evangelicals: $Y = a + b_1 X_{1evang} + b_2 X_{2attend}$

The two equations display how the interaction alters the slope for attendance, conditional on evangelical identification.

To explore the model in data, we start with *regmodel4* from Chapter 10. We create an indicator for a survey respondent's belonging to the evangelical Protestant religious tradition. After attaching the necessary packages:

```
library(tidyverse)
library(survey)
```

```
anes_20 <- anes_20 %>%
  mutate(evang =
           ifelse(relig_trad == "2. Evangelical Protestant", 1, 0))
```

The variable *evang* is scored 1 for respondent belonging to the evangelical Protestant tradition, 0 otherwise. To explore an interaction between evangelical Protestant religious belonging and service attendance, we will treat service attendance *attend* as a numeric measure. The numeric variable for religious service attendance is *attend_num*.

To simplify the model, we will treat formal educational attainment as a numeric variable, *edu*. The *edu* variable from *regmodel4* is converted to a numeric with `as.numeric()`:

```
anes_20<-anes_20 %>%
  mutate(edu=as.numeric(edu))
```

We will add one additional control, household income, as a numeric measure, to disentangle income from religious service attendance. (See the ANES codebook for scaling.)

```
anes_20 <- anes_20 %>%
  mutate(income = fct_recode(V202468x,
           NULL = " 9. Refused")) %>%
  mutate(income = as.numeric(income))
```

After creating new variables we have to rerun the `svydesign()` function:

```
anes2020_design<-svydesign(data = anes_20, ids = ~V200015c,
                  strata=~V200015d, weights=~V200015b, nest=TRUE)
```

Then we run the model, including both the *evang* and *service attendance* variables:

```
regmodel5<-svyglm(partyid_num ~  edu + sex +
           race_category + age + never_married + income+
           evang + attend_num, design=anes2020_design)
```

Observe the model output with `summary(regmodel5)`. In the model output, you will notice that the *evang* slope (0.529) is statistically significant, as is *attend_num* (0.355), both factors on average leading individuals to stronger Republican party identification.

The interaction between evangelical Protestant identity and attendance is a multiplicative interaction, the product of *evang* times *attend_num*. Interaction terms can be specified in a model in two ways. In the *anes_20* dataset, we could create a new variable as the product of the two terms: `anes_20<-anes_20 %>% mutate(evangXattend = evang * attend_num)`. Or simply include the product of the two individual variables within the `svyglm()` function, as the last independent variable *evangXattend_num* :

```
regmodel6<-svyglm(partyid_num ~  edu + sex +
                    race_category + age + never_married + income +
                    evang + attend_num +
                    evang*attend_num, design=anes2020_design)
```

To review the results and the differences between the two models, we will use the **memisc** (Elff, 2023) package:

```
library(memisc)
```

Storing each model results in `mtable()` :

```
reg_table <- mtable("Model 1" = regmodel5, "Model 2" = regmodel6,
                    sdigits = 2)
```

Printing the table:

```
print(reg_table)
```

```
##
## Calls:
## Model 1: svyglm(formula = partyid_num ~ edu + sex + race_category + age +
##     never_married + income + evang + attend_num, design = anes2020_design)
## Model 2: svyglm(formula = partyid_num ~ edu + sex + race_category + age +
##     never_married + income + evang + attend_num + evang * attend_num,
##     design = anes2020_design)
##
## ======================================================
##                              Model 1      Model 2
## ------------------------------------------------------
##   (Intercept)                 5.254***     5.141***
##                              (0.177)      (0.188)
##   edu                        -0.225***    -0.222***
##                              (0.038)      (0.038)
##   sex: 2. Female/1. Male     -0.375***    -0.377***
##                              (0.067)      (0.066)
##   race_category              -1.420***    -1.418***
##                              (0.101)      (0.102)
##   age                        -0.013***    -0.013***
##                              (0.003)      (0.003)
##   never_married              -0.589***    -0.575***
##                              (0.102)      (0.102)
##   income                      0.006        0.006
##                              (0.007)      (0.007)
##   evang                       0.529***     1.030***
```

```
##                              (0.101)       (0.228)
##    attend_num                 0.355***      0.404***
##                              (0.032)       (0.035)
##    evang x attend_num                      -0.224*
##                                            (0.094)
## ---------------------------------------------------------
##    Log-likelihood          -8918.29      -8897.44
##    N                        4356          4356
## =========================================================
##    Significance: *** = p < 0.001; ** = p < 0.01;
##                  * = p < 0.05
```

Notice that the demographic characteristics for Model 1 remain relatively stable in Model 2. In the interactive Model 2, the slopes are interpreted as follows:

Starting at the bottom of the table, note the statistically significant slope of -0.224 on *evang x attend_num*. The negative sign on the slope indicates the association between increasing religious attendance on party identification is weaker for evangelicals compared to non-evangelicals. Specifically, for every one-unit increase in *attend_num*, the slope of attend_num decreases by -0.224 for evangelicals relative to non-evangelicals, meaning that increased service attendance does not strengthen Republican party identification to the extent that it does for the electorate as a whole. The p-value on the slope for *evang X attend_num* , while arguably on the cusp of statistical significance ($* = p < 0.05$), nonetheless indicates that the difference slopes between evangelicals and non-evangelicals is statistically significant. The relationship between religious attendance and party identification depends, at least in part, on evangelical Protestant status.

The magnitude of the relationship between service attendance and party identification — for evangelicals – is the linear combination of the slopes for *attend_num* and the interactive term *evangXattend_num*: $0.404 + (--0.224) = 0.181$. While for non-evangelicals a one-unit increase in religious attendance is associated with a .404 point stronger Republican party identification; for evangelicals, the association is about half that, .18 points.

A useful test is whether this combined effect for evangelicals is likely distinguishable from 0 in the population. — It could be, for instance, that while weaker than for non-evangelicals, the relationship is null in the population. The *car* package provides a function `linearHypothesis()` to test linear combinations of slopes against the null hypothesis that in the population, the linear combination is 0. The argument specifies the linear combination, with variable names from the model specified in the model, in this case *regmodel6*: `attend_num + evang:attend_num = 0`:

```r
linearHypothesis(regmodel6,
    hypothesis.matrix = "attend_num + evang:attend_num = 0")
```

```
## Linear hypothesis test
##
## Hypothesis:
## attend_num  + evang:attend_num = 0
##
## Model 1: restricted model
## Model 2: partyid_num ~ edu + sex + race_category + age + never_married +
##     income + evang + attend_num + evang * attend_num
##
##   Res.Df Df Chisq Pr(>Chisq)
## 1     42
## 2     41  1  4.61      0.032 *
## ---
## Signif. codes:
## 0 '***' 0.001 '**' 0.01 '*' 0.05 '.' 0.1 ' ' 1
```

The linear hypothesis test evaluates whether the combined effect of religious service attendance plus the interaction with evangelical status equals zero, testing whether the slope for religious attendance among evangelicals is significantly different from zero while accounting for other predictors in the model. The test indicates the slope for evangelicals on religious service attendance is significantly different from 0 $p = 0.032$. The result indicates that religious attendance has a statistically significant, albeit weaker, relationship with party identification among evangelicals.

Interaction terms are symmetrical, and can be interpreted from the other perspective – considering nominal affiliation with evangelical Protestantism and how it relates to party identification as religious service attendance changes. The coefficient for *evang*, `1.02962`, represents the influence of being an evangelical on party identification when religious attendance *attend_num* is at 0, or not attending. Being a nominal evangelical (who never attends religious services) is associated with an increase of approximately 1.03 points on the party identification scale (shifting toward stronger Republican identification), compared to a non-evangelical who also never attends services.

The influence of being evangelical on party identification, as service attendance increases, depends on the combination of the slope and the interaction term. Among non-attending evangelicals the impact of 1.0296 is the result of $1.02962 + (- - 0.22359 * 0) = 1.02962$. Yet as attendance increases to 3, the impact reduces to $1.02962 + (- - 0.22359 * 3) = 1.02962 - -0.67077 = 0.35885$. At weekly service attendance levels, the effect of being evangelical decreases to about a third of the level for nomina,l (non-attending) evangelicals.

Model based predictions

Given the model, we can explore model-based predictions from the interaction term, via `predict()` . Like before, we create the prediction dataset, specifying 'average' values for the independent variables and particular combinations of the interaction terms. For example, we may be interested in calculating the model based prediction of a party identification for an evangelical who does not report attending religious services, versus one who attends services weekly. The prediction dataset contrasts service attendance values (0 and 3) and average characteristics, a male with some post high school education or training , 45 years old, and a household income of 65,000 to 69,999 USD (12th quantile), married or divorced:

```
predictiondata <- tribble(
  ~edu, ~sex, ~race_category, ~age, ~never_married, ~income, ~evang,
       ~attend_num,
  3, "1. Male", 0, 45, 0, 12, 1, 3,
  3, "1. Male", 0, 45, 0, 12, 1, 0)
```

Given the prediction dataset, the model based predictions and 95 percent confidence intervals:

```
predictions<-predict(regmodel6,predictiondata, interval = "confidence")

confint(predictions)
```

```
##    2.5 % 97.5 %
## 1 5.306  5.771
## 2 4.568  5.425
```

The model based predictions show that a weekly attending evangelical Protestant, given the other characteristics on average identifies as an Independent-leaning Republican (5.33), while a nominal evangelical identifies closely on the scale (4.79), but more Independent.

11.3 Testing and refining linear models

Presenting and interpreting regression results requires thoughtful model specification and a clear strategy for communicating findings. A standard approach involves presenting a series of nested models, starting with a parsimonious specification focused on the primary relationships of interest and gradually adding covariates as control variables. The models are combined into a table such as *reg_table* presented in this chapter. It is not uncommon to see three or four models specified this way. This stepwise approach, a sensitivity analysis, assesses the stability of the primary slopes of interest, as additional control variables are introduced. It also helps distinguish between potential sources of spuriousness (confounding variables) and mediation (variables that explain part of the relationship). Whether a variable serves as a confounder or mediator should be guided by prior theory and literature.

When interpreting model estimates, pay close attention to patterns in slope changes across model specifications; a decrease in magnitude when adding controls may indicate partial mediation or confounding effects. Remember, adding variables to a model effectively moves their relationship with the dependent variable out of the error term and into the model, isolating their contribution. A good linear model requires careful, theory drive model specification and clear sensitivity analysis to provide for drawing valid conclusions and effective communication about the sensitivity of the model to alternative specifications.

Resources

Further investigating the assumptions underlying linear models requires an intermediate-to-advanced study. See Urdinez and Cruz (2022) for specific statistical tests and remedies for common problems in regression models. See Fox and Weisberg (2019) for methods of applying the **car** (Fox et al., 2022a) package to graphically exploring interactive relationships in linear models. The *car* package includes two functions, `qqPlot()` and `residualPlot()` (note the capitalization) with additional nonlinear trend lines to assess departures from normality. There are also additional graphical plot types for more advanced study in the **car** package.

11.4 Exercises

Knit your **R** code into a document to answer the following questions:

(1) In the 2020 ANES dataset, find a 7 point issue attitude scale and construct a demographic linear model, with three independent variables such as educational attainment, gender, and race. (a) Create a residuals vs. fitted values plot. Interpret whether the residuals suggest any violations of linearity or homoskedasticity. (b) Generate a QQplot of the residuals and assess whether the residuals follow a normal distribution. In a paragraph, summarize your findings

and discuss whether the assumptions of the linear model appear reasonable.

(2) For the same model, calculate Cook's distances and identify any observations exceeding the rule-of-thumb threshold. Summarize whether these observations appear to unduly influence the model and suggest possible next steps.

(3) Build a regression model to explore the interaction between a religious tradition and religious service attendance on party identification, similar to the model presented in this chapter. Interpret the coefficient for the interaction term. What does it imply about how the relationship between attendance and party identification differs for the chosen religious group? Present and interpret model-based predictions to interpret the results.

12

Linear Models for Binary Outcomes

Chapter 12 introduces linear models for binary (two-level) dependent variables.

Learning objectives and chapter resources

By the end of Chapter 12 you should be able to (1) explain key concepts in forming binary logistic regression models; (2) interpret the summary of model output; and (3) generate tables and graphics of model based predicted probabilities. The material in the chapter requires the following packages: **jtools** (Long, 2023), **effects** (Fox et al., 2022b), and **sjPlot** (Lüdecke, 2024) which will need to be installed, as well as **survey** (Lumley, 2021), **car** (Fox et al., 2022a), and **dplyr** (Wickham et al., 2023b) from the **tidyverse** (Wickham, 2023b). The material uses *anes 2020 survey.rdata* from https://faculty.gvsu.edu/kilburnw/inpolr.html.

12.1 Introduction

In this chapter we review a particular type of linear model, binary logistic regression, for the analysis of variation in dichotomous (or binary) dependent variables. Dichotomous measures contrast presence and absence (or 'success' and 'failure'), such as in an election poll respondents reporting that they voted (or not), or supported (or opposed) a particular policy. Through a logistic regression model, just as in the OLS regression models from Chapters 10 and 11, we estimate slopes on independent variables to examine the multiple factors that explain the variation in these outcomes, such as why some people voted while others did not. This application of logistic regression is often referred to as "binary logistic regression", binary in reference to the scaling of the dependent variable.

Applying the same `lm()` type functions from Chapters 10 and 11 to dichotomous dependent variables is problematic. Instead, statisticians have developed "Generalized Linear Models", for which linear regressions with a continuously scaled, binary, or ordinal dependent variable are special cases. We first review key concepts in the estimation and interpretation of these models for binary logistic regression, followed by an example with R code for modeling Americans' support or opposition to the death penalty for capital crimes.

12.2 Conceptualizing binary logistic regression

As a starting point, recall that the dependent variable, having only two outcomes, could be scored 0 for a 'failure' or lack of an attribute and 1 for a 'success' or attribute. Ranging from 0 to 1, it could be interpreted in probabilistic terms. Applying a linear regression function (such as `lm()`) to model variation in this dependent variable results in the linear probability model. The term "probability" references the purpose of such a model – to examine how the probability of a value of 1 on the dependent variable changes as a linear function of the independent variables. For example, with data from the 2020 ANES, responses to the question of whether an individual supports the death penalty, considered as either "opposes" (0) or "supports" (1), we could construct a linear regression model of death penalty support as a function of various demographic characteristics. The primary problem with this approach, however, is that doing so violates some of the assumptions required for valid OLS regression model estimation: Because of the 0 or 1 variable scaling, errors are not normally distributed, the model induces heteroskedasticity, and perhaps most problematically the model-based predictions could easily be out of bounds, that is, a predicted probability that is less than 0 or greater than 1.

Researchers turn instead to "logistic regression". Like a multiple linear regression, the goal of a logistic regression is to estimate an intercept and slopes for independent variables. Instead of fitting a straight line through the data, a logistic regression fits an S-shaped curve to transformed values on the dependent variable, constrained at the probability boundaries of 0 and 1. The line guarantees that predictions never exceed these limits. Figure 12.1 displays this curve, fitting a curve of best fit to hypothetical data.

Rather than model the dependent variable directly, logistic regression first transforms the dependent variable (scored 0 or 1) into a quantity called 'log-odds.' Odds are ratios, the ratio of a probability p to $(1 - p)$:

$Odds = \frac{p}{(1-p)}$

For example, if the probability of an American voting age citizen supporting the death penalty is .65 or 65 percent, then the odds of supporting the death

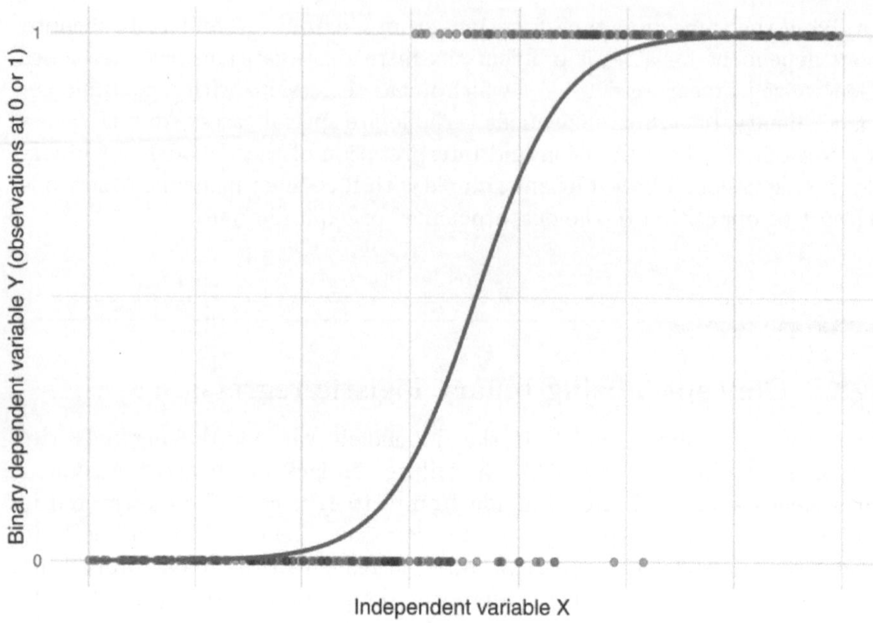

FIGURE 12.1 Hypothetical logistic regression S-shaped curve relating observations on a continuous X to a dichotomous Y.

penalty are $\frac{.65}{(1-.65)} = 1.857$.

The 'log-odds' is simply the natural logarithm of the odds, called a "logit":

$logit = ln(\frac{p}{(1-p)})$.

Given the probability of .65 (or odds of 1.857) the logged odds `log(1.857)` is .619. To further explain odds, consider this example from the 2020 ANES. We tabulate death penalty support (see question *V201345x* in the 2020 ANES data appendix) by respondent sex. (The code for creating the individual measures from the survey is presented further below.) For now, consider the cross-tabulation of death penalty attitude by sex:

```
##                sex
## death_penalty 1. Male 2. Female
##        Oppose     773       944
##        Favor     1487      1466
```

The tabulation displays the frequencies. So the table shows, for example, that 773 male respondents opposed the death penalty. A probability of an individual supporting the death penalty is the likelihood of it occurring, as a proportion between 0 (not occurring) and 1 (occurring with certainty).

Given the total number of males and females in the sample, we can calculate the probability of any given male or female supporting the death penalty. The total for males (773 + 1487) is 2260. The probability of a male supporting the death penalty $\frac{1487}{2260} \approx .66$. The total for females (944 + 1466) is 2410. The probability of a female supporting the death penalty is $\frac{1466}{2410} \approx .61$.

Odds are ratios of probabilities. The odds of an individual supporting the death penalty is the ratio of the probability of supporting it divided by the probability of opposing it. Odds can be expressed as

$Odds = \frac{P}{1-P}$,

where P is the probability of an event occuring.

The probability of a male supporting the death penalty is approximately .66; the probability of opposing is $1 - .66 \approx .34$. The odds of a male supporting the death penalty are $\frac{.66}{.34} = 1.94$. The odds of a male supporting the death penalty are about 1.94 to 1, meaning that for every male who opposes the death penalty, approximately 1.94 support it. For females, odds of supporting the death penalty are $\frac{.61}{.39} \approx 1.56$. For females the odds are about 1.56 to 1. Thus the odds are slightly higher that any given male will support the death penalty compared to females.

Comparing two odds yields the odds ratio, a measure of the relative odds between two groups. As a ratio, it is a division of one odds by another. The odds ratio comparing female support to male support is:

Odds ratio $\frac{Odds(female)}{Odds(male)}$ or $\frac{1.56}{1.94} \approx .80$.

An odds ratio of .80 means that the odds of a female supporting the death penalty are about 80 percent of the odds of a male supporting it. Probabilities, odds, and odds ratios all describe the same phenomenon under study, but are expressed in different units. Simply personal preference may determine which appears the most intuitive. It is helpful to remember that (a) if the probability of an event occurring is less than .5, then the odds are less than 1; and (b) if the probability of an event occurring has a probability greater than .5 then the odds are greater than 1.

In a logistic regression model, with the dependent variable (Y) converted to logits, and a set of independent (X) variables, the algorithm for logistic regression finds a best fitting y-intercept and slopes for each X. Note that the model is still a linear model — the logit is expressed as a linear function of the independent variables:

$logit(p) = log(\frac{p}{(1-p)}) = \beta_0 + \beta_1 x_1 + \beta_2 x_2 + ... + \beta_k x_k$\$

The parameters (coefficients) from the model are β_0 the estimated intercept, with the estimated slopes $\beta_1, \beta_2, ..., \beta_k$. And the independent variables are $x_1, x_2, x_3, ..., x_k$. This is where the 'linear' in linear model comes from: the model is linear with respect to the parameters and the predictors, even though the relationship between the predictors and the actual probability p (the s-curve) is nonlinear.

Modeling support versus opposition to the death penalty

Below we review a logistic regression model, followed by the **R** code, of Americans' support or opposition to the death penalty for capital crimes. The data are from the 2020 ANES. The first is a simple example to illustrate the concepts involved in interpreting a logistic regression model. While the R code presents a fuller model and interpretation.

Often a goal of logistic regression modeling is to estimate probabilities given observations with the same values on the X independent variables. For example, a model of attitudes toward the death penalty (support or oppose) based on demographic characteristics could estimate the probability of supporting the death penalty given a particular set of individual characteristics such as education level, age, and race. The logistic regression model is then intended to provide estimates of the probability of thc dependent variable being equal to 1 (or "support") given values on the independent variables.

Consider the estimates for a simplified model, death penalty support (1) or opposition (0) as a function of just one X variable, individual educational attainment, measured on a five-point scale from 1 (less than a high school diploma) to 5 (a graduate degree). The education variable records degree attainment on an ordinal scale. For ease of interpretation, the model estimates one constant slope parameter across the scale rather than multiple slopes for categorical 'dummy variables'. Imagine we run the R code and find the model estimates correspond to this equation:

$$logit(deathpenaltysupport) = 1.5 - .302 * x_{education}$$

This equation includes a y-intercept and a slope on education. The model estimates are in units of 'logit' or log-odds: 1.5 is the logit of supporting the death penalty for someone with a hypothetical '0' score on the education variable. The slope on education, -.302, means that for each unit increase in the education scale, the logit of supporting the death penalty decreases by .302 units. A logit is not an intuitive metric. Fortunately, we can convert the estimates to odds ratios and probabilities. By exponentiating the y-intercept and slope (raising e, Euler's number\$ $\approx$ 2.718\$ to the power of the estimate), we remove the log, resulting in the odds and odds ratios. For example, while corresponding to a hypothetical person with a "0" education, the $e^{1.5} = 4.482$. (In **R**, `exp(1.5)`). For this hypothetical person, the odds of supporting the death penalty are 4.48 to 1, or 4.48 times more likely to be supported than not. While the scale of odds is practically limitless and generally dependent on units of measurement, these odds are very high.

For education, $e^{-.302} = .739$. The result, .739, is the odds ratio associated with a one-unit increase in the education level. An odds ratio of 0.739 means that for each one-unit increase in education, the odds of supporting the death penalty decrease. The odds are multiplied by 0.739 with each additional unit increase in education. The decrease in odds is calculated as $(1 - .739)$, a decrease of about .261 or 26.1 percent. In this sense, the odds of supporting the death penalty decrease by about 26.1 percent for each one-unit increase in education level. An alternative interpretation is based on probabilities; because probabilities are scaled from 0 to 1, some researchers prefer working with probabilities rather than odds ratios.

Calculating the predicted probabilities requires some additional arithmetic, to pull out the predicted probabilities within the model equation of the log-odds: $ln(\frac{p}{(1-p)}) = 1.5 - .302 * x_{education}$. To calculate the predicted probability for a person with a high school diploma ($education = 2$), we substitute 2 into the equation:

$$ln(\frac{p}{(1-p)}) = 1.5 - .302 \times 2$$

$$ln(\frac{p}{(1-p)}) = 1.5 - .604 = .896.$$

To remove the natural logarithm (ln), we exponentiate both sides of the equation:

$$\frac{p}{(1-p)} = e^{.896}$$

Calculating $e^{.896}$, (in **R** `exp(.896)`), is 2.45. From $\frac{p}{(1-p)} = 2.45$, we solve for p by multiplying both sides by $(1 - p)$ resulting in

$$p = 2.45 \times (1 - p)$$

From $p = 2.45 - 2.45p$, we rearrange terms to get to $p + 2.45p = 2.45$.

The terms $p + 2.45p$ combine as $1 + 2.45$, since the coefficient of the first p is 1.

So $3.45p = 2.45$. And finally, $p = \frac{2.45}{3.45}$, which equals about .71, the probability of someone with a high school diploma supporting the death penalty. The model assumes all observations with the same combinations of values on independent variables have the same probability on the dependent variable.

Fortunately, as demonstrated below, much of this process of calculating and interpreting odds ratios and predicted probabilities is automated with various **R** packages.

In what follows we estimate and interpret a logistic regression model of support for the death penalty, as a function of individual demographic characteristics. The code begins with attaching packages, importing the 2020 ANES dataset, creating the variables for inclusion in the model, followed by the estimation and interpretation of it.

Attach packages:

```
library(tidyverse)
library(survey)
```

Download the Workspace. (Take care to either set a working directory, move the Workspace to an RStudio project file location, or include the file directory path in the `load()` function.)

```
load("anes 2020 survey data.rdata")
```

```
anes_design<-svydesign(data = anes_20, ids = ~V200015c,
                       strata=~V200015d,
                       weights=~V200015b, nest=TRUE)
```

Variable *V201345x* measures support toward the death penalty in capital crimes. (See the Appendix for question wording.)

Responses are measured on a four-point scale from "Favor strongly" to "Oppose strongly". In the sample, combining the two favorable attitudes sums to 63 percent, with 37 percent opposed.

```
prop.table(svytable(~V201345x, anes_design))*100
```

```
## V201345x
##       1. Favor strongly  2. Favor not strongly
##                   42.06                  21.10
## 3. Oppose not strongly     4. Oppose strongly
##                   19.36                  17.47
```

The strength of the attitude can be collapsed to contrast favoring or opposing the death penalty.

```
anes_20<-anes_20 %>%
  mutate(death_penalty = fct_recode(V201345x,
              "Favor" = "1. Favor strongly",
              "Favor" = "2. Favor not strongly",
              "Oppose" = "3. Oppose not strongly",
              "Oppose" = "4. Oppose strongly")) %>%
  mutate(death_penalty = fct_rev(death_penalty))
```

The `fct_rev()` function reverses the direction of the scale, so that `Oppose` is the first or lower category, followed by `Favor` as the second or higher category.

```
str(anes_20$death_penalty)
```

```
##  Factor w/ 2 levels "Oppose","Favor": 1 1 1 1 1 2 1 NA 2 1 ...
```

As a factor stored variable, *death_penalty*, has two levels "Oppose" and "Favor" associated with numeric scores 1 and 2, respectively. The function for estimating the logistic regression model, `glm()` , will identify as the reference category the factor labels first in alphabetical order. So in this case, because "F" precedes "O" in the alphabet, the function will model the probability of a respondent being in "Favor" of the death penalty, scored at the reference category.

In preparing data, for any dependent variable, verify the factor levels are in the intended order. (The `relevel()` function could be used to set a specific reference category.) In estimating the model, the algorithm converts the factor variable to a numeric, with the reference category stored as *1* and the other category as *0*. To avoid the use of factor levels and potential confusion over reference categories, an alternative scaling is to convert the variable to a numeric, with two values 0 and 1. In this example, with contrasting "Oppose" at 0 and "Favor" at 1:

```
anes_20<-anes_20 %>%
  mutate(death_penalty = as.numeric(death_penalty)-1)

str(anes_20$death_penalty)
```

```
##  num [1:4783] 0 0 0 0 0 1 0 NA 1 0 ...
```

We will model support or opposition to the death penalty as a function of

- age (in years),
- sex (dichotomous identity, 0 male and 1 female),
- racial identification (dichotomous categories, 0 for white non-Hispanic and 1 for all others),
- educational attainment (the 1 through 5 attainment scale).

In this example, a researcher could estimate this model to examine how educational attainment (as the focal variable) shapes support for the death penalty, given a robust set of controls for other stable characteristics.

We will contrast these results with an additional model controlling for religious affiliation (Roman Catholicism) and religious service attendance, along with the multiplicative interaction between the two. In this second model, our interest will be in investigating to what extent either of these aspects of religion affects death penalty support, along with their interaction — whether nominal Catholic identifiers have a different attitude toward the death penalty compared to frequent service-attending Catholics.

For age, we create a new variable that copies the age in years from *V201507x*.

```
anes_20$age<-anes_20$V201507x
```

Creating the sex measure, for simplicity contrast *"2. Female"* with *"1. Male"*, by recoding to missing the respondents refusing to answer the question.

```
anes_20<- anes_20 %>%
  mutate(sex = fct_recode( V201600,
              NULL = "-9. Refused"))
```

And create a measure contrasting whether a respondent selected *"White, non-Hispanic"* recorded as 0 or any other response as 1:

```
anes_20<-anes_20 %>%
    mutate(race_category=fct_recode(V201549x,
              NULL="-9. Refused",
              NULL="-8. Don't know")) %>%
  mutate(race_minority = ifelse(race_category ==
                    '1. White, non-Hispanic', 0, 1))
```

The `fct_recode()` function nulls out the refused and don't know, followed by the `ifelse()` function, which creates a numeric variable. Both the factor variable *sex* and the numeric *race_category* are valid storage types for inclusion in the model as independent variables. For estimating the models either storage type (numeric or factor) is valid. It only affects the formatting of the results in the model output.

```
str(anes_20$race_minority)
```

```
##  num [1:4783] 1 0 0 0 0 0 1 1 0 0 ...
```

Education remains the five-point scale, from "1. Less than high school credential" to "5. Graduate degree". (Set the survey design and tabulate *education* to observe the proportions of the sample at each level.)

```
anes_20<-anes_20 %>%
  mutate(education = V201511x) %>%
  mutate(education = fct_recode(education,
                    NULL = "-9. Refused",
                    NULL = "-8. Don't know"))
```

Because we want to estimate a single (constant) slope across each level of education, the *education* variable must be stored or entered into the model as a numeric rather than factor (labeled) variable. Otherwise, as a factor, the model will treat the variable as nominal categories, estimating a separate slope for each education level, relative to a baseline. Rather than include a `as.numeric()` wrapper in the model function, we will convert education to a numeric storage type:

```
anes_20$education<-as.numeric(anes_20$education)
```

Check the storage (`str(anes_20$education)`) or tabulate the (unweighted) frequencies to observe the numeric storage categories, from 1 through 5.

12.3 Estimating the binary logistic regression model

To estimate the logistic regression model, the base R function is `glm()`. In relies upon the familiar formula based model specification, followed by the argument `family=quasibinomial(link="logit")` specifying the "logit" link function between the dependent and independent variables. Since the data are from a survey sample where observations must be weighted, weights are specified with a `weight=` argument. (Otherwise, the `weight=` argument is not needed.) The `glm()` function can account for the weighted observations in the ANES, but not the uncertainty accompanying the complex sampling design, which is reviewed with the `svyglm()` function from the **survey** package further below.

```
model1<-glm(death_penalty ~ sex + age + education + race_minority,
            family=binomial(link="logit"),
            data=anes_20, weight=V200015b)

summary(model1)
```

```
##
## Call:
## glm(formula = death_penalty ~ sex + age + education + race_minority,
##     family = binomial(link = "logit"), data = anes_20, weights = V200015b)
##
## Deviance Residuals:
##    Min      1Q  Median      3Q     Max
## -4.579  -0.948   0.532   0.903   2.765
##
## Coefficients:
##               Estimate Std. Error z value Pr(>|z|)
## (Intercept)    1.66159    0.13813   12.03  < 2e-16 ***
## sex2. Female  -0.21467    0.06414   -3.35  0.00082 ***
## age            0.00571    0.00188    3.03  0.00242 **
```

```
## education      -0.34993    0.02877  -12.16  < 2e-16 ***
## race_minority -0.58360    0.06837   -8.54  < 2e-16 ***
## ---
## Signif. codes:
## 0 '***' 0.001 '**' 0.01 '*' 0.05 '.' 0.1 ' ' 1
##
## (Dispersion parameter for binomial family taken to be 1)
##
##     Null deviance: 5821.4  on 4396  degrees of freedom
## Residual deviance: 5589.0  on 4392  degrees of freedom
##   (386 observations deleted due to missingness)
## AIC: 5458
##
## Number of Fisher Scoring iterations: 4
```

Interpreting the logistic regression model summary

The summary of the model (`summary(model1)`) resembles the summary output for `lm()`, read row-wise, although with a few differences. The coefficient estimates appear under `Estimate`, next to the standard error. The ratio of `Estimate` divided by `Std. Error` yields a z score, the `z value`. The `z value` corresponds to the probability under `Pr(>|z|)`, the p-value measure of statistical significance of each entry from `Estimate`.

A smaller p-value indicates a smaller probability of having observed in the sample an `Estimate` at least as large as the observed value, if the null hypothesis of no population level relationship is true. Using a conventional p-value of .05 to decide whether there is sufficient evidence within the sample to reject the null hypothesis, we would conclude based on the `Signif. codes` that sex, age, education, and race of the survey respondent are all statistically significant factors shaping their attitudes toward the death penalty. The signs on the slopes under `Estimate` indicate that being female, a member of a racial minority, and higher educational attainment are associated with opposition to the death penalty, while being older is associated with support.

Interpreting the magnitude of each slope coefficient, however, is not as straightforward as with the linear regression models reviewed in Chapters 10 and 11. Each entry under `Estimate` is on a scale of log-odds, that is the natural logarithm of the odds of supporting the death penalty. For example, the coefficient for sex (*sex2. Female*) is -0.21467. The negative sign indicates that being female is associated with a decrease in the log-odds of supporting the death penalty relative to being male.

In logistic regression, each coefficient represents the change in log-odds of the outcome associated with a one-unit increase in each corresponding independent variable. Equivalently, if we exponentiate the coefficient, it yields an odds ratio, which reflects the multiplicative change in the odds of the outcome. In this example, the coefficient compares females to males, and the exponentiated value represents the odds ratio of supporting the death penalty for females compared to males. A 'log-odds' scale is not an intuitively meaningful metric. So prior to interpretation, usually the estimates are transformed to odds ratios by exponentiation (i.e., applying the inverse of the natural logarithm). This transformation yields odds ratios, which are typically easier to interpret.

Odds are central to interpreting coefficients in a logistic regression model. Exponentiating each coefficient using `exp()` removes the log transformation and yields a more interpretable quantity: the odds ratio associated with a one-unit increase in the independent variable (holding other variables constant).

```
exp(model1$coefficients)
```

```
##   (Intercept)  sex2. Female            age
##        5.2677        0.8068         1.0057
##     education race_minority
##        0.7047        0.5579
```

Just like the linear regression models of Chapters 10 and 11, the intercept in a logistic regression model represents the expected value of the outcome when all independent variables are set to zero. Interpretation depends on whether those zero values are meaningful.

In *model1*, the intercept estimate is 1.66, which is on the log-odds scale. Exponentiating this value `exp(1.66159)`, results in 5.2677, which would mean the odds of supporting the death penalty are approximately 5.26 to 1 for a respondent who is male ($sex = 0$), non-Hispanic White ($race\_minority = 0$), and has values of zero on both *education* and *age*. The last two characteristics are out-of-sample or illogical, so the intercept is not meaningful to interpret.

For *sex2. Female* the exponentiated coefficient is .8068: the odds of supporting the death penalty for females are 0.8068 times the odds for males, holding other factors (education, racial minority status) constant. Note the comparison between the two categories of the *sex* variable; .8068 is an odds ratio, comparing the odds for females to the odds for males. Because the odds ratio is less than 1, it indicates that females have lower odds of supporting the death penalty relative to males. Proportional or percentage difference helps to clarify this gap in odds. Substracting the odds ratio from 1 and multiplying by 100 yields the percentage change in odds from males to females: $(1 - .8068) \times 100 = 19.32$ percent. The odds of supporting the death penalty are 19.32% lower for females than for males, controlling for the other covariates in the model.

For *age* the exponentiated coefficient is 1.0057, or just barely above 1. In the model summary, the coefficient is just slightly above 0. While statistically significant, the `Estimate` and exponentiated coefficient indicate that a one-year increase in age is associated with only a small increase in the odds of supporting the death penalty, controlling for other factors. A slope of 0 in the log-odds corresponds to an odds ratio of 1, meaning no change in odds. The coefficient on *age* is not exactly zero, showing there is a slight positive association between age and support for the death penalty. In percentage change terms $(1.0057 - -1) \times 100 = .57$percent, meaning that the odds of supporting the death penalty increase by approximately .57 percent for each additional year of age. While the model treats age as a continuous linear predictor, it is possible that the relationship between age and support for the death penalty is nonlinear — increasing, then at a certain age decreasing. To explore this possibility age could be grouped into age categories and modeled with reference to an older or younger age group.

For *as.numeric(education)* the exponentiated coefficient is 0.7047. Each additional unit increase on the education scale is associated with a decrease in the odds of supporting the death penalty by a factor of about 0.7047, holding other factors constant. Because the odds ratio is less than 1, higher levels of education are associated with lower odds of support. As a percentage change, we subtract the odds ratio from 1 and multiply by 100: $(1 - 0.7407) \times 100 = 29.53$ percent. Each unit increase in the education scale is associated with an approximate 29.5% decrease in the odds of supporting the death penalty, relative to the previous level of education.

For *race_minority* the exponentiated coefficient is 0.5579. Being a member of a racial or ethnic minority is associated with 0.5579 times the odds of supporting the death penalty compared to being non-Hispanic white, holding other factors constant. Because the odds ratio is less than 1, minority respondents have lower odds of supporting the death penalty relative to non-Hispanic Whites. The percentage decrease in odds is $(1 - .5579) \times 100 = 44.2$ percent. All else equal, the odds of supporting the death penalty are approximately 44.2% lower for members of racial or ethnic minority groups than for non-Hispanic Whites.

Using the `confint()` function, we can generate 95 percent confidence intervals for the model coefficients. The confidence intervals are regenerated rather than calculated directly on the existing model standard errors, so we apply the function to the model object:

```
confint(model1)
```

```
##                   2.5 %    97.5 %
## (Intercept)    1.392267  1.933840
## sex2. Female  -0.340497 -0.089034
```

```
## age            0.002025  0.009411
## education     -0.406541 -0.293761
## race_minority -0.717751 -0.449725
```

The output displays the lower and upper bounds for each coefficient on the log-odds scale. For example, the 95% confidence interval for `sex2. Female` is (-0.3405, -0.0890). The interval lies entirely below zero, reflecting the fact that the coefficient is statistically significant. To interpret the confidence interval on the odds ratio scale, we exponentiate the lower and upper bounds: $exp(-.3405) = .7114$ and $exp(-.0890) = .9148$. In percentage change terms, the population odds of supporting the death penalty for females are between 29% and 8.5% lower than for males, holding other factors constant.

Estimating the model with the complex survey design

Compare the estimates (and confidence intervals) from *model1* to the logistic regression estimates accounting for the complex survey design. Since we created additional variables, we need to re-specify the complex survey design:

```
anes_design<-svydesign(data = anes_20, ids = ~V200015c,
                       strata=~V200015d,
                       weights=~V200015b, nest=TRUE)
```

The model function to incorporate the survey design is simply `svyglm()`, with the same arguments.

Saving model results as *model2*:

```
model2<-svyglm(death_penalty ~ sex + age+ education + race_minority,
               family=quasibinomial(link="logit"),
               design=anes_design)

summary(model2)
```

```
##
## Call:
## svyglm(formula = death_penalty ~ sex + age + education + race_minority,
##     design = anes_design, family = quasibinomial(link = "logit"))
##
## Survey design:
## svydesign(data = anes_20, ids = ~V200015c, strata = ~V200015d,
##     weights = ~V200015b, nest = TRUE)
##
## Coefficients:
##               Estimate Std. Error t value Pr(>|t|)
```

```
## (Intercept)     1.66159    0.19315    8.60  3.9e-11 ***
## sex2. Female   -0.21467    0.09703   -2.21    0.032 *
## age             0.00571    0.00230    2.49    0.017 *
## education      -0.34993    0.04283   -8.17  1.7e-10 ***
## race_minority -0.58360    0.09621   -6.07  2.3e-07 ***
## ---
## Signif. codes:
## 0 '***' 0.001 '**' 0.01 '*' 0.05 '.' 0.1 ' ' 1
##
## (Dispersion parameter for quasibinomial family taken to be 1.004)
##
## Number of Fisher Scoring iterations: 4
```

Notice the coefficient estimates in *model2* are identical to those in *model1*, because the estimates are based on the same underlying sample. The standard errors, however, are larger given the adjustments for the greater uncertainty of the complex sample. Despite the larger standard errors, the coefficients for sex and age remain statistically significant at the conventional 0.05 level. Still, note how the p-values and standard errors increase when the survey design is correctly specified. As in the linear model examples from earlier chapters, these results demonstrate the importance of incorporating the complex survey design where applicable.

12.4 Generating model-based predicted probabilities

Researchers may prefer probabilities, rather than odds ratios, to present and interpret model estimates. The probabilities, on the outcome (1) of the dependent variable, are calculated on particular combination of values on the independent variables, in tabular or graphical form. For example, given *model2*, perhaps what would be of interest are the probabilities of males and females (non-racial minority status) as education varies from lowest to highest. Below we will calculate these probabilities via the `predict()` function and graphing the probabilities in **ggplot2**. There are, of course, several packages that will calculate and visualize predicted probabilities more automatically, which are reviewed briefly.

One method to calculate probabilities is to first specify a table of covariate (independent variable) values as the basis for generating the predicted probabilities:

```
predictiondata<-tribble(
~education, ~sex, ~race_minority, ~age,
```

```
1, "2. Female", 0, 47.5,
2, "2. Female", 0, 47.5,
3, "2. Female", 0, 47.5,
4, "2. Female", 0, 47.5,
5, "2. Female", 0, 47.5)
```

The *predictiondata* is a data table created by the `tribble()` function. The first line displays the names of the three independent variables specified within *model2*. Notice that although we entered *education* into the `svyglm()` function as `as.numeric(education)`, for creating the predicted probabilities the name of the dataset variable is sufficient. The subsequent five lines, however, display the combination of values on the covariates as entered in the model – starting with females scored as "2. Female" in the *sex* variable, and varying over levels of education recorded numerically, from 1 to 5, and a 0 for non racial minority status.

We use the `predict()` function to generate the predicted probabilities. Note that the argument `type="response"` is necessary; without it, `predict()` will generate predictions on the logistic regression model log-odds scale.

```
predict(model2, predictiondata, type="response")
```

```
##    response   SE
## 1     0.797 0.02
## 2     0.735 0.02
## 3     0.661 0.01
## 4     0.579 0.02
## 5     0.492 0.02
```

The predicted probability appears under the heading *response*. Each row corresponds to the prediction for an average person with the characteristics in the *predictiondata* table. For example, the first row (coincidentally labelled *1* for the lowest score on the education variable) shows a predicted probability of .802 for White, non-Hispanic females with less than a high school diploma. Each additional row shows the corresponding probability as educational attainment increases to *5*, a graduate degree. Among White, non-Hispanic females with a graduate (post four year university) degree the probability of supporting the death penalty is '50-50', or .501. Next to the probabilities under the heading *SE* is the standard error for each probability.

An equivalent set of predictions for men would begin with another table of values on the covariates:

```
predictiondata2<-tribble(
~education, ~sex, ~race_minority, ~age,
1, "1. Male", 0, 47.5,
2, "1. Male", 0, 47.5,
3, "1. Male", 0, 47.5,
4, "1. Male", 0, 47.5,
5, "1. Male", 0, 47.5 )
```

Followed by

```
predict(model2, predictiondata2, type="response")
```

```
##    response   SE
## 1     0.830 0.02
## 2     0.774 0.02
## 3     0.707 0.02
## 4     0.630 0.02
## 5     0.546 0.02
```

Researchers often present tables of probabilities such as these, and given sampling uncertainty, accompanied by 95 percent confidence intervals. To add confidence intervals, store each of the predictions as an object. Predictions for males will be stored as *preds_m*, and for females *preds_f*:

```
preds_m<-predict(model2, predictiondata, type="response")
preds_f<-predict(model2, predictiondata2, type="response")
```

Converting each to a dataframe:

```
preds_m<-as.data.frame(preds_m)
preds_f<-as.data.frame(preds_f)
```

Then we use specific columns from the dataframe to create the confidence intervals.

Storing the probabilities for males in an object *predm*:

```
predm <- preds_m$response
```

Then creating the lower and upper bounds for a 95 percent confidence interval on the predicted probability:

```
lowerm <- preds_m$response - (1.96*preds_m$SE)
upperm <- preds_m$response + (1.96*preds_m$SE)
```

Then repeating these steps for females:

```
predf<-preds_f$response
lowerf <- preds_f$response - (1.96*preds_f$SE)
upperf <- preds_f$response + (1.96*preds_f$SE)
```

The predicted probabilities form a probability 'profile', probabilities for individuals with specific characteristics. Often these probabilities, along with confidence intervals, are presented in tabular form. Of course, we can graph the predicted probabilities and confidence intervals. We need to create a dataset for visualization, the predicted probabilities and boundaries for the confidence intervals. We could create the dataset as a tribble, or pull the data directly from the different objects with the `data.frame()` function:

```
logit_male_predictions<-data.frame(
education = 1:5,
predicted=predm,
lower=lowerm,
upper=upperm
)
```

```
logit_male_predictions
```

```
##   education predicted  lower  upper
## 1         1    0.7971 0.7581 0.8362
## 2         2    0.7347 0.7008 0.7686
## 3         3    0.6612 0.6326 0.6897
## 4         4    0.5790 0.5478 0.6102
## 5         5    0.4922 0.4485 0.5359
```

This *logit_male_predictions* could be the basis of a table, or a graph across each series across the range of education levels. Figure 12.2 displays the predicted probability of supporting the death penalty.

```
ggplot(logit_male_predictions, aes(x = education, y = predicted)) +
  geom_point(color = "blue", size = 3) +
  geom_errorbar(aes(ymin = lower, ymax = upper),
                width = .1, color="blue") +
  labs(x = "Education (1 less than high school, 5 graduate degree)",
```

```
    y = "Probability of supporting the death penalty",
    subtitle="'white non-hispanic' male respondents") +
  theme_minimal() +
  scale_x_continuous(breaks = 1:5) +
  scale_y_continuous(limits = c(0, 1))
```

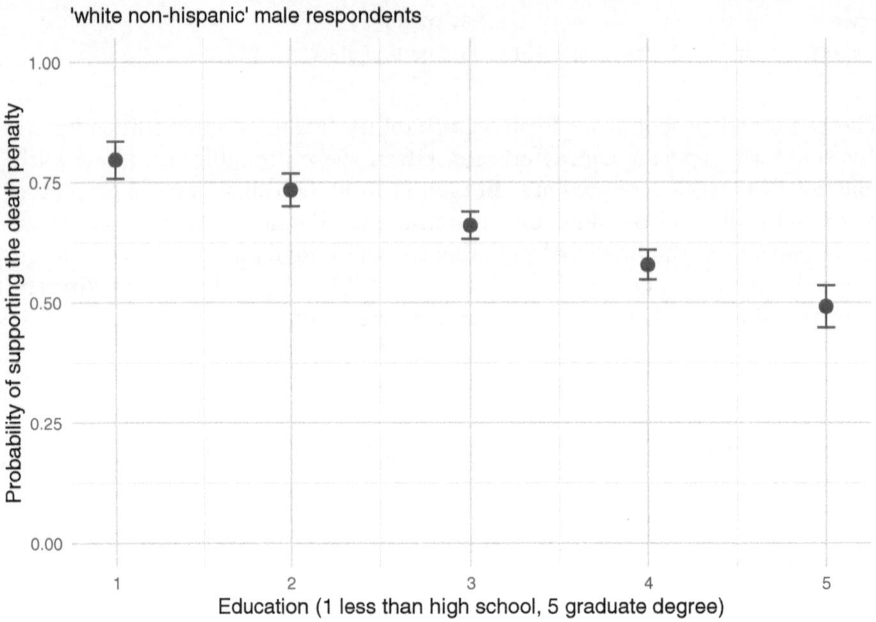

FIGURE 12.2 Predicted probability of supporting the death penalty across levels of education for White, non-Hispanic male respondents, with 95 percent confidence intervals.

Graphing female and male respondents together requires creating the probabilities dataset in long format. The values for female respondents should be added, or row bound, to the male respondents, with an additional variable to record sex.

```
logit_male_predictions<-logit_male_predictions %>%
  mutate(sex="male")
```

```
logit_female_predictions<-data.frame(education = 1:5,predicted=predf,
    lower=lowerf, upper=upperf, sex="female")
```

Using `rbind()` to bind the rows for female respondents below males.

```
logit_predictions<- rbind(logit_male_predictions,
                          logit_female_predictions)
```

```
logit_predictions
```

```
##    education predicted  lower  upper    sex
## 1          1    0.7971 0.7581 0.8362   male
## 2          2    0.7347 0.7008 0.7686   male
## 3          3    0.6612 0.6326 0.6897   male
## 4          4    0.5790 0.5478 0.6102   male
## 5          5    0.4922 0.4485 0.5359   male
## 6          1    0.8296 0.7919 0.8674 female
## 7          2    0.7744 0.7397 0.8090 female
## 8          3    0.7075 0.6772 0.7378 female
## 9          4    0.6303 0.5992 0.6613 female
## 10         5    0.5457 0.5042 0.5873 female
```

Then plotting the data from *logit_predictions* in Figure 12.3:

```
ggplot(logit_predictions, ) +
  geom_point(aes(x = education, y = predicted, color = sex), size = 3) +
  geom_errorbar(aes(x = education, y = predicted, color = sex,
                    ymin = lower, ymax = upper), width = .1) +
  labs(x = "Education Level", y = "Predicted Probability",
       color = "Sex") +
  theme_minimal() +
  scale_x_continuous(breaks = 1:5) +
  scale_y_continuous(limits = c(0, 1))
```

Plotting sex across education implies interest in testing an interaction between sex and educational attainment in support for the death penalty. The inclusion and graphing of multiplicative interaction terms (such as *sex* multiplied by *education*) is discussed further below.

Quick package functions for model-based predictions

Of course, several packages will automatically create tables and figures of odds ratios and probabilities. The Resources section at the end of the chapter mentions a few. The core function in the **effects** (Fox et al., 2022b)package creates a graphic of predicted probabilities across levels of predictors. In the **effects** package, the function `predictorEffects()` will plot a panel of graphics displaying predicted probabilities.

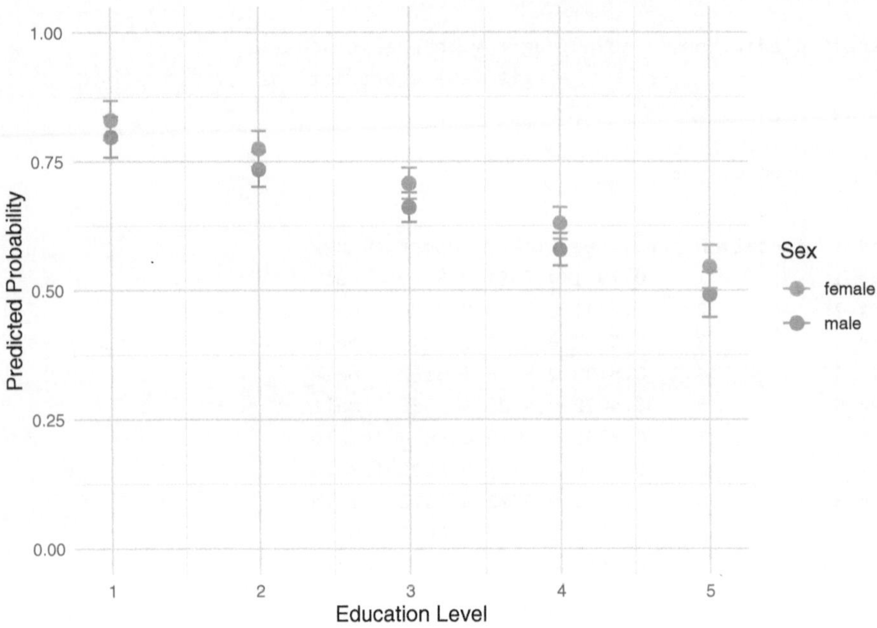

FIGURE 12.3 Predicted probabilities of supporting the death penalty by sex and educational attainment (White, non-Hispanic), from less than high school (1) to graduate degree (5). The error bars (95 percent confidence intervals) overlap, displaying the lack of statistically significant difference between sexes at each level of educational attainment.

With the name of the model object as the sole argument, we wrap this function inside `print()`:

```
plot(predictorEffects(model2))
```

The result is a default graphic of predicted probabilities and confidence intervals across the range of all predictors included in the model. Typically researchers will create such graphics for focal predictors, rather than across all covariates. For example, if testing a hypothesis that death penalty support decreases as educational attainment increases, with education as the focal predictor it would be preferable to examine predicted probabilities over the range of education. The `Effect()` function will calculate the predicted probabilities and confidence intervals. For example, to print a set of predicted probabilities for education, the function would be `print(Effect("education", model2))`.[1]

[1] A simple way to add labels for axes and a title would be to save the predictions as an object and print the object with arguments for labels `education_effect <- Effect("education", model2)`, then `plot(education_effect, type = "response", main = "Effect of Education on Support for Death Penalty", xlab = "Education Level", ylab = "Predicted Probability of`

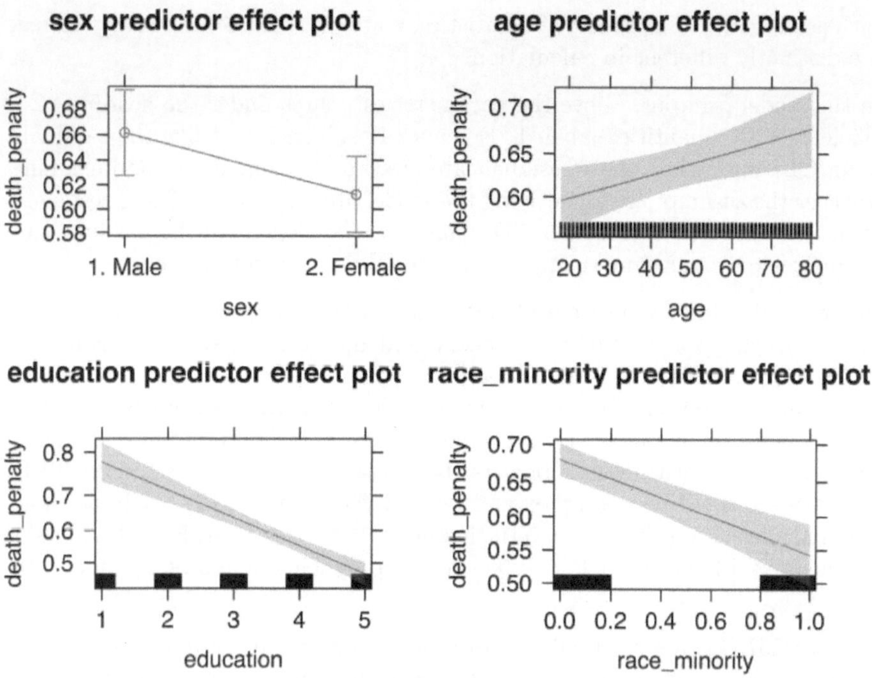

FIGURE 12.4 Panel of predicted probabilities from the **effects** package. The `predictorEffects()` function creates a graphic displaying predicted probabilities across each range of each independent variable. For each panel, all other independent variables are held at a mean (for numeric) or reference category (for factor) variables.

Another package for interpreting model summaries is **jtools**. The `effect_plot()` function in **jtools** for a logistic regression will plot predicted probabilities across levels of a predictor. For example, to plot predicted probabilities given age: `effect_plot(model1, pred = age, interval=TRUE)`.

12.5 Assessing model fit

Assessment of model fit for a binary logistic regression, like the OLS models of Chapters 10 and 11, is measured both in absolute terms and relative to a nested model. Just as in those chapters, it is complicated by a survey analysis function

`Supporting Death Penalty").`

such as svyglm(). The specific fit statistics, however, while conceptually related, are distinctly different in calculation.

In the model printout, above the parameter estimates under the heading "Coefficients", are quantities labeled "Deviance Residuals". Just like the residuals in an OLS regression, these residuals are used to assess model fit and examine whether the assumptions underlying the model are reasonable. These residuals, sometimes simply referred to as "Deviance", are analogous to the residual sum of squares (or error sum of squares) in the OLS regression model.

Before explaining Deviance and the relationship to model fit, we must first briefly touch upon a statistical theory and method, Maximum Likelihood Estimation (MLE). The logistic regression models are estimated through an algorithm that powers the glm() function. The algorithm is iterative, like the algorithm for scaling analyses in Chapters 7 and 8, but different in one crucial respect. The algorithm that powers the glm() function is based upon MLE. The subject of MLE is complex and beyond the scope of this textbook. (See Fox and Weisberg (2011).) An intuition behind the idea of MLE in the context of generalized linear models is sufficient to apply logistic and other models to data.

In brief, MLE is a statistical criterion used to find the parameter estimates in the model. A likelihood is simply the idea of measuring how typical or probable an observation is given a sample, or how typical a particular sample is given a population. In the context of a logistic regression model, the MLE algorithm queries how likely an individual is to have a particular value on a dependent variable Y (such as supporting or opposing the death penalty), given the values of the independent variables (e.g., gender, age) and the estimated model parameters (the slopes and y-intercept). Basically, the MLE solution provides the model estimates that are most consistent with the observed data.

The MLE algorithm operates iteratively, step by step, attempting to find the best solution based on a 'convergence criterion' — a standard for evaluating whether a different solution would be better than the previous one. When the algorithm stops, the MLE results are said to have 'converged' on a solution. However, this process can sometimes fail to work, leading to 'convergence failure', due to errors in the data, multicollinearity among predictors, or even the problem of "separability" in which at least one predictor is too neatly correlated with the dependent variable. For a more detailed explanation of model estimation and potential pitfalls to estimation, see Long (1997).

The starting point for assessing model fit begins with the quantity used to evaluate the MLE solution, the measured value of this maximum likelihood algorithm itself. Every model, once the algorithm has converged on a solution, has a maximum likelihood (ML) value. Comparing these values to MLs under different model specifications is one way to measure fit. And it involves a second quantity, termed "deviance", to assess model fit to the data.

Information about deviance appears in the model printouts, as in `summary(model1)`. The deviance is somewhat analogous to sums of squares that appear in an OLS regression model printout. Deviances can be measured for individual observations used to estimate a model, or the deviances can be combined to calculate an overall summary deviance measure for the model. It is used to quantify fit in generalized linear models such as logistic regression. Like a comparison of sums of squares, such as a ratio comparing residual to total sums of squares to construct an r^2 model fit statistic in OLS regression, the Deviance is often interpreted in ratio form as well.(For a more in-depth explanation of deviance and ML values, see Fox and Weisberg (2019).)

Let us start with the deviance reported in `summary(model1)`. The "Null deviance" is a sort of baseline deviance, the deviance corresponding to a measure with no independent variables (no predictors) at all. Any model with predictors (such as *model1*) can be compared against it, to observe if the model improves upon the null deviance. The "Residual deviance" refers to the deviance score for the estimated model, in this case for *model1* the model with four independent variables. If the deviance for this model is significantly lower than the deviance for the null model — analogous to the idea of having smaller residuals in one model compared to another — then the fuller model fits the data better than the null model, or the model nested within it. For a more technical explanation of deviance calculations and comparisons, see Long (Long, 1997).

The summary of the model reports a "Null deviance" of 5821.4 "on 4396 degrees of freedom". The 'degrees of freedom' are equal to the effective number of observations N minus k parameters. The null model parameter is 1 (for the intercept), so the degrees of freedom are $N - 1$. The "Residual deviance" of 5589 on $N - 5$ degrees of freedom (accounting for four slopes and the intercept). To what extent is the smaller Residual deviance enough of an improvement that we can conclude the proposed model fits the data better than the null model? The likelihood ratio test (LRT) is used to compare the goodness of fit of two nested models, in this case comparing a null to the proposed (fitted) model.

Comparing a fitted to a null model

Do the estimated slopes have any 'value added' to the model? The two deviances can be compared in a test statistic, often abbreviated as G for a 'goodness of fit' test (Cohen et al., 2003). The test, under the null hypothesis of no improvement in the fitted model compared to the null model, is distributed as chi-squared (χ^2) with degrees of freedom equal to the number of k predictors in the fitted model (or alternatively the difference in degrees of freedom between fitted and null deviances). The test statistic is the difference $G = \textit{Null deviance} - \textit{Residual deviance}$. A larger value of G provides evidence that the fitted model offers a better fit than the null model.

It is important to keep in mind that complex survey designs — as opposed to data collected through a simple random sample — can complicate the exact calculation of the test statistic. Nonetheless, for a general model comparison, we can still calculate an approximate goodness-of-fit test using the χ^2 `pchisq()` function in **R**. We begin by calculating the test statistic: $G =$ Null deviance - Residual deviance $= 5821.4 - -5589.0 = 232.4$.

The degrees of freedom correspond to the number of additional parameters in the fitted model, or equivalently, the difference in degrees of freedom between the null and residual models:

$$df = df_{\text{null}} - df_{\text{residual}} = 4396 - -4392 = 4$$

We then compute the p-value associated with the test statistic through the `pchisq()` function, which returns the cumulative probability up to a specified value under the χ^2 distribution.

Since we want the upper-tail probability (the area beyond our observed G value), we subtract this value from 1: `1 - pchisq(232.4, 4)`. Running this line will return a p-value of zero, a quantity too small for the standard precision limits in **R**. The extremely low p-value, given any reasonable threshold such as ($p < .05$ or $p < .01$), leads to the rejection of the null hypothesis of no difference in model fit between the null and fitted model. In short, the inclusion of the predictors significantly improves model fit, even under this conservative test.

Base **R** functions and various packages perform model comparison tests automatically, and in some cases, apply adjustments for the survey design. For example, applying the `anova()` function to the fitted model and null (intercept only model) yields the results of a more specific, design-adjusted test comparing the two model likelihoods (Fox and Weisberg, 2011). First we estimate the null model and save it as `null_model`:

```
null_model<-svyglm(death_penalty ~ 1,
              family=quasibinomial(link="logit"),
              design=anes_design)
```

The `~ 1` indicates the model should be 'intercept only'. Comparing the two models *model2* and *null_model* (including the argument `type="ChiSq"` for the χ^2 test):

```
anova(null_model, model2, test="Chisq")
```

```
## Working (Rao-Scott) LRT for sex age education race_minority
##  in svyglm(formula = death_penalty ~ sex + age + education + race_minority,
##     design = anes_design, family = quasibinomial(link = "logit"))
```

```
## Working 2logLR =  283.8 p= <2e-16
## (scale factors:  1.4 1.1 0.87 0.62 )
```

The output reports a modified test, the "Working (Rao-Scot)" likelihood ratio test, which adjusts the standard likelihood ratio test to account for the complex survey design. In this example, the test statistic is 283.8, reflecting the improvement in model fit from adding predictors. The associated p-value is effectively 0, indicating that such a large test statistic would be extremely unlikely under the null hypothesis of no improvement in fit.

Given a standard significance cutoff such as ($\alpha = .05$) we reject the null hypothesis and conclude that the fitted model significantly improves upon the null model. The reported scale factors are internal adjustment values used to account for the survey design and are not directly used when interpreting the test result.

Another approach to comparing nested models is the **car** package `Anova()` function — notice the capital `A` — that compares improvements in model fit for individual predictors. That is, the function returns a likelihood ratio test for the improvement in model fit for each predictor versus excluding that predictor (Fox and Weisberg, 2011). Including *model2* as an argument:

```
library(car)
Anova(model2)
```

```
## Analysis of Deviance Table (Type II tests)
##
## Response: death_penalty
##                Df Chisq Pr(>Chisq)
## sex             1  4.89      0.027 *
## age             1  6.19      0.013 *
## education       1 66.74    3.1e-16 ***
## race_minority   1 36.79    1.3e-09 ***
## ---
## Signif. codes:
## 0 '***' 0.001 '**' 0.01 '*' 0.05 '.' 0.1 ' ' 1
```

The "Analysis of Deviance Table" should be read line by line. Each line shows likelihood ratio test statistics for the exclusion of each predictor. For example, in the line `sex`, the `DF` stands for the degrees of freedom in the test, which is 1 for each predictor excluded in the test model. The `Chisq` value measures how much the inclusion of each predictor improves the fit of the model compared to the model excluding the predictor — along with the p-value `Pr(>Chisq)` reporting the probability of observing the `Chisq` test statistic at least that large if the null hypothesis of no model improvement is true. In the case of `sex` and each additional row, the small p-value shows that we can reject the null of no improvement for each predictor. The inclusion of each predictor, relative to a model excluding that individual predictor, improves the model.

Pseudo R^2 statistics

The model fit statistic R^2 from OLS regression has no direct analog in logistic regression, but typically researchers construct "pseudo-R^2" statistics to assess model fit. The simplest "pseudo-R^2" is to compare the ratio of fitted (or residual) model deviance to null (or total) deviance. This ratio, subtracted from 1, could be interpreted as the proportion of the total deviance explained by the model.

"Pseudo-R^2" $= 1 - \frac{fittedmodeldeviance}{nulldeviance}$

This statistic is often referred to as a "McFadden's Pseudo R^2" (Long, 1997). The quantity does not have the same interpretation as an R^2 in OLS regression, in terms of 'percent of variation explained in the dependent variable'. Many different modifications to it have been proposed; often these modified pseudo-R^2 statistics are reported instead.

Recall in the model summary for `glm()` reports two deviance values: Residual deviance and Null deviance. These two quantities are used to calculate "McFadden's Pseudo R^2". For example, we can calculate the pseudo-R^2 for *model1* as:

```
1 - (model1$deviance / model1$null.deviance)
```

```
## [1] 0.03993
```

While this value may appear low compared to the the typical linear regression R^2 values, pseudo-R^2 values in logistic regression are typically much smaller and should not be interpreted as "percent of variance explained". Still, a value of .03993 suggests that the model provides a modest improvement in fit over the null model.

The same components are stored in models estimated with `svyglm()`. For *model2*, accessing `model2$deviance` and `model2$null.deviance` and calculating the pseudo-R^2 in the same way yields the same result. One common variation is an adjusted pseudo-R^2, sometimes referred to as "McFadden's adjusted Pseudo R^2". This statistic penalizes model complexity by substracting the number of estimated parameters k from the residual deviance:

"Adjusted Pseudo-R^2" $= 1 - \frac{fittedmodeldeviance - k}{nulldeviance}$

For further discussion of pseudo-R^2 measures and relevant **R** packages, see Urdinez (2022). The **DescTools** package (Signorell and many others, 2024) calculates various measures of fit and data visualizations, as will **jtools** (Long, 2023), which supports the `svyglm()` function. The section further below describing the tabular and graphical presentation of model estimates presents how to find some of these quantities with **jtools**.

Akaike Information Criterion (AIC)

The Akaike Information Criterion (AIC) is another model fit statistic appears in the output of `glm()` models, such as *model1*. It is based on the model's log-likelihood function and provides a measure of model quality that balances goodness of fit with model complexity.

The AIC is most often used as a relative fit statistic, useful for comparing models that differ in specification, including both nested and non-nested models. A smaller AIC statistic indicates better fit.[2]. See Urdinez and Cruz (2022) and Fox and Weisberg (2019) for a further discussion of assessing the AIC in R.

12.6 Model summary tables and coefficient graphics

Several packages format model summaries for neat tables and graphical summaries. The **jtools** package supports both `glm()` and `svyglm()` estimators. The `summ()` function in **jtools**, such as `summ(model1)`, produces a summary table of model estimates and fit statistics[3]

```
library(jtools)
```

Adding an argument `exp=TRUE` will print the exponentiated coefficients:

```
summ(model2, exp = TRUE)
```

Observations	4397
Dependent variable	death_penalty
Type	Survey-weighted generalized linear model
Family	quasibinomial
Link	logit

Pseudo-R^2 (Cragg-Uhler)	0.06
Pseudo-R^2 (McFadden)	0.04
AIC	NA

[2]The AIC is calculated as $2k - 2ln(L)$, with k the number of slopes plus the intercept in the model, and L the model likelihood function

[3]See https://jtools.jacob-long.com/ for additional examples, particularly https://cran.r-project.org/web/packages/jtools/vignettes/summ.html for exporting tables to Word or PDF format.

	exp(Est.)	S.E.	t val.	p
(Intercept)	5.27	0.19	8.60	0.00
sex2. Female	0.81	0.10	-2.21	0.03
age	1.01	0.00	2.49	0.02
education	0.70	0.04	-8.17	0.00
race_minority	0.56	0.10	-6.07	0.00

Standard errors: Robust

The result is the above summary table, along with fit statistics, displaying exponentiated coefficients and corresponding components of a null hypothesis test. Researchers frequently publish graphical depictions of model estimates, and save coefficient tables for appendices. The `plot_summs()` function will plot a figure displaying model estimates and confidence intervals.

12.7 Checking model assumptions and diagnostics

For any independent variables entered into the model as numeric measures — that is, treated as continuous measures, not categorical (nominal or ordinal) — it is necessary to check whether the relationship between continuous measures and the dependent variable could be reasonably summarized with a linear slope. For example, in the models of death penalty support, if there were evidence of a nonlinear relationship between age or education and support for the death penalty, perhaps because support increases with age but at some point decreases over additional years, this nonlinear relationship would need to be accounted for in the model. Just as with OLS regression, checking residual plots can help identify any potential nonlinearities.

Researchers usually check the "deviance" residuals rather than the Pearson's residuals for a logistic regression, although both tend to lead to the same conclusions. (See Long (1997) for a further explanation and formula for calculation.) A standard scatterplot of residuals versus fitted values — and trend line — helps to identify potential problems. Remember the key is to look for any unusual systematic pattern. The deviance residuals should be 'randomly' distributed around zero. And the trend line – showing how residuals (on the Y-axis) vary over fitted values – should be more or less flat at 0 and straight across the X-axis. A distinct U-shape in residuals and the trend line would evidence of potential nonlinearity.

We construct a scatterplot for *model2*, from the deviance residuals stored from the `residuals()` function and `type="deviance")`. Fitted values are extracted from one of the *model2* components:

```
deviance_residuals <- residuals(model2, type = "deviance")
```

The *deviance_residuals* are stored as a vector, which we need to combine with the fitted values to create a dataframe. One way to do so is to combine the two with the `data.frame()` function, defining the two columns: `data.frame(fitted = fitted(model2), residuals = deviance_residuals)`. The data source and aesthetic are included in the `ggplot()` function, since we will create two geometries, the points and the smoothed line. Adding some labels:

```
ggplot(data.frame(fitted = fitted(model2),
                  residuals = deviance_residuals),
  aes(x = fitted, y = residuals)) +
  geom_point() +
  geom_smooth(method = "loess", se = FALSE, color = "red") +
  labs(x = "fitted values", y = "deviance residuals") +
  theme_minimal()
```

Figure 12.5 displays the residuals (on Y) versus fitted values (on X) for *model2*. (The figure based on *model1* would be identical.) The residuals are centered around 0 in a pattern of two bands, which is not unusual given the categorical variables included in the model. If the model is 'well behaved', there would be no obvious unusual pattern in the figure and the trend line would be straight. The pattern of the residuals, however, especially in the lower right quadrant displays a fanning problem that potentially signals a problem – the model fits worse at higher predicted values. Of course, detecting a potential problem does not reveal the precise source of the problem (Fox and Weisberg, 2011). In this example, it is likely that assuming a constant increase in the probability of supporting the death penalty for additional steps in educational attainment is a source – necessitating a series of categorical variables to model these changes relative to a baseline. Generally, the diagnostic tools from the **car** package reviewed in Chapter 11 will help, such as plotting residuals against individual predictors, or using metrics to identify outliers, or observations with unusual leverage or influence. Variance Inflation Factors (VIFs) can be inspected to detect potential multicollinearity.

Yet the diagnostic 'rules of thumb' in OLS regression do not uniformly apply to logistic regression. Thus if leverage or influence is a potential concern for particular observations, for example, consult a source with a more in-depth review of the potential pitfalls of using these tools in logistic regression, such as Long (1997).

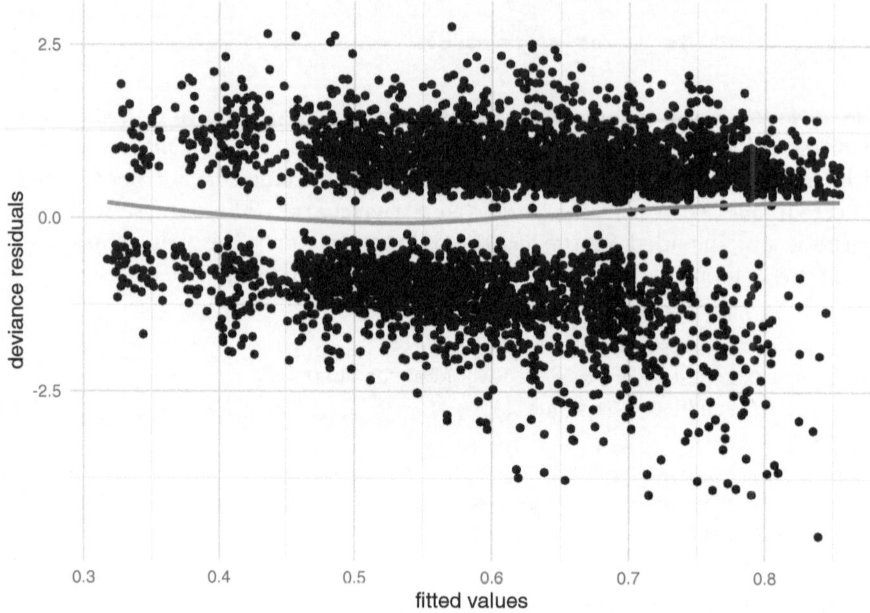

FIGURE 12.5 Plot of residuals versus fitted values from *model2*. While the two blobs of residuals centered around 0 on the Y-axis are not unusual for a model with categorical X-variables, the fanning pattern (and curvy trend line) from top-left to bottom-right reveal potential problems in model specification, particularly the problem of heteroskedasticity. The larger spread of residuals at higher fitted values shows that the model fit varies — fit is worse at higher predicted values.

12.8 Modeling and interpreting interactions

Since predicted probabilities can be calculated for any combination of independent variables, a researcher could implicitly test an interaction – for example, observing how the probability of supporting the death penalty over levels of educational attainment is different for men versus women. Calculating the probability acknowledges an interest in testing an interaction — whether sex alters the strength and/or direction of the relationship between education and support for the death penalty. A multiplicative interaction term should be specified within the model to test this intuition.

To compare *model2* with a fuller model (investigating the interaction between *sex* and *education* is left for the exercises), consider an interaction between Catholic religious identity and frequency of religious service attendance, reported in ordinal categories such as "monthly" and "weekly". The interaction between the two could be viewed from two perspectives. Nominal, non-attending Catholic identifiers may differ from Catholic identifiers attending services frequently in support of the death penalty, while Catholics, over varying levels of religious service attendance, may differ in support from non-Catholic identifiers.

We re-estimate the model after creating a *Catholic* identifier and a numeric version of the *attend* variable. This interaction – a qualitative category compared against a rank order of service attendance treated as a numeric, continuous measure — is known as a 'categorical by continuous' interaction.

```
anes_20 <- anes_20 %>%
  mutate(Catholic=ifelse(relig_trad=="4. Roman Catholic", 1, 0))

anes_20$attend_num<-as.numeric(anes_20$attend)-1
```

For the interaction, service attendance *attend_num* must be saved as a numeric measure; for *Catholic*, it could be saved either as a labeled factor or numeric. With `ifelse()`, a numeric is simply faster to create. Subtracting 1 from the resulting numeric measure means the scale ranges from 0 for 'never' attendees to 3 for weekly attendees.

To use the `svyglm()` estimator, since we created two new variables in the *anes_20* dataframe we rerun the `svy_design()` function.

```
anes_design<-svydesign(data = anes_20, ids = ~V200015c,
                       strata=~V200015d,
                       weights=~V200015b, nest=TRUE)
```

Then we include the interaction between *Catholic* and *attend_num* in the `svyglm()` function as `Catholic*attend_num`, along with the individual components of the interaction. This interaction could be saved directly into the dataset as the product of the two variables, saving it as `anes_20 %>% mutate(CatholicXattend_num = Catholic*attend_num)`. The model is saved as *model3*:

```
model3<-svyglm(death_penalty ~ sex + age+ education + race_minority +
                 Catholic + attend_num + Catholic*attend_num,
               family=quasibinomial(link="logit"),
               design=anes_design)

summary(model3)
```

```
##
## Call:
```

```
## svyglm(formula = death_penalty ~ sex + age + education + race_minority +
##     Catholic + attend_num + Catholic * attend_num, design = anes_design,
##     family = quasibinomial(link = "logit"))
##
## Survey design:
## svydesign(data = anes_20, ids = ~V200015c, strata = ~V200015d,
##     weights = ~V200015b, nest = TRUE)
##
## Coefficients:
##                     Estimate Std. Error t value
## (Intercept)          1.62055    0.18814    8.61
## sex2. Female        -0.22467    0.09678   -2.32
## age                  0.00539    0.00240    2.25
## education           -0.35463    0.04400   -8.06
## race_minority       -0.60556    0.09967   -6.08
## Catholic             0.40203    0.14161    2.84
## attend_num           0.09007    0.04427    2.03
## Catholic:attend_num -0.34877    0.10414   -3.35
##                     Pr(>|t|)
## (Intercept)          6.5e-11 ***
## sex2. Female          0.0251 *
## age                   0.0298 *
## education            3.9e-10 ***
## race_minority        2.8e-07 ***
## Catholic              0.0069 **
## attend_num            0.0481 *
## Catholic:attend_num   0.0017 **
## ---
## Signif. codes:
## 0 '***' 0.001 '**' 0.01 '*' 0.05 '.' 0.1 ' ' 1
##
## (Dispersion parameter for quasibinomial family taken to be 1.007)
##
## Number of Fisher Scoring iterations: 4
```

The interpretation of the interactive term (and components) follows the same logic as the multiplicative interaction discussed in Chapter 10. Recall that these are no longer 'independent' marginal effects. In brief:

The negative sign on the interaction term `Catholic:attend_num` slope means that, for Catholics, increased service attendance is associated with decreased support for the death penalty. As religious service attendance increases, support for the death penalty decreases — the quantity `-0.34877` represents the additional effect of service attendance on the log-odds of supporting the death penalty for Catholics.

The slopes for the constituent parts of the interaction are:

For `Catholic`, the slope `0.40203` represents the effect of being Catholic on the log-odds of supporting the death penalty when attendance is at the baseline of non-attendance, the category corresponding to 0 on the attendance scale. The positive sign on the slope suggests that Catholics who never attend services are more likely to support the death penalty compared to non-Catholics.

For `attend_num`, the slope `0.09007` represents the effect of religious service attendance on the log-odds of supporting the death penalty for non-Catholics, the category corresponding to 0 on the `Catholic` variable scale. The positive sign on the slope suggests that among non-Catholics, higher attendance is associated with increased support for the death penalty.

Taking into account the interactions and component parts, we could interpret the model results as follows:

The total effect of attendance on the log-odds of supporting the death penalty for Catholics is the sum of the main effect of attendance `attend_num` and the interaction term: $0.09007 + (-0.34877) = -0.25870$. This negative combined effect means that, for Catholics, as attendance increases, the support for the death penalty decreases.

For non-Catholics, the log-odds of supporting the death penalty increase by 0.09007 for each unit increase in attendance. Among Catholics, the log-odds of supporting the death penalty decrease by 0.2587 for each unit increase in attendance, reflecting a reversal in the direction of the relationship between attendance and support depending on religious affiliation.

Converting any of these quantities into exponentiated coefficients – as odds ratios – proceeds in the usual fashion. The combined or total interaction effect is summed on the log-odds scale, then exponentiated to interpret it as an odds ratio. The attendance total effect for Catholics is $0.09007 - -0.34877 = - - 0.2587$, then $exp(-0.2587) \approx 0.772$. Thus each unit increase in attendance decreases the odds of supporting the death penalty by approximately 22.8 percent.

Predicted probabilities

To visualize the predicted probabilities, we could repeat the same steps reviewed earlier for creating a dataset of values on the right-hand side of the model. For the interaction, to compare Catholics versus non-Catholics across levels of service attendance (like comparing females and males on levels of educational attainment), the dataset should have the appropriate 1 or 0 for religious group and service attendance score in the interaction term to match that of the individual variable components. The **sjPlot** package will graph this interaction with the `plot_model()` function on *model3*:

```
library(sjPlot)
plot_model(model3, type = "pred", terms=c("attend_num", "Catholic"))
```

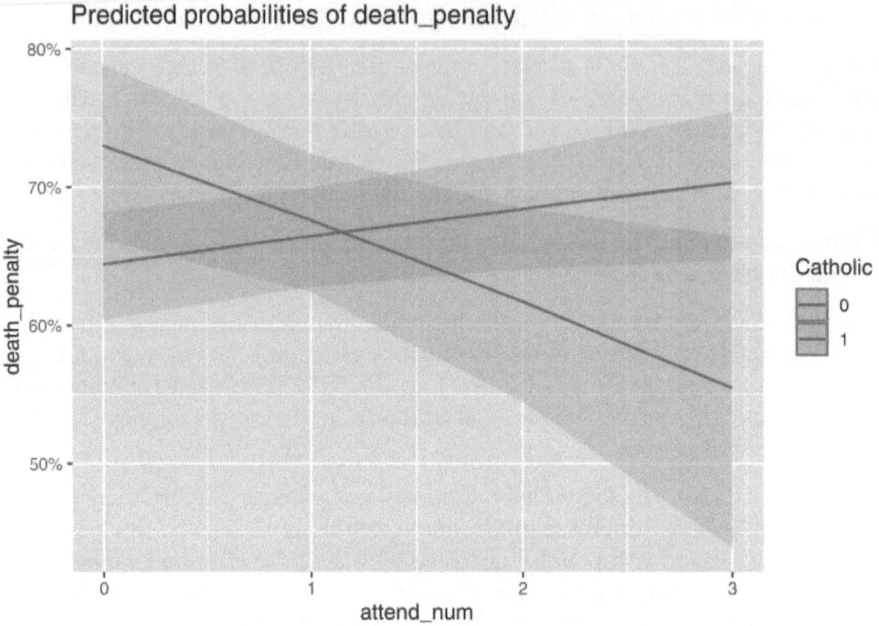

FIGURE 12.6 Predicted probabilities produced with the **sjPlot** package. The lines display predicted probabilities of supporting the death penalty at each of the levels of religious service attendance, treated as a numeric measure. The two different slopes of the lines for Catholic and non-Catholic show that for Catholic identifiers, increased service attendance lowers the predicted probabilty of supporting the death penalty, controlling for other factors, while it appears to slightly increase for non-Catholics.

In `plot_model()` for `model3` the argument `terms=c("attend_num", "Catholic")` with the two variables in that order and `type="pred"` will produce the predicted probabilities over the levels of religious service attendance for each level of *Catholic*. Figure 12.6 displays this graphic of predicted probabilities. The graphic displays a 'crossover' interaction – while for non-Catholics (an amorphous large group) support for the death penalty increases from approximately 65 percent at the lowest (or non-attendance) of services but increases to a little over 70 percent at weekly service. Among Catholics there is a steep decline, roughly 18 percent from non-attendance (73 percent support) to weekly attendance (about 55 percent) support. Keep in mind that although the attendance scale consists of ordered categories, in the model the measure is treated as a continuous measure, thus the `plot_model()` function calculates the probabilities at multiple points along the scale, connecting each one in a line. Still, the pattern of crossover interaction holds.

The predicted probabilities do show confidence intervals that overlap — yet the interaction term in the model estimates is statistically significant. The apparent contradiction between a statistically significant interaction term and overlapping probability confidence intervals can occur for a few different reasons (see Long (1997)). In short, the logistic regression model operates on the 'logit' (log-odds) scale, where the effects are linear. When transforming log-odds to probabilities, small differences in log-odds can translate to overlapping probabilities. The statistical significance of interaction does mean that there appears to be a meaningful interaction effect between Catholic identity and religious service attendance on the log-odds scale – the direction or magnitude of the relationship between service attendance and death penalty support differs for Catholics versus non-Catholics. It is just that in predicted probabilities, the level of uncertainty in the predictions is such that the differences are not clearly distinguishable at all levels of attendance. Overlapping confidence intervals do not imply that there is no interactive relationship. They suggest that the effect is not clearly distinguishable given the level of uncertainty. The interaction term's statistical significance indicates that, on average, there is an interaction effect, but the variability in the data means that we cannot be as confident about the predicted probabilities at each level of attendance.

Key considerations in data preparation, model specification, and interpretation

In preparing data, formulating a model, estimating it, and interpreting the results, there are a few key considerations to keep in mind.

The linearity assumption in the slopes relating X to log-odds of Y means that for severely skewed independent variables, prior to estimation it may be necessary to transform variables – such as the log transformations in right-skewed measures reviewed in Chapter 5. Different R packages provide tools for selecting appropriate transformations; see Fox and Weisberg (2011) for a review.

Beyond being clear about whether the data at hand require estimation with a complex survey estimator such as `svyglm()` versus `glm()`, potential variables for inclusion in the model need to be scaled appropriately. Determine whether the variables are scaled as continuous (numeric) measures or categorical (factor) measures. For model specification, avoid reinventing the wheel and base the model off the literature as closely as possible. Existing theory provides a guide to the inclusion of specific variables, ensuring that the model is based on established knowledge. Once estimated, check the linearity between continuous predictors and the dependent variable by inspecting residuals. If multicollinearity is a potential concern, check the VIF. Always examine residual plots, leverage plots, and influence measures to identify any potential issues with the model.

Resources

The **blorr** package (Hebbali, 2020) provides many functions for validating models. The **sjPlot** (Lüdecke, 2024) and **ggeffects** (Lüdecke, 2018) packages produce different visualizations of model estimates.

12.9 Exercises

Include R code and relevant output in an .Rmd file knitted to Word or PDF.

(1) Start with the code for estimating **model2**, but include educational attainment as an ordinal measure with obtaining a high school diploma as the reference category. Relative to a high school diploma, how does obtaining a four-year college degree change support for the death penalty? Print out a model summary of the exponentiated coefficients and interpret the odds ratios from the model estimates.

(2) Start with the code for estimating **model2**, but include an additional demographic characteristic, marital status, contrasting individuals who are currently married or widowed with all those who are not. Is marital status associated with death penalty support? How or how not?

(3) Reestimate *model2*, including a multiplicative interaction between *sex* and *education*, `sex*education` in the right-hand side of the equation in `svyglm()`. Does the interaction term indicate there is a significant interaction between the two? How or how not?

(4) Compare **model2** with **model3** controlling for Catholic religious affiliation and service attendance, as well as the interaction between the two. Is **model3** a 'better' model than **model2**? Base your answer in an analysis of the results and model fit statistics, including a likelihood ratio test.

(5) With the ANES2020 data, estimate a logistic regression model of participation in "political meetings, rallies, speeches, dinners", variable

V202014, using the same demographic characteristics in **model2**. Interpret the results of the model, converting the coefficients into odds ratios. Which demographic characteristics appear to be associated with this mode of political participation?

(6) Using the US States dataset, construct a model of Trump or Biden winning States (and DC) in either 2016 or 2020. Include three to four independent variables, at least one independent variable should be treated as a numeric or continuous measure. In two paragraphs, explain the model and present a graphic illustrating how variation in one of the independent variables is associated with the probability of support for either candidate.

(7) Using the model of US States, investigate linearity assumptions for interval scaled measures, and whether any observations exercise an extraordinary influence on the parameter estimates.

A

Data Codebooks

The following describes the contents of datasets available from the text websites, https://github.com/whittkilburn/inpolr and https://faculty.gvsu.edu/kilburnw/inpolr.

A.1 Fundamentals of US presidential elections

prez.csv

The file *prez.csv* contains annual observations for US presidential elections.

year Election year

incprez Name of incumbent president

vote Incumbent president share of the two-party vote percentage

approval Gallup Poll presidential approval for late June to July of election year

qgdprate GDP growth rate for the second quarter of each election year

incstatus An indicator (1) if the incumbent party has served two consecutive terms and (0) otherwise.

Sources: The American Presidency Project (Woolley and Peters, 2024), US Bureau of Economic Analysis (2024b), Gallup (Gallup, 2024).

A.2 Countries

covid_g7.csv

The file *covid_g7.csv* contains COVID-19 related daily observations on seven countries from 2020 to 2021.

iso_code and *country* Country identifier

date Date in year-month-day

total_cases COVID-19 cases by date

new_cases COVID-19 cases

cases_per_100K New COVID-19 cases per 100,000 people

cases_smoothed_per_100k Seven-day rolling average of case counts per 100,000 people

new_cases_smoothed Seven-day rolling average of case counts

total_deaths COVID-19 attributed deaths by date

new_deaths COVID-19 attributed deaths

new_deaths_smoothed Seven-day rolling average of COVID-19 attributed deaths

total_vaccinations Total COVID-19 vaccine doses administered by date.

people_vaccinated Total number of people receiving at least one vaccine dose by date

people_fully_vaccinated Total number of people receiving all prescribed vaccine doses by date.

new_vaccinations Vaccine doses administered

new_vaccinations_smoothed Seven-day rolling average of daily COVID-19 vaccine doses administered

population Total country population

Source: Our World in Data (2020).

gdpfert.csv

The file *gdpfert.csv* contains country-level characteristics for the year 2010.

iso2c Country abbreviation

country Name of country

year Year 2010

NY.GDP.PCAP.KD Gross domestic product (GDP) per capita in 2010 U.S. dollars

SP.DYN.TFRT.IN Average number of births per woman, 2010

SP.DYN.LE00.IN Life expectancy at birth in years

region World Bank region

longitude Geographic longitude centroid of country

latitude Geographic latitude centroid of country

income World Bank income group

gdp Gross domestic Product (GDP) per capita in 2010 U.S. dollars

fert Average number of births per woman 2010

Source: The World Bank (The World Bank, 2024).

countries.csv

The file *countries.csv* contains various country-level characteristics for the year 2019 and COVID related measures for 2020.

country Name of country

lending World Bank lending status

iso2c, **iso3c** Two and three letter country code

region Continent of country

income World Bank country level income level

total_population Total population

gdp_constant GDP (constant 2015 USD)

gdp_growth GDP growth (annual percent)

gdp_constant_int_pp GDP, PPP (constant 2017 international USD)

gdp_pc_constant GDP per capita (constant 2015 USD)

gdp_pc_growth GDP per capita growth (annual percent)

gdp_pc_pp GDP per capita, PPP (constant 2017 USD)

fert_rate Fertility rate (average births per female)

life_expect_female Life expectancy at birth in 2019, females

life_expect Life expectancy at birth in 2019, total

life_expect_male Life expectancy at birth in 2019, males

ag_forest Agriculture, forestry, and fishing, value added (percent of GDP)

schl_enroll School enrollment secondary (percent gross)

gdp_import Imports of goods and services (percent of GDP)

labor_fem Labor force, female (percent of total labor force)

fuel_exports Fuel exports (percent of merchandise exports)

net_migrate Net migration

fert_ado Adolescent fertility rate (births per 1,000 women ages 15-19)

gdp_export Exports of goods and services (percent of GDP)

co2_per_gdp Carbon dioxide emissions (kg per 2010 USD of GDP)

co2_emit Carbon dioxide emissions (kt)

co2_per_cap Carbon dioxide emissions (metric tons per capita)

co2_pp_gdp Carbon dioxide emissions (kg per PPP USD of GDP)

gas_total Total greenhouse gas emissions (kt of carbon dioxide equivalent)

pop_0_14 Population ages 0-14 (percent of total population)

pop_15_64 Population ages 15-64 (percent of total population)

pop_65_over Population ages 65 and above (percent of total population)

pop_dens Number of people divided by land area, in square kilometers

infant_mort Infant Mortality Rate (per 1,000 live births)

women_par Percentage of women in the national parliament

Source: The World Bank (2024).

polyarchy Electoral democracy index

libdem Liberal democracy index

partipdem Participatory democracy index

delibdem Deliberative democracy index

egaldem Egalitarian democracy index

Source: Varieties of Democracy (Coppedge et al., 2022b).

total_cases Total COVID cases in 2020

total_deaths Total COVID deaths in 2020

total_vaccinations Total number of COVID-19 vaccination doses administered in 2020

Source: Our World in Data (Mathieu et al., 2020).

hdi Human Development Index value

life_expect_un Life expectancy at birth in years

expect_school Expected years of schooling in years

mean_school Mean years of schooling in years

gni_pc_2017ppp Gross National Income Per Capita (2017 PPP USD)

gender_ineq Gender Inequality Index value

matern_mort Maternal Mortality Ratio (deaths per 100,000 live births)

ado_birth Adolescent Birth Rate (births per 1,000 women ages 15-19)

pop_sec_f Population with at least some secondary education, female (percent ages 25 and older)

pop_sec_m Population with at least some secondary education, male (percent ages 25 and older)

labor_par_f Labor force participation rate, female (percent ages 15 and older)

labor_par_m Labor force participation rate, male (percent ages 15 and older)

plant_adj_hdi Planetary pressures–adjusted Human Development Index value

co2_per_capita Carbon dioxide emissions per capita in tonnes

mat_foot Material footprint per capita in tonnes

ineq_ad_hdi Inequality-adjusted Human Development Index value

ineq_life Inequality in life expectancy

ineq_educ Inequality in education

ineq_inc Inequality in income

Source: United Nations Development Programme (United Nations Development Programme, 2022).

countries_tsxs.csv

The file *countries_tsxs.csv* consists of annual observations from 2000 to 2019 on countries around the world, as a subset of the variables in *countries.csv*. The dataset is structured in wide format, with approximately 4,500 country-year observations and 46 columns.

hdi_2021.csv

The file *hdi_2021.csv* contains observations on the Human Development Index and components for countries around the world, measured in 2021 (United Nations Development Programme, 2022).

Country Country name

hdi Human Development Index value

life_expect Average life expectancy at birth

expect_school Expected number of years of schooling for a school entrance age child

mean_school Average number of years of education for people ages 25 and older

gni_pc Gross national income per capita

A.3 US States and counties

The file *us_states.csv* consists of observations on 50 States plus the District of Columbia on the following measures.

state Name of US State

trump_votes_2020, **biden_votes_2020**, **other_votes_2020**, Vote totals for each candidate or all others combined, separate columns for 2016, 2012, and 2008.

Source: Federal Election Commission (Federal Election Commission, 2025)

cases_2020 COVID-19 total cases for 2020

cases_jan_2020 to **cases_oct_2021** Cumulative COVID-19 case counts by month

vax_dec_2020 to **vax_oct_2021** Cumulative vaccination counts per month (all ages)

deaths_jan_2020 to **deaths_oct_2021** Cumulative COVID-19 attributed deaths by month

deaths_pc_jun_2020 Per capita (10,000 people) COVID-19 deaths by June 2020

Source: *The New York Times* (2021).

stay_begin and **stay_end** Date of beginning and end of COVID-19 stay at home order

shutdown_date_range Date length of State COVID-19 stay at home order

shutdown_length Length of State COVID-19 stay at home order in numeric number of days

Source: Ballotopedia.org (2024).

population_density Persons per square mile by census tract, 2020 decennial census

total_population_2010 State population from 2010 decennial US Census, US Census Bureau

total_population_est_2019 Estimated population in 2019, 1-year estimate, US Census Bureau

census_fips_code Numeric code for State

pop Total population, 5-year American Community Survey (ACS) 2018 estimate

white_pop black_pop native_pop asian_pop latino_pop, First mention ACS estimate

median_income ACS 2018 median household income

gini ACS 2018 Gini coefficient of inequality

Source: US Census Bureau (2018).

bachelors_pop Persons 25 years or older with a bachelor's degree

q1_gdp_2020 to **q4_gdp_2020** Quarterly percentage change in GDP from the previous quarter.

pc_income_2020 State per capita income in 2020

oct_2020_unemployed Proportion of population unemployed in October 2020

Source: Bureau of Economic Analysis (U.S. Bureau of Economic Analysis, 2024a).

drug_alc_deaths_2019 Number of deaths attributed to drug or alcohol overdose in 2019.

expectancy Average life expectancy at birth.

Source: Centers for Disease Control and Prevention (Centers for Disease Control and Prevention, 2024).

state_social_liberalism_policy Index of social policy liberalism 2014, higher scores more liberal

state_econ_liberalism_policy Index of economic policy liberalism 2014, higher scores more liberal

mass_social_liberalism_policy Index of mass social liberalism 2014, higher scores more liberal

mass_econ_liberalism_policy Index of mass economic liberalism 2014, higher scores more liberal

Source: Caughey and Warshaw (2018).

percent_women_legislature Percent of women in the State legislature

Source: Center for American Women and Politics, Eagleton Institute of Politics, & Rutgers, The State University of New Jersey (Center for American Women and Politics, 2021).

vehicle_fatality_2019 to **vehicle_fatality_2021** Human fatalities involving motor vehicle crashes

Source: National Highway Traffic Safety Administration (Stewart, 2022)

Michigan County level presidential votes, 2012 and 2016

The file *michigan12_16.csv* contains Michigan County level presidential votes for election years 2012 and 2016.

fips Within Michigan FIPS code for County

Id2 FIPS code with Michigan (2600) identifier plus County FIPS

geography County name

county County name

clinton_vote Votes for Hillary Clinton in 2016

other2016_vote Votes for all other presidentialcandidates in 2016

trump_vote Votes for Donald Trump in 2016

trump_percent Percentage votes for Trump

clinton_percent Percentage votes for Clinton

total_votes Total votes cast in 2016

vote_contrib Contribution of vote to the total vote count

dem_index Index of County electoral importance to Democratic party in 2016 (calculated from Gimpel and Schuknecht (2004))

rep_index Index of County electoral importance to Republican party (Gimpel and Schuknecht (2004))

obama_vote Votes for Barack Obama in 2012

other2012_vote Votes for all other presidential candidates in 2012

romney_vote Votes for Mitt Romney in 2012

romney_percent Percentage votes for Romney

obama_percent Percentage votes for Obama

total_votes2012 Total votes cast for president in 2012

dem_index2012 Index of County electoral importance to Democratic party in 2012 (calculated from Gimpel and Schuknecht (2004))

rep_index2012 Index of County electoral importance to Republican party in 2012 (Gimpel and Schuknecht (2004))

The file *2016GEN_MI_CENR_BY_COUNTY.xls* contains elections results for all races appearing on the ballot in 2016 in Michigan.

Source: Michigan Department of State (Michigan Secretary of State, 2024).

A.4 Political geography

The file *Michigan_State_Senate_Districts_2011_plan.geojson* describes the geography of Michigan Senate districts for the 2011 apportionment plan.

The file *Counties___v17a.shp* describes the geography of Michigan counties.

Source: Michigan Geographic Information Systems (GIS) Open Data Portal (2024).

The file *ZAF_ADM4.geojson* describes South African administrative boundaries.

Source: Goodman et al. (2019).

The file *Crime_Incidents_in_2024.geojson* describes the geography of crime reported in Washington, DC, from January to September 2024.

Source: DC Open Data Portal (District of Columbia Government, 2024).

A.5 Votes and ideology in the United States Congress

The file *co_na_h_votes_117_wide.csv* records in wide format the roll call votes for Representatives from Colorado to the U.S. House in the 117th U.S. Congress.

The file *s_memb_votes_117_wide.csv* records in wide format the roll call votes for the entire U.S. Senate in the 117th U.S. Congress.

name Last name, party caucus and State identifier.

1 to **996** Yea (1) or Nay (0) on roll call vote.

The file *co_117_euc_distances.csv* contains a matrix of pairwise Euclidean distances on roll call votes for Representatives from Colorado.

The files *86_house_dw.csv* and *house_memb_117.csv* record for each representative (voting and non-voting) to the 86th and 117th US Congresses, respectively.

chamber "House" for the House of Representatives

icpsr Numeric identifier for each Representative

district_code Number of district within State

state_abbrev Two letter State abbreviation

party_code Name of party affiliation

bioname Full name of Representative

born Year born

nominate_dim1 NOMINATE dimension 1 score

nominate_dim2 NOMINATE dimension 2 score

age Age in years

Source: Voteview.com (Lewis et al., 2023).

A.6 Text of *The Federalist*

The file *federalist.zip* contains individual 85 text (.txt) documents. Each document is the full text of each essay, given the number and identified author from the Unitd States Library of Congress.

Source: U.S. Library of Congress (Library of Congress, 2024).

A.7 Comments on Title IX regulatory interpretation

The file *reg_comments_sampled.csv* contains random sampled comments (n=1000) on a proposed rule affecting Title IX enforcement.

id Identification key

city Location of submitter

state Location of submitter

name First name of submitter

date Date comment submitted

comment Full text of the submitted comment

Source: *The Federal Register* (U.S. Department of Education, 2023).

A.8 2020 American National Election Studies

The **R** Workspace *anes 2020 survey data.rdata* is a subset of the 2020 ANES survey (American National Election Studies, 2021), an election poll of U.S. citizens 18 years of age or older by the November 2020 election date. The dataset is saved within the Workspace as a dataframe *anes_20*. (Load the data with `load(file="anes 2020 survey data.rdata")`).

In the following list of variables, `pre` refers to survey questions administered in the fall 2020 pre-election survey, while `post` refers to the winter 2020 post-election survey. Pre-election interviews occurred between August 18 to election day November 3; post-election interviews occurred between November 8 until January 4, 2021. For most variables, missing responses due to non-completion of the survey or (in very few cases) technical errors have been recoded to a missing value of `NA`; for some questions, survey responses originally coded "Refused" or "Don't know" remain as in the original dataset from the ANES. Otherwise, responses appear solely on the intended measurement scale for each question. Some variables with an **x** at the end of the name include the responses for a two-part question. For example, variable **V201126x** on approval of the U.S. Congress is the response scale combined from two separate questions, "Do you approve or disapprove of the way the U.S. Congress has been handling its job?" and then "Do you [approve/disapprove] strongly or not strongly?". In the variable label, `[respondent]` refers to the person answering the question, the survey respondent.

Variable list

version Version of ANES 2020 Time Series Release

V200001 2020 Case ID

V200015b Post-election weight

V200015c Variance unit

V200015d Variance stratum

V201005 pre: How often does [respondent] pay attention to politics and elections

V201006 pre: How interested in following campaigns

V201020 pre: Did [respondent] vote in a Presidential primary or caucus

V201033 pre: For whom does [respondent] intend to vote for President

V201100 pre: How likely is it that [respondent] will vote in November

V201115 pre: How hopeful [respondent] feels about how things are going in the country

V201116 pre: How afraid [respondent] feels about how things are going in the country

V201118 pre: How angry [respondent] feels about how things are going in the country

V201119 pre: How happy [respondent] feels about how things are going in the country

V201126x pre: Approval of Congress handling its job

V201129x pre: Approve or disapprove President handling job

V201151 pre: Feeling Thermometer: Joe Biden, Democratic Presidential candidate

V201152 pre: Feeling Thermometer: Donald Trump, Republican Presidential candidate

V201153 pre: Feeling Thermometer: Kamala Harris, Democratic Vice-Presidential candidate

V201154 pre: Feeling Thermometer: Mike Pence, Republican Vice-Presidential candidate

V201200 pre: 7pt scale liberal-conservative self-placement

V201225x pre: Voting as duty or choice

V201228 pre: Party ID: Does [respondent] think of self as Democrat, Republican, or Independent

V201231x pre: Party ID [respondent] 7-point scale

V201232 pre: Party identity importance

V201234 pre: Government run by a few big interests or for benefit of all

V201235 pre: Does government waste much tax money

V201236 pre: How many in government are corrupt

V201237 pre: How often can people be trusted

V201240 pre: Which party better: handling health care

V201241 pre: Which party better: handling immigration

V201244 pre: Which party better: handling COVID-19

V201246 pre: 7pt scale spending & services: self-placement

V201249 pre: 7pt scale defense spending: self-placement

V201247 pre: 7pt scale spending & services: Democratic Presidential candidate

V201248 pre: 7pt scale spending & services: Republican Presidential candidate

V201250 pre: 7pt scale defense spending: Democratic Presidential candidate

V201251 pre: 7pt scale defense spending: Republican Presidential candidate

V201305x pre: Federal Budget Spending: public schools

V201308x pre: Federal Budget Spending: Tightening border security

V201327x pre: National economy better or worse in last year

V201330x pre: Economy better or worse in next 12 months

V201336 pre: Abortion self-placement

V201345x pre: [respondent] favor/oppose death penalty

V201346 pre: During last year, US position in world weaker or stronger

V201349x pre: Country would be better off if we just stayed home

V201350 pre: Force to solve international problems

V201351 pre: Votes counted accurately

V201352 pre: Trust election officials

V201353 pre: How often people denied right to vote

V201356x pre: Favor/oppose vote by mail

V201359x pre: Favor/oppose requiring ID when voting

V201362x pre: Favor/oppose allowing felons to vote

V201372x pre: Helpful/harmful if Pres didn't have to worry about congress/courts

V201375x pre: Favor or oppose restricting journalist access

V201377 pre: How much trust in news media

V201411x pre: Transgender policy

V201414x pre: Favor/oppose laws protect gays lesbians against job discrimination

V201415 pre: Should gay and lesbian couples be allowed to adopt

V201416 pre: [respondent] position on gay marriage

V201423x pre: Should children brought illegally be sent back or allowed to stay

V201426x pre: Favor or oppose building a wall on border with Mexico

V201433 pre: Is religion important part of [respondent] life

V201434 pre: Is Bible word of God or men

V201452 pre: Ever attend church or religious services

V201453 pre: Attend religious services how often

V201456 pre: Does Christian [respondent] consider self born again

V201458x pre: [respondent] Major group religion

relig_trad pre: reduced set of major group religion categories, from V201458x

secular pre: [respondent] identity as Agnostic or Atheist

attend pre: Religious service attendance

V201502 pre: [respondent] how much better or worse off financially than 1 year ago

V201507x pre: [respondent] age

V201508 pre: Marital status

V201511x pre: [respondent] 5 Category level of education

V201549x pre: [respondent] self-identified race/ethnicity

V201600 pre: What is [respondent] sex

V201628 pre: How many guns owned

V201645 pre: On which program does Federal government spend the least

V201646 pre: Party with most members in House before election

V201647 pre: Party with most members in Senate before election

V202471 post: [respondent] has family/neighbors/coworkers/friends who are gay, lesbian or bisexual

V202013 post: [respondent] attend online political meetings, rallies, speeches, fundraisers

V202014 post: [respondent] go to any political meetings, rallies, speeches, dinners

V202015 post: [respondent] wear campaign button or post sign or bumper sticker

V202016 post: [respondent] do any (other) work for party or candidate

V202017 post: [respondent] contribute money to individual candidate running for public office

V202019 post: [respondent] contribute money to political party during this election year

V202021 post: [respondent] contribute to any other group that supported or opposed candidates

V202022 post: [respondent] ever discuss politics with family or friends

V202023 post: How many days in past week discussed politics with family or friends

V202066 post: Did [respondent] vote in November 2020 election

V202072 post: Did [respondent] vote for President

V202073 post: For whom did [respondent] vote for President

V202143 post: Feeling thermometer: Democratic presidential candidate: Joe Biden

V202144 post: Feeling thermometer: Republican presidential candidate: Donald Trump

V202156 post: Feeling thermometer: Democratic Vice Presidential candidate: Kamala Harris

V202157 post: Feeling thermometer: Republican Vice Presidential candidate: Mike Pence

V202158 post: Feeling thermometer: Dr. Anthony Fauci

V202159 post: Feeling thermometer: Christian fundamentalists

V202160 post: Feeling thermometer: feminists

V202161 post: Feeling thermometer: liberals

V202162 post: Feeling thermometer: labor unions

V202163 post: Feeling thermometer: big business

V202164 post: Feeling thermometer: conservatives

V202165 post: Feeling thermometer: U.S. Supreme Court

V202166 post: Feeling thermometer: gay men and lesbians

V202167 post: Feeling thermometer: congress

V202168 post: Feeling thermometer: Muslims

V202169 post: Feeling thermometer: Christians

V202170 post: Feeling thermometer: Jews

V202171 post: Feeling thermometer: police

V202172 post: Feeling thermometer: transgender people

V202173 post: Feeling thermometer: scientists

V202174 post: Feeling thermometer: Black Lives Matter

V202175 post: Feeling thermometer: journalists

V202176 post: Feeling thermometer: North Atlantic Treaty Organization (NATO)

V202177 post: Feeling thermometer: United Nations (UN)

V202178 post: Feeling thermometer: National Rifle Association (NRA)

V202179 post: Feeling thermometer: socialists

V202180 post: Feeling thermometer: capitalists

V202181 post: Feeling thermometer: Federal Bureau of Investigation (FBI)

V202182 post: Feeling thermometer: Immigration and Customs Enforcement (ICE) agency

V202183 post: Feeling thermometer: MeToo movement

V202184 post: Feeling thermometer: rural Americans

V202185 post: Feeling thermometer: Planned Parenthood

V202186 post: Feeling thermometer: World Health Organization (WHO)

V202187 post: Feeling thermometer: Centers for Disease Control (CDC)

V202225 post: Limits on campaign spending

V202231x post: Favor/oppose new limits on imports

V202232 post: What should immigration levels be

V202233 post: How likely immigration will take away jobs

V202236x post: Favor/oppose allowing refugees to come to US

V202239x post: Effect of illegal immigration on crime rate

V202242x post: Favor/oppose providing path to citizenship

V202245x post: Favor/oppose returning unauthorize immigrants to native country

V202248x post: Favor/oppose separating children of detained immigrants

V202259x post: Favor/oppose government trying to reduce income inequality

V202260 post: Society should make sure everyone has equal opportunity

V202261 post: We'd be better off if worried less about equality

V202262 post: Not a big problem if some have more chance in life

V202263 post: If people were treated more fairly we would have fewer problems

V202264 post: The world is changing and we should adjust view of moral behavior

V202265 post: Fewer problems if there was more emphasis on traditional family values

egal_index post: Additive index of **V202261** to **V202265**

V202266 post: Which child trait more important: independence or respect

V202267 post: Which child trait more important: curiosity or good manners

V202268 post: Which child trait more important: obedience or self-reliance

V202269 post: Which child trait more important: considerate or well-behaved

auth_index post: Authoritarianism index, average of **V202266** to **V202269**

V202286x post: Easier/harder for working mother to bond with child

V202290x post: Better/worse if man works and woman takes care of home

V202291 post: Do women demanding equality seek special favors

V202292 post: Do women complaining about discrimination cause more problems

sexism_index post: Modern sexism index, average of **V202286x** to **V202292**

V202300 post: Agree/disagree: blacks should work their way up without special favors

V202301 post: Agree/disagree: past slavery and discrimination make it difficult for blacks

V202302 post: Agree/disagree: blacks have gotten less than they deserve

V202303 post: Agree/disagree: if blacks tried harder they'd be as well off as whites

V202304 post: Our political system only works for insiders with money and power

V202305 post: Because of rich and powerful it's difficult for the rest to get ahead

V202308x post: Trust ordinary people/experts for public policy

V202309 post: How much do people need help from experts to understand science

V202310 post: How important should science be for decisions about COVID-19

V202311 post: Business and politics controlled by few powerful people

V202312 post: Much of what people hear in schools and media are lies by those in power

V202313 post: Post-materialism most important 1A

V202314 post: Post-materialism next most important 1B

V202315 post: Post-materialism most important 2A

V202316 post: Post-materialism next most important 2B

V202325 post: Favor or oppose tax on millionaires

V202331x post: Favor or oppose requiring vaccines in schools

V202236x post: Favor or oppose allowing refugees to come to US

V202337 post: Should federal government make it more difficult or easier to buy a gun

V202341x post: Favor or oppose background checks for gun purchases

V202344x post: Favor or oppose banning 'assault-style' rifles

V202347x post: Favor or oppose government buy back of 'assault-style' rifles

V202351 post: How often do police officers use more force than necessary

V202355 post: Does [respondent] currently live in a rural or urban area

V202361x post: Favor or oppose free trade agreements

V202367x post: International trade increased/decreased jobs in US

V202373x post: Increasing diversity made US better/worse place to live

V202376x post: Favor or oppose federal program giving citizens $12K/year

V202377 post: Should the minimum wage be raised, kept the same, or lowered

V202383x post: Health benefits of vaccinations outweigh risks

V202387x post: Attention to sexual harrassment as gone too far/not far enough

V202390x post: Favor/oppose transender people serve in military

V202456 post: During past 12 months, [respondent] or any family members stopped by police

V202457 post: Have you ever been arrested, or has that never happened to you?

V202468x post: Summary: total family income

V202562 post: Life experience: does [respondent] currently owe money on student loans

V202475 post: Does [respondent] consider themself a feminist or anti-feminist

V202541a post: Which social media platforms [respondent] visited - Facebook

V202541b post: Which social media platforms [respondent] visited - Twitter

V202541c post: Which social media platforms [respondent] visited - Instagram

V202541d post: Which social media platforms [respondent] visited - Reddit

V202541e post: Which social media platforms [respondent] visited - YouTube

V202541f post: Which social media platforms [respondent] visited - SnapChat

V202541g post: Which social media platforms [respondent] visited - TikTok

V202559 post: Evidence that hydroxychloroquine is effective treatment for COVID-19 or not

V202410 post: Attitudes about elites: politicians do not care about people

V202411 post: Attitudes about elites: most politicians are trustworthy

V202412 post: Attitudes about elites: politicians are main problem in US

V202413 post: Attitudes about elites: strong leader in government is good

V202414 post: Attitudes about elites: people should make policy decisions

Question wording

version to **V200015d** data and sampling meta variables

V201005 "How often do you pay attention to what's going on in government and politics?" 1. Always 2. Most of the time 3. About half the time 4. Some of the time 5. Never

V201006 "Some people don't pay much attention to political campaigns. How about you? Would you say that you have been very much interested, somewhat interested or not much interested in the political campaigns so far this year?" 1. Very much interested 2. Somewhat interested 3. Not much interested

V201020 "Did you vote in a Presidential primary election or caucus this year?" 1. Yes, voted in primary or caucus 2. No, didn't vote in primary or caucus

V201033 "Who do you think you will vote for? [Joe Biden, Donald Trump/Donald Trump, Joe Biden], Jo Jorgensen, Howie Hawkins, or someone else?" 1. Joe Biden 2. Donald Trump 3. Jo Jorgensen 4. Howie Hawkins 5. Other candidate

pre_vote_choice simplified categories of vote choice from V201033

V201100 "How likely is it that you will vote in the general election this November?" 1. Extremely likely 2. Very likely 3. Moderately likely 4. Slightly likely 5. Not likely at all

V201115 "The next few questions are about how you feel about how things are going in the country these days.How hopeful do you feel about how things are going in the country these days?" 1. Not at all 2. A little 3. Somewhat 4. Very 5. Extremely

V201116 "How afraid do you feel about how things are going in the country?" Response scale **V201115**

V201118 "How angry do you feel about how things are going in the country?" Response scale **V201115**

V201119 "How happy do you feel about how things are going in the country?" For responses see **V201115**

V201126x "Do you approve or disapprove of the way the U.S. Congress has been handling its job? Do you [approve/disapprove] strongly or not strongly?" 1. Approve strongly 2. Approve not strongly 3. Disapprove not strongly 4. Disapprove strongly

V201129x "Do you approve or disapprove of the way Donald Trump is handling his job as President? Do you [approve/disapprove] strongly or not strongly?" Response scale **V201126x**

V201151 "I'd like to get your feelings toward some of our political leaders and other people who are in the news these days. I'll read the name of a person and I'd like you to rate that person using something we call the feeling thermometer. Ratings between 50 degrees and 100 degrees mean that you feel favorable and warm toward the person. Ratings between 0 degrees and 50 degrees mean that you don't feel favorable toward the person and that you don't care too much for that person. You would rate the person at the 50 degree mark if you don't feel particularly warm or cold toward the person. If we come to a person whose name you don't recognize, you don't need to rate that person. Just tell me and we'll move on to the next one." "How would you rate: Joe Biden"

V201152 "How would you rate: Donald Trump"

V201153 "How would you rate: Kamala Harris"

V201154 "How would you rate: Mike Pence"

V201200 "We hear a lot of talk these days about liberals and conservatives. Here is a seven-point scale on which the political views that people might hold are arranged from extremely liberal to extremely conservative.[Scale shown to respondents.] Where would you place yourself on this scale, or haven't you thought much about this?" 1. Extremely liberal 2. Liberal 3. Slightly liberal 4. Moderate; middle of the road 5. Slightly conservative 6. Conservative 7. Extremely conservative 99. Haven't thought much about this

V201225x "Different people feel differently about voting. For some, voting is a choice - they feel free to vote or not to vote, depending on how they feel about the candidates and parties. For others voting is a duty - they feel they should vote in every election no matter how they feel about the candidates and parties. For you personally, is voting mainly a choice, mainly a duty, or neither a choice nor a duty?" 1. Very strongly a duty 2. Moderately strongly a duty 3. A little strongly a duty 4. Neither a duty nor a choice 5. A little strongly a choice 6. Moderately strongly a choice 7. Very strongly a choice

V201228 "Generally speaking, do you usually think of yourself as Democrat, a Republican, an independent, or what?" 1. Democrat 2. Republican 3. Independent 5. Other party {SPECIFY}

V201231x "... Would you call yourself a strong [Democrat / Republican] or a not very strong [Democrat / Republican]? Do you think of yourself as closer to the Republican Party or to the Democratic Party?" 1. Strong Democrat 2. Not very strong Democrat 3. Independent-Democrat 4. Independent 5. Independent-Republican 6. Not very strong Republican 7. Strong Republican

V201232 "How important is being [a Democrat/a Republican/an Independent] to your identity?" 1. Extremely important 2. Very important 3. Moderately important 4. A little important 5. Not at all important

V201234 "Would you say the government is pretty much run by a few big interests looking out for themselves or that it is run for the benefit of all the people?" 1. Run by a few big interests 2. For the benefit of all the people

V201235 "Do you think that people in government waste a lot of the moneywe pay in taxes, waste some of it, or don't waste very much of it?" 1. Waste a lot 2. Waste some 3. Don't waste very much

V201236 "How many of the people running the government are corrupt?" 1. All 2. Most 3. About half 4. A few 5. None

V201237 "Generally speaking, how often can you trust other people?" 1. Always 2. Most of the time 3. About half the time 4. Some of the time 5. Never

V201240 "Which party do you think would do a better job of handling health care?" 1. Democrats would do a much better job 2. Democrats would do a somewhat better job 3. Not much difference between them 4. Republicans would do a somewhat better job 5. Republicans would do a much better job

V201241 "Which party do you think would do a better job of handling immigration?" Response scale **V201240**

V201244 "Which party do you think would do a better job of handling the COVID-19 pandemic?" Response scale **V201240**

V201246 "Some people think the government should provide fewer services even in areas such as health and education in order to reduce spending. Suppose these people are at one end of a scale, at point 1. Other people feel it is important for the government to provide many more services even if it means an increase in spending. Suppose these people are at the other end, at point 7. And, of course, some other people have opinions somewhere in between, at points 2, 3, 4, 5 or 6. Where would you place yourself on this scale, or haven't you thought much about this?" 1. Government should provide many fewer services 2. 3. 4. 5. 6. 7. Government should provide many more services

V201247 "Where would you place Joe Biden on this issue?" Response scale in **V201246**

V201248 "Where would you place Donald Trump on this issue?" Response scale in **V201246**

V201249 "Some people believe that we should spend much less money for defense. Suppose these people are at one end of a scale, at point 1. Others feel that defense spending should be greatly increased. Suppose these people are at the other end, at point 7. And, of course, some other people have opinions somewhere in between, at points 2, 3, 4, 5 or 6. Where would you place yourself on this scale, or haven't you thought much about this?" 1. Greatly decrease defense spending 2. 3. 4. 5. 6. 7. Greatly increase defense spending

V201250 "Where would you place Joe Biden on this issue?" Response scale in **V201250**

V201251 "Where would you place Donald Trump on this issue?" Response scale in **V201250**

V201305x "Next I am going to read you a list of federal programs. For each one, I would like you to tell me whether you would like to see spending increased, decreased, or kept the same....What about public schools? Should federal spending on public schools be increased, decreased, or kept the same?" 1. Increased a lot 2. Increased a little 3. Kept the same 4. Decreased a little 5. Decreased a lot

V201308x "What about tightening border security to prevent illegal immigration?Should federal spending on tightening border security to prevent illegal immigration be increased, decreased, or kept the same?" Response scale **V201305x**

V201327x "Now thinking about the economy in the country as a whole, would you say that over the past year the nation's economy has gotten better, stayed about the same, or gotten worse?" 1. Gotten much better 2. Gotten somewhat better 3. Stayed about the same 4. Gotten somewhat worse 5. Gotten much worse

V201330x "What about the next 12 months? Do you expect the economy, in the country as a whole, to get better, stay about the same, or get worse?" Response scale in **V201330x**

V201336 "There has been some discussion about abortion during recent years. Which one of the opinions on this page best agrees with your view? You can just tell me the number of the opinion you choose." 1. "By law, abortion should never be permitted." 2. "The law should permit abortion only in case of rape, incest, or when the woman's life is in danger." 3. "The law should permit abortion other than for rape/incest/danger to woman but only after need clearly established." 4. "By law, a woman should always be able to obtain an abortion as a matter of personal choice."

V201345x "Do you favor or oppose the death penalty for persons convicted of murder? Do you [favor / oppose] the death penalty for persons convicted of murder strongly or not strongly?" 1. Favor strongly 2. Favor not strongly 3. Oppose not strongly 4. Oppose strongly

V201346 "Turning to some questions about America's role in the world: During the past year, would you say that the United States' position in the world has grown weaker, stayed about the same, or has it grown stronger?" 1. Weaker 2. Stayed about the same 3. Stronger

V201349x "Do you agree or disagree with this statement: 'This country would be better off if we just stayed home and did not concern ourselves with problems in other parts of the world.' Do you [agree / disagree] strongly or [agree / disagree] somewhat with this statement?" 1. Agree strongly 2. Agree somewhat 3. Disagree somewhat 4. Disagree strongly

V201350 "How willing should the United States be to use military force to solve international problems?" 1. Extremely willing 2. Very willing 3. Moderately willing 4. A little willing 5. Not at all willing

V201351 "In the November 2020 general election, how accurately do you think the votes will be counted?" 1. Not at all accurately 2. A little accurately 3. Moderately accurately 4. Very accurately 5. Completely accurately

V201352 "How much do you trust the officials who oversee elections where you live?" 1. Not at all 2. A little 3. A moderate amount 4. A lot 5. A great deal

V201353 "How often are people who are eligible to vote denied the right to vote?" 1. Never 2. Rarely 3. Occasionally 4. Fairly often 5. Very often

V201356x "Do you favor, oppose, or neither favor nor oppose conducting all elections by mail, instead of people voting in-person? Do you [favor / oppose]

that a great deal, moderately, or a little?" 1. Favor a great deal 2. Favor moderately 3. Favor a little 4. Neither favor nor oppose 5. Oppose a little 6. Oppose moderately 7. Oppose a great deal

V201359x "Do you favor, oppose, or neither favor nor oppose requiring all people to show a government issued photo ID when they vote?" Response scale **V201356x**

V201362x "Do you favor, oppose, or neither favor nor oppose allowing convicted felons to vote once they complete their sentence?" Response scale **V201356x**

V201372x "Would it be helpful, harmful, or neither helpful nor harmful if U.S. presidents could work on the country's problems without paying attention to what Congress and the courts say? How [helpful / harmful]?" 1. Extremely helpful 2. Moderately helpful 3. A little helpful 4. Neither helpful nor harmful 5. A little harmful 6. Moderately harmful 7. Extremely harmful

V201375x "Do you favor, oppose, or neither favor nor oppose elected officials restricting journalists' access to information about government decision-making? Do you [favor / oppose] that a great deal, moderately, or a little?" Response scale **V201356x**

V201377 "In general, how much trust and confidence do you have in the news media when it comes to reporting the news fully, accurately, and fairly?" 1. None 2. A little 3. A moderate amount 4. A lot 5. A great deal

V201411x "Should transgender people - that is, people who identify themselves as the sex or gender different from the one they were born as - have to use the bathrooms of the gender they were born as, or should they be allowed to use the bathrooms of their identified gender? How strongly do you feel that way?" 1. Feels very strongly transgender people should use bathroom of birth gender 2. Feels moderately strongly... 3. Feels a little strongly... 4. Feels a little strongly transgender people be allowed to use bathroom of identified gender 5. Feels moderately strongly... 6. Feels very strongly transgender people be allowed to use bathroom of identified gender

V201414x "Do you favor or oppose laws to protect gays and lesbians against job discrimination? Do you [favor / oppose] such laws strongly or not strongly?" 1. Favor strongly 2. Favor not strongly 3. Oppose not strongly 4. Oppose strongly

V201415 "Do you think gay or lesbian couples should be legally permitted to adopt children?" 1. Yes 2. No

V201416 "Which comes closest to your view? You can just tell me the number of your choice." 1. "Gay and lesbian couples should be allowed to legally marry." 2. "Gay and lesbian couples should be allowed to form civil unions but not legally marry." 3. "There should be no legal recognition of gay or lesbian couples' relationship."

V201423x "What should happen to immigrants who were brought to the U.S. illegally as children and have lived here for at least 10 years and graduated high school here? Should they be sent back where they came from, or should they be allowed to live and work in the United States?...Do you favor that a great deal, a moderate amount, or a little?" 1. Favors a great deal they should be sent back 2. Favors a moderate amount they should be sent back 3. Favors a little they should be sent back 4. Favors a little they should be allowed to live and work in US 5. Favors a moderate amount they should be allowed to live and work in US 6. Favors a great deal they should be allowed to live and work in US

V201426x "Do you favor, oppose, or neither favor nor oppose building a wall on the U.S. border with Mexico? Do you [favor / oppose] that a great deal, a moderate amount, or a little?" Response scale **V201356x**

V201433 "How important is religion in your life?" 1. Extremely important 2. Very important 3. Moderately important 4. A little important 5. Not important at all

V201434 "Which of these statements comes closest to describing your feelings about the Bible? You can just give me the number of your choice." 1. "The Bible is the actual word of God and is to be taken literally, word for word." 2. "The Bible is the word of God but not everything in it should be taken literally, word for word." 3. "The Bible is a book written by men and is not the word of God."

V201452 "Lots of things come up that keep people from attending religious services even if they want to. Thinking about your life these days, do you ever attend religious services, apart from occasional weddings, baptisms or funerals?" 1. Yes 2. No

V201453 "Do you go to religious services every week, almost every week, once or twice a month, a few times a year, or never?" 1. Every week 2. Almost every week 3. Once or twice a month 4. A few times a year 5. Never

V201456 "Would you call yourself a born-again Christian, that is, have you personally had a conversion experience related to Jesus Christ?" 1. Yes 2. No

V201458x Major group religion 1. Mainline Protestant 2. Evangelical Protestant 3. Black Protestant 4. Undifferentiated Protestant 5. Roman Catholic 6. Other Christian 7. Jewish 8. Other religion 9. Not religious

relig_trad reduced categories of religious traditions (from **V201458x**) 1. Mainline Protestant 2. Evangelical Protestant 3. Other Christian 4. Roman Catholic 5. Jewish 6. Other religion 7. Not religious

secular Whether an individual reported religious identification as "Atheist" or "Secular" in variables V201435 or V201436 0. Not Secular 1. Secular

attend Religious service attendance, from V201452 and V201453 1. Never 2. few times a year, month 3. almost every week 4. every week

V201502 "We are interested in how people are getting along financially thesedays. Would you say that [you / you and your family living here] are much better off financially, somewhat better off, about the same, somewhat worse off, or much worse off than you were a year ago?" 1. Much better off 2. Somewhat better off 3. About the same 4. Somewhat worse off 5. Much worse off

V201507x Respondent age in years

V201508 "Are you now married, widowed, divorced, separated or never married?" 1. Married: spouse present 3. Widowed 4. Divorced 5. Separated 6. Never married

V201511x "What is the highest level of school you have completed or the highest degree you have received?" 1. Less than high school credential 2. High school credential 3. Some post-high school, no bachelor's degree 4. Bachelor's degree 5. Graduate degree

V201549x "I am going to read you a list of five race categories. You may choose one or more races. For this survey, Hispanic origin is not a race. Are you White; Black or African American; American Indian or Alaska Native; Asian; or Native Hawaiian or Other Pacific Islander?" 1. White, non-Hispanic 2. Black, non-Hispanic 3. Hispanic 4. Asian or Native Hawaiian/other Pacific Islander, non-Hispanic alone 5. Native American/Alaska Native or other race, non-Hispanic alone 6. Multiple races, non-Hispanic

V201600 "What is your sex?" 1. Male 2. Female

V201628 "How many guns do you or anyone else living here own? Type the number."

V201645 "On which of the following does the U.S. federal government currently spend the least?" 1. Foreign aid 2. Medicare 3. National defense 4. Social Security

V201646 "Do you happen to know which party currently has the most members in the U.S. House of Representatives in Washington?" 1. Democrats 2. Republicans

V201647 "Do you happen to know which party currently has the most members in the U.S. Senate?" 1. Democrats 2. Republicans

V202471 "Among your immediate family members, relatives, neighbors, co-workers, or close friends, are any of them gay, lesbian, or bisexual as far as you know?" 1. Yes 2. No

V202013 "Did you participate in any online political meetings, rallies, speeches, fundraisers, or things like that in support of a particular candidate?" (for

V202013 through **V202022** response scale 1. Yes 2. No)

V202014 "Did you go to any political meetings, rallies, speeches, dinners, or things like that in support of a particular candidate?"

V202015 "Did you wear a campaign button, put a campaign sticker on your car, or place a sign in your window or in front of your house?"

V202016 "Did you do any other work for one of the parties or candidates?"

V202017 "During an election year people are often asked to make a contribution to support campaigns. Did you give money to an individual candidate running for public office?"

V202019 "Did you give money to a political party during this election year?"

V202021 "Did you give any money to any other group that supported or opposed candidates?"

V202022 "Do you ever discuss politics with your family or friends?"

V202023 "How many days in the past week did you talk about politics with family or friends?" 0. Zero days 1. One day 2. Two days 3. Three days 4. Four days 5. Five days 6. Six days 7. Seven days

V202066 "In talking to people about elections, we often find that a lot of peoplewere not able to vote because they weren't registered, they were sick, or they just didn't have time. Which of the following statements best describes you:" 1. "I did not vote in the election this November." 2. "I thought about voting this time, but didn't." 3. "I usually vote, but didn't this time." 4. "I am sure I voted."

V202072 "How about the election for President? Did you vote for a candidate for President?" 1. Yes, voted for President 2. No, didn't vote for President

V202073 "Who did you vote for? [Joe Biden, Donald Trump/Donald Trump, Joe Biden], Jo Jorgensen, Howie Hawkins, or someone else?" 1. Joe Biden 2. Donald Trump 3. Jo Jorgensen 4. Howie Hawkins 5. Other candidate

post_vote_choice categories of vote choice from **V202073** 1. Joe Biden 2. Donald Trump 3. Other

V202143 "I'd like to get your feelings toward some of our political leaders and other people who are in the news these days.I'll read the name of a person and I'd like you to rate that person using something we call the feeling thermometer. Ratings between 50 degrees and 100 degrees mean that you feel favorable and warm toward the person. Ratings between 0 degrees and 50 degrees mean that you don't feel favorable toward the person and that you don't care too much for that person. You would rate the person at the 50 degree mark if you don't feel particularly warm or cold toward the person. If we come to a person whose name you don't recognize, youdon't need to rate that person. Just tell me and we'll move on to the next one. On the feeling thermometer scale from 0 to 100, how would you rate Joe Biden?" [All feeling thermometers are recorded as numeric scores 0 to 100.]

V202144 "How would you rate: Donald Trump"

V202156 "How would you rate: Kamala Harris"

V202157 "How would you rate: Mike Pence"

V202158 "How would you rate: Dr. Anthony Fauci"

V202159 "Still using the feeling thermometer scale from 0 to 100, how would you rate the following groups:" "How would you rate: Christian Fundamentalists"

V202160 "How would you rate: Feminists"

V202161 "How would you rate: Liberals"

V202162 "How would you rate: Labor unions"

V202163 "How would you rate: Big business"

V202164 "How would you rate: Conservatives"

V202165 "How would you rate: The U.S. Supreme Court"

V202166 "How would you rate: Gay men and lesbians"

V202167 "How would you rate: Congress"

V202168 "How would you rate: Muslims"

V202169 "How would you rate: Christians"

V202170 "How would you rate: Jews"

V202171 "How would you rate: Police"

V202172 "How would you rate: Transgender people"

V202173 "How would you rate: Scientists"

V202174 "How would you rate: Black Lives Matter movement"

V202175 "How would you rate: Journalists"

V202176 "How would you rate: NATO"

V202177 "How would you rate: The United Nations (UN)"

V202178 "How would you rate: The National Rifle Association (NRA)"

V202179 "How would you rate: Socialists"

V202180 "How would you rate: Capitalists"

V202181 "How would you rate: The Federal Bureau of Investigation (FBI)"

V202182 "How would you rate: The Immigration and Customs Enforcement (ICE) agency"

V202183 "How would you rate: The #MeToo movement"

V202184 "How would you rate: Rural Americans"

V202185 "How would you rate: Planned Parenthood"

V202186 "How would you rate: The World Health Organization (WHO)"

V202187 "How would you rate: The Centers for Disease Control (CDC)"

V202225 "Do you favor, oppose, or neither favor nor oppose placing limits on political campaign spending?" 1. Favor 2. Oppose 3. Neither favor nor oppose

V202231x "Some people have suggested placing new limits on foreign imports in order to protect American jobs. Others say that such limits would raise consumer prices and hurt American exports. Do you favor or oppose placing new limits on imports? Do you [favor/oppose] placing new limits on imports strongly or not strongly?" Response scale **V201356x**

V202232 "Do you think the number of immigrants from foreign countries who are permitted to come to the United States to live should be increased a lot, increased a little, left the same as it is now, decreased a little, or decreased a lot?" 1. Increased a lot 2. Increased a little 3. Left the same as it is now 4. Decreased a little 5. Decreased a lot

V202233 "Now I'd like to ask you about immigration in recent years. How likely is it that recent immigration levels will take jobs away from people already here – extremely likely, very likely, somewhat likely, or not at all likely?" 1. Extremely likely 2. Very likely 3. Somewhat likely 4. Not at all likely

V202236x "Do you favor, oppose, or neither favor nor oppose allowing refugees who are fleeing war, persecution, or natural disasters in other countries to come to live in the U.S.? Do you [favor/oppose] that a great deal, a moderate amount, or a little?" Response scale **V201356x**

V202239x "Does illegal immigration increase, decrease, or have no effect on the crime rate in the U.S.? Does it [increase/decrease] the crime rate a lot, a moderate amount, or a little?" Response scale **V202232**

V202242x "Do you favor, oppose, or neither favor nor oppose providing a path to citizenship for unauthorized immigrants who obey the law, pay a fine, and pass security checks? Do you [favor/oppose] that a great deal, a moderate amount, or a little?" Response scale **V201356x**

V202245x "Do you favor, oppose, or neither favor nor oppose returning all unauthorized immigrants to their native countries? Do you [favor/oppose] that a great deal, a moderate amount, or a little?" Response scale **V202232**

V202248x "Do you favor, oppose, or neither favor nor oppose separating the children of detained immigrants, rather than keeping them with their parents in adult detention centers? Do you [favor/oppose] that a great deal, a moderate amount, or a little?" Response scale **V202232**

V202259x "Next, do you favor, oppose, or neither favor nor oppose the government trying to reduce the difference in incomes between the richest and poorest households? Do you [favor/oppose] that a great deal, a moderate amount, or a little?" Response scale **V202232**

V202260 "I am going to read several more statements. After each one, I would like you to tell me how strongly you agree or disagree. The first statement is:'Our society should do whatever is necessary to make sure that everyone has an equal opportunity to succeed.' Do you agree strongly, agree somewhat, neither agree nor disagree, disagree somewhat, or disagree strongly with this statement?" 1. Agree strongly 2. Agree somewhat 3. Neither agree nor disagree 4. Disagree somewhat 5. Disagree strongly

V202261 'This country would be better off if we worried less about how equal people are.' Response scale **V202260**

V202262 'It is not really that big a problem if some people have more of a chance in life than others.' Response scale **V202260**

V202263 'If people were treated more equally in this country we would have many fewer problems.' Response scale **V202260**

V202264 'The world is always changing and we should adjust our view of moral behavior to those changes.' Response scale **V202260**

V202265 'This country would have many fewer problems if there were more emphasis on traditional family ties.' Response scale **V202260**

egal_index post: additive index of **V202261** to **V202265**

V202266 "On another topic. Although there are a number of qualities that people feel that children should have, every person thinks that some are more important than others. I am going to read you pairs of desirable qualities.Please tell me which one you think is more important for a child to have: Independence or respect for elders" 1. Independence 2. Respect for elders 3. Both

V202267 "Which one is more important for a child to have: Curiosity or good manners" 1. Curiosity 2. Good manners 3. Both

V202268 "Which one is more important for a child to have: Obedience or self-reliance" 1. Obedience 2. Self-reliance 3. Both

V202269 "Which one is more important for a child to have: Being considerate or well behaved" 1. Being considerate 2. Well behaved 3. Both

auth_index post: additive index of **V202266** to **V202269**

V202286x "Do you think it is easier, harder, or neither easier nor harder for mothers who work outside the home to establish a warm and secure relationship with their children than it is for mothers who stay at home? Is it a great deal easier, somewhat easier, or slightly easier for mothers who work outside the home to establish a warm and secure relationship with their children? Is it a great deal harder, somewhat harder, or slightly harder for mothers who work outside the home to establish a warm and secure relationship with their children?" 1. A great deal easier 2. Somewhat easier 3. Slightly easier 4. Neither easier nor harder 5. Slightly harder 6. Somewhat harder 7. A great deal harder

V202290x "Do you think it is better, worse, or makes no difference for the family as a whole if the man works outside the home and the woman takes care of the home and family? Is it much better, somewhat better, or slightly better? Is it much worse, somewhat worse, or slightly worse?" 1. Much better 2. Somewhat better 3. Slightly better 4. Makes no difference 5. Slightly worse 6. Somewhat worse 7. Much worse

V202291 "When women demand equality these days, how often are they actually seeking special favors?" 1. Always 2. Most of the time 3. About half the time 4. Some of the time 5. Never

V202292 "When women complain about discrimination, how often do they cause more problems than they solve?" Response scale **V202291**

sexism_index post: additive index of **V202286x** to **V202292**

V202300 " 'Irish, Italian, Jewish and many other minorities overcame prejudice and worked their way up. Blacks should do the same without any special favors.' Do you agree strongly, agree somewhat, neither agree nor disagree,disagree somewhat, or disagree strongly with this statement?" Response scale **V202260**

V202301 " 'Generations of slavery and discrimination have created conditions that make it difficult for blacks to work their way out of the lower class.' Do you agree strongly, agree somewhat, neither agree nor disagree, disagree somewhat, or disagree strongly with this statement?" Response scale **V202260**

V202302 " 'Over the past few years, blacks have gotten less than they deserve.' Do you agree strongly, agree somewhat, neither agree nor disagree, disagree somewhat, or disagree strongly with this statement?" Response scale **V202260**

V202303 " 'It's really a matter of some people not trying hard enough; if blacks would only try harder they could be just as well off as whites.' Do you agree strongly, agree somewhat, neither agree nor disagree, disagree somewhat, or disagree strongly with this statement?"Response scale **V202260**

V202304 "How well does the following statement describe your view? 'Our political system only works for the insiders with money and power.' "1. Not at all well 2. Not very well 3. Somewhat well 4. Very well 5. Extremely well

V202305 "How well does the following statement describe your view? 'Because of the rich and powerful, it becomes difficult for the rest of us to get ahead.' " Response scale **V202304**

V202308x "When it comes to public policy decisions, whom do you tend to trust more: ordinary people, experts, or trust both the same?Do you trust [ordinary people/experts] much more or somewhat more?" 1. Trust ordinary people much more 2. Trust ordinary people somewhat more 3. Trust both the same 4. Trust experts somewhat more 5. Trust experts much more

V202309 "How much do ordinary people need the help of experts to understand complicated things like science and health?" 1. Not at all 2. A little 3. A moderate amount 4. A lot 5. A great deal

V202310 "In general, how important should science be for making government decisions about COVID-19?" 1. Not at all important 2. A little important 3. Moderately important 4. Very important 5. Extremely important

V202311 "How well does the following statement describe your view? Most business and politics in this country are secretly controlled by the same few powerful people." Response scale **V202304**

V202312 "How well does the following statement describe your view?"Much of what people hear in schools and the media are lies designed to keep people from learning the real truth about those in power." Response scale **V202304**

V202313 "Which of these do you consider to be the most important? You can just tell me the number of the option you choose." 1. A high level of economic growth 2. Making sure this country has strong defense forces 3. Seeing that people have more say about how things are done at their jobs/in their communities 4. Trying to make our cities and countryside more beautiful

V202314 "And which would be the next most important? You can just tell me the number ofthe option you choose." Response scale **V202313**

V202315 "Here is another list. Which one of these do you consider most important? You can just tell me the number of the option you choose." 1. Maintaining order in the nation 2. Giving people more say in important government decisions 3. Fighting rising prices 4. Protecting freedom of speech

V202316 "And which would be the next most important? You can just tell me the number of the option you choose." Response scale **V202315**

V202325 "Do you favor, oppose, or neither favor nor oppose increasing incometaxes on people making over one million dollars per year?" 1. Favor 2. Oppose 3. Neither favor nor oppose

V202331x "Do you favor, oppose, or neither favor nor oppose requiring children to be vaccinated in order to attend public schools? Do you [favor/oppose] that a great deal, a moderate amount, or a little?" Response scale **V201356x**

V202236x "Do you favor, oppose, or neither favor nor oppose allowing refugees who are fleeing war, persecution, or natural disasters in other countries to come to live in the U.S.? Do you [favor/oppose] that a great deal, a moderate amount, or a little?" Response scale **V201356x**

V202337 "Do you think the federal government should make it more difficult for people to buy a gun than it is now, make it easier for people to buy a gun, or keep these rules about the same as they are now?" 1. More difficult 2. Easier 3. Keep these rules about the same

V202341x "Do you favor, oppose, or neither favor nor oppose requiring background checks for gun purchases at gun shows or other private sales? Do you [favor/oppose] that a great deal, a moderate amount, or a little?" Response scale **V201356x**

V202344x "Do you favor, oppose, or neither favor nor oppose banning the sale of semi-automatic "assault-style" rifles? Do you [favor/oppose] that a great deal, a moderate amount, or a little?" Response scale **V201356x**

V202347x "Do you favor, oppose, or neither favor nor oppose a mandatory program where the government would buy back semi-automatic assault-style rifles from citizens who currently own them? Do you [favor/oppose] that a great deal, a moderate amount, or a little?" Response scale **V201356x**

V202351 "How often do you think police officers use more force than is necessary?" 1. Never 2. Rarely 3. About half the time 4. Most of the time 5. All the time

V202355 "Do you currently live in a rural area, small town, suburb, or a city?" 1. Rural area 2. Small town 3. Suburb 4. City

V202361x "Do you favor, oppose, or neither favor nor oppose the U.S. making free tradeagreements with other countries? How strongly do you [favor/oppose] it?" Response scale **V201356x**

V202367x "Has international trade increased, decreased, or neither increased nor decreased the number of jobs available in the United States?Has it [increased/decreased] the number of jobs in the U.S. a lot or somewhat?" 1. Increased a lot 2. Increased somewhat 3. Neiter increased nor decreased 4. Decreased somewhat 5. Increased a lot

V202373x "Does the increasing number of people of many different races and ethnic groups in the United States make this country a better place to live, a worse place to live, or does it make no difference?Does it make it a lot [better/worse] or a little [better/worse]?" 1. A lot better 2. A little better 3. Makes no difference 4. A little worse 5. A lot worse

V202376x "Do you favor, oppose, or neither favor nor oppose establishing a federal program that gives all citizens $12,000 per year, provided they meet certain conditions? This program would be paid for with higher taxes. Do you [favor/oppose] that a great deal, a moderate amount, or a little?" Response scale **V201356x**

V202377 "Should the federal minimum wage be raised, kept the same, lowered but not eliminated, or eliminated altogether?" 1. Raised 2. Kept the same 3. Lowered 4. Eliminated

V202383x "Now for questions on different topics. Do the health benefits of vaccinations generally outweigh the risks, do the risks outweigh the benefits, or is there no difference? Are the [health benefits/risks] of vaccinations much greater, moderately greater, or slightly greater?" 1. Health benefits much greater than risks 2. Health benefits moderately greater than risks 3. Health benefits slightly greater than risks 4. No difference 5. Risks slightly greather than health benefits 6. Risks moderately greater than health benefits 7. Risks much greater than health benefits

V202387x "Do you think attention to sexual harassment has gone too far, has not gone far enough, or has been about right? [Has it gone much too far or a little too far?] [Has it not gone nearly far enough or not gone quite far enough?]" 1. Has gone much too far 2. Has gone a little too far 3. Has been about right 4. Has not gone quite far enough 5. Has not gone nearly far enough

V202390x "Do you favor, oppose, or neither favor nor oppose allowing transgender people to serve in the United States Armed Forces? Do you [favor/oppose] that a great deal, a moderate amount, or a little?" Response scale **V201356x**

V202410 "Please tell me whether you agree strongly, agree somewhat, neither agree nor disagree, disagree somewhat, or disagree strongly with each of the following statements: Most politicians do not care about the people. Do you agree strongly, agree somewhat, neither agree nor disagree, disagree somewhat, or disagree strongly?" 1. Agree strongly 2. Agree somewhat 3. Neither agree nor disagree 4. Disagree somewhat 5. Disagree strongly

V202411 " 'Most politicians are trustworthy.' " Response scale **V202410**

V202412 " 'Politicians are the main problem in the United States.' " Response scale **V202410**

V202413 " 'Having a strong leader in government is good for the United States even if the leader bends the rules to get things done.' " Response scale **V202410**

V202414 " 'The people, and not politicians, should make our most important policy decisions.' " Response scale **V202410**

V202456 "During the past 12 months, were you or any of your family members stopped or questioned by a police officer, or did this not happen in the past 12 months?" 1. Was stopped or questioned in the past 12 months 2. Was not stopped or questioned in the past 12 months

V202457 "Have you ever been arrested, or has that never happened to you?" 1. Have been arrested 2. Never arrested

V202468x "The next question is about [the total combined income of all members of your family / your total income during the past 12 months. This includes money from jobs, net income from business, farm or rent, pensions, dividends, interest, Social Security payments and any other money income received by members of your family who are 15 years of age or older. What was the total income of your family during the past 12 months?"

V202475 "Do you consider yourself a feminist, an anti-feminist, or neither of these?" 1. Feminist 2. Anti-feminist 3. Neither

V202541a Which social media platforms have you visited in the past year? Please tell me any that apply. (**V202541a** through **V202541g** 0. Not mentioned 1. Mentioned)

V202541b Which social media platforms [respondent] visited - Twitter

V202541c Which social media platforms [respondent] visited - Instagram

V202541d Which social media platforms [respondent] visited - Reddit

V202541e Which social media platforms [respondent] visited - YouTube

V202541f Which social media platforms [respondent] visited - SnapChat

V202541g Which social media platforms [respondent] visited - TikTok

V202559 "Which of these two statements do you think is most likely to be true?" 1. Scientific evidence that hydroxychloroquine is a safe and effective treatment for COVID-19 2. No scientific evidence that hydroxychloroquine is a safe and effective treatment for COVID-19

V202562 "Do you currently owe money on student loans, or not?" 1. Yes 2. No

Bibliography

Abramowitz, A. I. (2016). Will time for change mean time for trump? *PS: Political Science & Politics*, 49(4):659–660.

American National Election Studies (2021). Anes 2020 time series study full release [dataset and documentation]. https://www.electionstudies.org. Accessed July 1, 2024.

Arel-Bundock, V. (2022). *WDI: World Development Indicators and Other World Bank Data*. R package version 2.7.8.

Armstrong, D. A., Bakker, R., Carroll, R., Hare, C., Poole, K. T., and Rosenthal, H. (2020). *Analyzing Spatial Models of Choice and Judgment*. Statistics in the Social and Behavioral Sciences. Chapman & Hall/CRC, Boca Raton, FL, 2nd edition.

Bache, S. M. and Wickham, H. (2020). *magrittr: A Forward-Pipe Operator for R*. R package version 2.0.1.

Ballotpedia (2024). Ballotpedia. https://ballotpedia.org/Main_Page. Accessed April 19, 2024.

Bartholomew, D. J., Steele, F., Moustaki, I., and Galbraith, J. I. (2008). *Analysis of Multivariate Social Science Data*. Statistics in the Social and Behavioral Sciences. Chapman & Hall/CRC, Boca Raton, FL, 2nd edition.

Baumer, B. S., Kaplan, D. T., and Horton, N. J. (2017). *Modern Data Science with R*. Texts in Statistical Science. Chapman & Hall/CRC, Boca Raton, FL, 1st edition.

Benoit, K. (2020). Text as data: An overview. In Curini, L. and Franzese, R., editors, *The SAGE Handbook of Research Methods in Political Science and International Relations*, pages 461–497. SAGE Publications Ltd, London.

Benoit, K., Muhr, D., and Watanabe, K. (2021). *stopwords: Multilingual Stopword Lists*. R package version 2.3.

Benoit, K. and Müller, S. (2024). Text analysis using r. https://quanteda.github.io/Text-Analysis-Using-R/. Accessed August 1, 2024.

Benoit, K. and Obeng, A. (2023). *readtext: Import and Handling for Plain and Formatted Text Files*. R package version 0.90.

Benoit, K., Watanabe, K., Wang, H., Nulty, P., Obeng, A., Müller, S., and Matsuo, A. (2018). quanteda: An r package for the quantitative analysis of textual data. *Journal of Open Source Software*, 3(30):774.

Benoit, K., Watanabe, K., Wang, H., Nulty, P., Obeng, A., Müller, S., Matsuo, A., and Lowe, W. (2023). *quanteda: Quantitative Analysis of Textual Data*. R package version 3.3.1.

Brady, H. E. (2011). The art of political science: Spatial diagrams as iconic and revelatory. *Perspectives on Politics*, 9(2):311–331.

Brady, H. E., Collier, D., and Box-Steffensmeier, J. M. (2011). Overview of political methodology: Post-behavioral movements and trends. In Goodin, R. E., editor, *The Oxford Handbook of Political Science*, pages 775–802. Oxford University Press.

Brauer, M., Freedman, G., Frostad, J., Van Donkelaar, A., Martin, R. V., Dentener, F., Dingenen, R. v., Estep, K., Amini, H., Apte, J. S., et al. (2016). Ambient air pollution exposure estimation for the global burden of disease 2013. *Environmental science & technology*, 50(1):79–88.

Cairo, A. (2019). *How Charts Lie: Getting Smarter about Visual Information*. W.W. Norton & Company, New York.

Campbell, A. and Kahn, R. L. (1948). Codebook introduction file: 1948 pre-post study (1948.t). https://electionstudies.org/data-center/1948-time-series-study/. Accessed April 20, 2025.

Caughey, D. and Warshaw, C. (2018). Policy preferences and policy change: Dynamic responsiveness in the american states, 1936–2014. *American Political Science Review*, 112(2):249–266.

Center for American Women and Politics (2021). Women in state legislatures 2021. https://cawp.rutgers.edu/facts/levels-office/state-legislature/women-state-legislatures-2021. Eagleton Institute of Politics, Rutgers University. Accessed April 19, 2024.

Centers for Disease Control and Prevention (2024). CDC WONDER (Wide-ranging Online Data for Epidemiologic Research). http://wonder.cdc.gov. National Center for Health Statistics. Accessed April 20, 2024.

Chang, W. (2018). *R Graphics Cookbook: Practical Recipes for Visualizing Data*. O'Reilly Media, Sebastopol, CA, 2nd edition.

Cheng, J., Karambelkar, B., and Xie, Y. (2023). *leaflet: Create Interactive Web Maps with the JavaScript Leaflet Library*. R package version 2.1.2.

Cleveland, W. S. and McGill, R. (1984). Graphical perception: Theory, experimentation, and application to the development of graphical methods. *Journal of the American Statistical Association*, 79(387):531–554.

Cohen, J., Cohen, P., West, S. G., and Aiken, L. S. (2003). *Applied Multiple Regression / Correlation Analysis for the Behavioral Sciences.* Routledge, New York, 3rd edition.

Coppedge, M., Gerring, J., Knutsen, C. H., Lindberg, S. I., Teorell, J., Alizada, N., Altman, D., Bernhard, M., Cornell, A., Fish, M. S., Gastaldi, L., Gjerløw, H., Glynn, A., Grahn, S., Hicken, A., Hindle, G., Ilchenko, N., Kinzelbach, K., Krusell, J., Marquardt, K. L., McMann, K., Mechkova, V., Medzihorsky, J., Paxton, P., Pemstein, D., Pernes, J., Rydén, O., von Römer, J., Seim, B., Sigman, R., Skaaning, S.-E., Staton, J., Sundström, A., Tzelgov, E., Wang, Y.-t., Wig, T., Wilson, S., and Ziblatt, D. (2022a). V-Dem Country-Year Dataset v12. https://www.v-dem.net/data/the-v-dem-dataset/. Accessed April 20, 2024.

Coppedge, M., Gerring, J., Knutsen, C. H., Lindberg, S. I., Teorell, J., Alizada, N., Gastaldi, L., Grahn, S., Hindle, G., Ilchenko, N., Natsika, N., Pernes, J., and von Römer, J. (2022b). V-Dem Organization and Management v12. https://www.v-dem.net. Accessed April 20, 2024.

Dasu, T. and Johnson, T. (2003). *Exploratory Data Mining and Data Cleaning.* Wiley Series on Methods and Applications in Data Mining. John Wiley & Sons, Hoboken, NJ.

De Sola Pool, I. and Abelson, R. (1961). The simulmatics project. *Public Opinion Quarterly*, 25(2):167–183.

de Vries, A. and Ripley, B. D. (2021). *ggdendro: Create Dendrograms and Tree Diagrams Using 'ggplot2'.* R package version 0.1.23.

District of Columbia Government (2024). Crime incidents in 2024. https://opendata.dc.gov/datasets/DCGIS::crime-incidents-in-2024. DC Open Data Portal. Accessed February 1, 2025.

Donaldson, G. A. (2007). *The First Modern Campaign: Kennedy, Nixon, and the Election of 1960.* Rowman and Littlefield, New York.

Donoho, D. (2017). 50 years of data science. *Journal of Computational and Graphical Statistics*, 26(4):745–766.

Dunnington, D. (2023). *ggspatial: Spatial Data Framework for ggplot2.* R package version 1.1.8.

Eder, M., Rybicki, J., and Kestemont, M. (2023). *stylo: Text Analysis and Stylometry with R.* R package version 0.7.5.

Elff, M. (2023). *memisc: Management of Survey Data and Presentation of Analysis Results.* R package version 0.99.31.7.

Erikson, R. S. and Tedin, K. L. (2015). *American Public Opinion.* Pearson, Boston, 9th edition.

Everson, P., Valelly, R., Vishwanath, A., and Wiseman, J. (2016). Nominate and american political development: a primer. *Studies in American Political Development*, 30(2):97–115.

Federal Election Commission (2025). Election results and voting information. https://www.fec.gov/introduction-campaign-finance/election-results-and-voting-information/. Accessed June 1, 2023.

Fox, J. (2016). *Applied Regression Analysis & Generalized Linear Models*. SAGE Publications, Inc., Thousand Oaks, CA, 3rd edition.

Fox, J. and Weisberg, S. (2011). *An R Companion to Applied Regression*. SAGE Publications, Inc., Thousand Oaks, CA, 2nd edition.

Fox, J. and Weisberg, S. (2019). *An R Companion to Applied Regression*. SAGE Publications, Inc., Thousand Oaks, CA, 3rd edition.

Fox, J., Weisberg, S., and Price, B. (2022a). *car: Companion to Applied Regression*. R package version 3.1-1.

Fox, J., Weisberg, S., Price, B., Friendly, M., and Hong, J. (2022b). *effects: Effect Displays for Linear, Generalized Linear, and Other Models*. R package version 4.2-2.

Galili, T. (2015). dendextend: an r package for visualizing, adjusting, and comparing trees of hierarchical clustering. *Bioinformatics*, 31(22):3718–3720.

Gallup (2024). Presidential job approval center. https://news.gallup.com/interactives/507569/presidential-job-approval-center.aspx. Accessed April 19, 2024.

Gapminder Foundation (2024). Gapminder. https://www.gapminder.org/. An educational resource providing data visualizations and insights on global development trends. Accessed July 1, 2024.

Gatscha, S., Karambelkar, B., and Schloerke, B. (2024). *leaflet.extras: Extra Functionality for 'leaflet' Package*. R package version 2.0.1.

Gelman, A. (2008). Scaling regression inputs by dividing by two standard deviations. *Statistics in Medicine*, 27(15):2865–2873.

Gelman, A., Pasarica, C., and Dodhia, R. (2002). Let's practice what we preach: turning tables into graphs. *The American Statistician*, 56(2):121–130.

Gelman, A. and Unwin, A. (2013). Infovis and statistical graphics: Different goals, different looks. *Journal of Computational and Graphical Statistics*, 22(1):2–28.

Gimpel, J. G. and Schuknecht, J. E. (2004). *Patchwork Nation: Sectionalism and Political Change in America*. The University of Michigan Press, Ann Arbor, MI.

Goodman, S., BenYishay, A., Lv, Z., and Runfola, D. (2019). Geoquery: Integrating hpc systems and public web-based geospatial data tools. *Computers and Geosciences*, (122):103–112.

Groves, R. M., Fowler, F. J. J., Couper, M. P., Lepkowski, J. M., Singer, E., and Tourangeau, R. (2011). *Survey Methodology.* John Wiley & Sons, Hoboken, NJ, 2nd edition.

HathiTrust Digital Library (2024). Hathitrust digital library. https://www.hathitrust.org. Accessed November 14, 2024.

Healy, K. (2019). *Data Visualization: A Practical Introduction.* Princeton University Press, Princeton, NJ.

Hebbali, A. (2020). *blorr: Tools for Developing Binary Logistic Regression Models.* R package version 0.3.0.

Ignatow, G. and Mihalcea, R. (2017). *Text Mining: A Guidebook for the Social Sciences.* SAGE Publications, Inc., Thousand Oaks, CA.

Jacoby, W. G. (1991). *Data Theory and Dimensional Analysis.* SAGE Publications, Inc., Newbury Park, CA.

Jacoby, W. G. and Armstrong, D. A. (2014). Bootstrap confidence regions for multidimensional scaling solutions. *American Journal of Political Science*, 58(1):264–278.

Jacoby, W. G. and Ciuk, D. J. (2018). Multidimensional scaling: An introduction. In Irwing, P., Booth, D. A., and Hughes, D. J., editors, *The Wiley Handbook of Psychometric Testing: A Multidisciplinary Reference on Survey, Scale and Test Development*, volume 1, pages 375–412. Wiley, Chichester, UK.

James, G., Witten, D., Hastie, T., and Tibshirani, R. (2013). *An Introduction to Statistical Learning: With Applications in R.* Springer, New York, 1st edition.

Jeworutzki, S. (2023). *cartogram: Create Cartograms with R.* R package version 0.3.0.

Jones, T. (2020). Dewey defeats truman: The most famous wrong call in electoral history. *Chicago Tribune.* https://www.chicagotribune.com/2020/10/31/dewey-defeats-truman-the-most-famous-wrong-call-in-electoral-history/. Accessed May 3, 2023.

Kastellec, J. P. and Leoni, E. L. (2007). Using graphs instead of tables in political science. *Perspectives on Politics*, 5(4):755–771.

Lepore, J. (2021). *If Then: How the Simulmatics Corporation Invented the Future.* Liveright Publishing Corporation, New York.

Lewis, J. B., Poole, K., Rosenthal, H., Boche, A., Rudkin, A., and Sonnet, L. (2023). Voteview: Congressional roll-call votes database. https://voteview.com. Accessed December 5, 2023.

Lewis-Beck, M. S. (2005). Election forecasting: Principles and practice. *The British Journal of Politics and International Relations*, 7(2):145–164.

Library of Congress (2024). The federalist papers: Primary documents in american history. https://guides.loc.gov/federalist-papers/. Research guide. Accessed April 20, 2024.

Long, J. A. (2023). *jtools: Analysis and Presentation of Social Scientific Data.* R package version 2.2.2.

Long, J. S. (1997). *Regression Models for Categorical and Limited Dependent Variables.* SAGE Publications, Inc., Thousand Oaks, CA.

Lovelace, R., Nowosad, J., and Muenchow, J. (2019). *Geocomputation with R.* Chapman & Hall/CRC Press, Boca Raton, FL.

Lumley, T. (2011). *Complex Surveys: A Guide to Analysis Using R.* John Wiley & Sons, Hoboken, NJ.

Lumley, T. (2021). *survey: Analysis of Complex Survey Samples.* R package version 4.1-1.

Lutosławski, W. (1897). *The Origin and Growth of Plato's Logic: With an Account of Plato's Style and of the Chronology of His Writings.* Longmans, Green and Company, London. Digitized by HathiTrust. Accessed April 20, 2024.

Lüdecke, D. (2018). ggeffects: Tidy data frames of marginal effects from regression models. *Journal of Open Source Software*, 3(26):772.

Lüdecke, D. (2024). *sjPlot: Data Visualization for Statistics in Social Science.* R package version 2.8.16.

Machlis, S. (2019). *Practical R for Mass Communication and Journalism.* The R Series. Chapman & Hall/CRC, Boca Raton, FL.

Mair, P., De Leeuw, J., and Groenen, P. J. F. (2022). *smacof: Multidimensional Scaling.* R package version 2.1-5.

Manning, C. D., Raghavan, P., and Schütze, H. (2009). *An Introduction to Information Retreival.* Cambridge University Press.

Mathieu, E., Ritchie, H., Rodés-Guirao, L., Appel, C., Giattino, C., Hasell, J., Macdonald, B., Dattani, S., Beltekian, D., Ortiz-Ospina, E., and Roser, M. (2020). Coronavirus pandemic (covid-19). https://ourworldindata.org/coronavirus. Accessed October 15, 2020.

McGhee, E. (2020). Partisan gerrymandering and political science. *Annual Review of Political Science*, 23:171–185.

Mervosh, S., Tumin, R., and Sasani, A. (2023). Trans athletes facing limits in biden plan. *The New York Times*, page A1. New York edition, Section A.

Michigan Secretary of State (2024). Election results and data. https://www.michigan.gov/sos/elections/election-results-and-data. Michigan Secretary of State website. Accessed November 20, 2024.

Mosteller, F. and Wallace, D. L. (1963). Inference in an authorship problem. *Journal of the American Statistical Association*, 58(302):275–309.

Müller, K. and Wickham, H. (2023). *tibble: Simple Data Frames*. R package version 3.2.0.

Ooms, J. (2014). The jsonlite package: A practical and consistent mapping between json data and r objects. *arXiv:1403.2805 [stat.CO]*.

Open Data DC (2024). Crime incidents in 2024. https://catalog.data.gov/dataset/crime-incidents-in-2024. Accessed October 31, 2024.

Pebesma, E. (2018). Simple Features for R: Standardized Support for Spatial Vector Data. *The R Journal*, 10(1):439–446.

Pebesma, E. and Bivand, R. (2023). *Spatial Data Science: With Applications in R*. Chapman & Hall/CRC The R Series. Chapman and Hall/CRC, Boca Raton, FL.

Pew Research Center (2021). Beyond red vs. blue: The political typology. https://www.pewresearch.org/topic/politics-policy/political-parties-polarization/political-typology/. Accessed December 1, 2024.

Polsby, D. D. and Popper, R. D. (1991). The third criterion: Compactness as a procedural safeguard against partisan gerrymandering. *Yale Law and Policy Review*, 9:301–353.

Poole, K., Lewis, J., Lo, J., and Carroll, R. (2024). *wnominate: Roll Call Analysis Software*. R package version 1.5.

Poole, K. T. (2005). *Spatial Models of Parliamentary Voting*. Cambridge University Press.

Poole, K. T. and Rosenthal, H. (2006). *Ideology and Congress: A Political Economic History of Roll Call Voting*. Transaction Publishers, New Brunswick, NJ, 2nd edition.

Project Gutenberg (2024). https://www.gutenberg.org. Accessed November 14, 2024.

R-core (2023). *foreign: Read Data Stored by 'Minitab', 'S', 'SAS', 'SPSS', 'Stata', 'Systat', 'Weka', 'dBase', ...* R package version 0.8-84.

R Core Team (2022). *R: A Language and Environment for Statistical Computing.* R Foundation for Statistical Computing, Vienna, Austria.

Rabinowitz, G. B. (1975). An introduction to nonmetric multidimensional scaling. *American Journal of Political Science*, 19(2):343–390.

Robinson, D. (2021). *unvotes: United Nations General Assembly Voting Data.* R package version 0.3.0.

Rorabaugh, W. J. (2009). *The Real Making of the President: Kennedy, Nixon, and the 1960 Election.* University Press of Kansas, Lawrence.

Rosling, H., Rönnlund, A. R., and Rosling, O. (2018). *Factfulness: Ten Reasons We're Wrong About the World — and Why Things Are Better Than You Think.* Flatiron Books, New York.

Runfola, D., Anderson, W., Baier, H., Kerckhoff, C., Gross, L., Manlove, K., Smith, E., Xu, S., and Samuels, C. (2020). geoboundaries: A global database of political administrative boundaries. *PLoS ONE*, 15(4):e0231866.

Rusk, J. G. and Weisberg, H. F. (1972). Perceptions of presidential sandidates: Implications for electoral change. *Midwest Journal of Political Science*, 16(3):388–410.

Shepard, R. N. (1962). The analysis of proximities: Multidimensional scaling with an unknown distance function. *Psychometrika*, 27(2):125–140.

Sievert, C. (2020). *Interactive Web-Based Data Visualization with R, Plotly, and Shiny.* The R Series. Chapman & Hall/CRC Press, Boca Raton, FL.

Signorell, A. and many others (2024). *DescTools: Tools for Descriptive Statistics.* R package version 0.99.58.

Silge, J. and Robinson, D. (2017). *Text Mining with R: A Tidy Approach.* O'Reilly Media, Sebastopol, CA.

Slowikowski, K. (2023). *ggrepel: Automatically Position Non-Overlapping Text Labels with ggplot2.* R package version 0.9.3.

State of Michigan (2024). Michigan geographic information systems (gis) open data portal. https://gis-michigan.opendata.arcgis.com/. Accessed April 20, 2025.

Steensland, B., Robinson, L. D., Wilcox, W. B., Park, J. Z., Regnerus, M. D., and Woodberry, R. D. (2000). The measure of american religion: Toward improving the state of the art. *Social forces*, 79(1):291–318.

Stewart, T. (2022). Overview of motor vehicle crashes in 2020 (report no. dot hs 813 266). Technical Report DOT HS 813 266, National Highway Traffic Safety Administration. Accessed April 19, 2024.

The New York Times (2021). Coronavirus (covid-19) data in the united states. https://github.com/nytimes/covid-19-data. Data based on reports from state and local health agencies. Accessed February 1, 2023.

The World Bank (2024). World development indicators. https://databank.worldbank.org/source/world-development-indicators. A comprehensive database of global development statistics, covering economic, social, and environmental indicators. Accessed February 1, 2024.

Tufte, E. R. (2001). *The Visual Display of Quantitative Information.* Graphics Press, Cheshire, CT, 2nd edition.

United Nations Development Programme (2022). *Human Development Report 2021/2022: Uncertain Times, Unsettled Lives: Shaping Our Future in a Transforming World.* United Nations Development Programme, New York. https://hdr.undp.org/content/human-development-report-2021-22, Accessed June 20, 2023.

Unwin, A. (2018). *Graphical Data Analysis with R.* The R Series. Chapman & Hall/CRC Press, Boca Raton, FL.

Urdinez, F. and Cruz, A., editors (2022). *R for Political Data Science: A Practical Guide.* The R Series. Chapman & Hall/CRC Press, Boca Raton, FL.

U.S. Bureau of Economic Analysis (2024a). Annual gdp, personal income, and employment by state. https://apps.bea.gov/itable/iTable.cfm?ReqID=70&step=1. Accessed July 5, 2024.

U.S. Bureau of Economic Analysis (2024b). Gross domestic product (gdp). https://www.bea.gov/data/gdp. Accessed April 19, 2024.

U.S. Census Bureau (2018). U.S. Census Bureau Data. https://www.census.gov/data.html. Population and income source: U.S. Census Bureau decennial census and American Community Survey 5-year estimate, 2018. Accessed July 1, 2024.

U.S. Department of Education (2023). Public comment on proposed rule: Nondiscrimination on the basis of sex in education programs or activities receiving federal financial assistance: Sex-related eligibility criteria for male and female athletic teams. https://www.regulations.gov/document/ED-2022-OCR-0143-0001/comment. Docket ID: ED-2022-OCR-0143-0001. Accessed January 15, 2024.

Walker, K. (2023). *Analyzing US Census Data: Methods, Maps, and Models in R.* The R Series. Chapman & Hall/CRC Press, Boca Raton, FL.

Walker, K. and Herman, M. (2024). *tidycensus: Load US Census Boundary and Attribute Data as tidyverse and sf-Ready Data Frames.* R package version 1.6.

Walker, K. and Rudis, B. (2024). *tigris: Load Census TIGER/Line Shapefiles and Extract and Use Census Data in R.* R package version 2.1.

Watanabe, K. and Baturo, A. (2023). Seeded sequential lda: A semi-supervised algorithm for topic-specific analysis of sentences. *Social Science Computer Review*, 42(1):224–248.

Watanabe, K. and Müller, S. (2023). Quanteda tutorials. https://tutorials.quanteda.io. Accessed July 13, 2024.

Watanabe, K. and Xuan-Hieu, P. (2023). *seededlda: Seeded Sequential LDA for Topic Modeling.* R package version 1.1.0.

Weisberg, H. F., Krosnick, J. A., and Bowen, B. D. (1996). *An Introduction to Survey Research, Polling, and Data Analysis.* SAGE Publications, Inc., Thousand Oaks, CA, 3rd edition.

Wickham, H. (2010). A layered grammar of graphics. *Journal of Computational and Graphical Statistics*, 19(1):3–28.

Wickham, H. (2022). *stringr: Simple, Consistent Wrappers for Common String Operations.* R package version 1.5.0.

Wickham, H. (2023a). *forcats: Tools for Working with Categorical Variables (Factors).* R package version 1.0.0.

Wickham, H. (2023b). *tidyverse: Easily Install and Load the Tidyverse.* R package version 2.0.0.

Wickham, H., Averick, M., Bryan, J., Chang, W., McGowan, L. D., François, R., Grolemund, G., Hayes, A., Henry, L., Hester, J., Kuhn, M., Pedersen, T. L., Miller, E., Bache, S. M., Müller, K., Ooms, J., Robinson, D., Seidel, D. P., Spinu, V., Takahashi, K., Vaughan, D., Wilke, C., Woo, K., and Yutani, H. (2019). Welcome to the tidyverse. *Journal of Open Source Software*, 4(43):1686.

Wickham, H. and Bryan, J. (2023). *readxl: Read Excel Files.* R package version 1.4.2.

Wickham, H., Chang, W., Henry, L., Pedersen, T. L., Takahashi, K., Wilke, C., Woo, K., Yutani, H., and Dunnington, D. (2023a). *ggplot2: Create Elegant Data Visualisations Using the Grammar of Graphics.* R package version 3.4.1.

Wickham, H., François, R., Henry, L., Müller, K., and Vaughan, D. (2023b). *dplyr: A Grammar of Data Manipulation.* R package version 1.1.0.

Wickham, H. and Grolemund, G. (2016). *R for Data Science: Import, Tidy, Transform, Visualize, and Model Data.* O'Reilly Media, Sebastopol, CA.

Wickham, H., Hester, J., and Bryan, J. (2023c). *readr: Read Rectangular Text Data.* R package version 2.1.4.

Wickham, H. and Miller, E. (2023). *haven: Import and Export 'SPSS', 'Stata' and 'SAS' Files.* R package version 2.5.3.

Wilkinson, L. (1999). *The Grammar of Graphics.* Statistics and Computing. Springer, New York.

Woolley, J. T. and Peters, G. (2024). The american presidency project. https://www.presidency.ucsb.edu/statistics/elections. University of California, Santa Barbara.

Xie, Y. (2023a). *bookdown: Authoring Books and Technical Documents with R Markdown.* R package version 0.33.

Xie, Y. (2023b). *knitr: A General-Purpose Package for Dynamic Report Generation in R.* R package version 1.42.

Zhu, H. (2024). *kableExtra: Construct Complex Table with kable and Pipe Syntax.* R package version 1.4.0.

Index